Synthesis and Operability Strategies for Computer-Aided Modular Process Intensification

Synthesis and Operability Strategies for Computer-Aided Modular Process Intensification

Efstratios N. Pistikopoulos
Texas A&M Energy Institute; Artie McFerrin Department of Chemical Engineering
Texas A&M University
College Station, TX, United States

Yuhe Tian
Department of Chemical and Biomedical Engineering
West Virginia University
Morgantown, WV, United States

Elsevier
Radarweg 29, PO Box 211, 1000 AE Amsterdam, Netherlands
The Boulevard, Langford Lane, Kidlington, Oxford OX5 1GB, United Kingdom
50 Hampshire Street, 5th Floor, Cambridge, MA 02139, United States

Copyright © 2022 Elsevier Inc. All rights reserved.

MATLAB® is a trademark of The MathWorks, Inc. and is used with permission.
The MathWorks does not warrant the accuracy of the text or exercises in this book.
This book's use or discussion of MATLAB® software or related products does not constitute endorsement or sponsorship by The MathWorks of a particular pedagogical approach or particular use of the MATLAB® software.

No part of this publication may be reproduced or transmitted in any form or by any means, electronic or mechanical, including photocopying, recording, or any information storage and retrieval system, without permission in writing from the publisher. Details on how to seek permission, further information about the Publisher's permissions policies and our arrangements with organizations such as the Copyright Clearance Center and the Copyright Licensing Agency, can be found at our website: www.elsevier.com/permissions.

This book and the individual contributions contained in it are protected under copyright by the Publisher (other than as may be noted herein).

Notices

Knowledge and best practice in this field are constantly changing. As new research and experience broaden our understanding, changes in research methods, professional practices, or medical treatment may become necessary.

Practitioners and researchers must always rely on their own experience and knowledge in evaluating and using any information, methods, compounds, or experiments described herein. In using such information or methods they should be mindful of their own safety and the safety of others, including parties for whom they have a professional responsibility.

To the fullest extent of the law, neither the Publisher nor the authors, contributors, or editors, assume any liability for any injury and/or damage to persons or property as a matter of products liability, negligence or otherwise, or from any use or operation of any methods, products, instructions, or ideas contained in the material herein.

ISBN: 978-0-323-85587-7

For information on all Elsevier publications
visit our website at https://www.elsevier.com/books-and-journals

Publisher: Susan Dennis
Acquisitions Editor: Anita Koch
Editorial Project Manager: Lena Sparks
Production Project Manager: Bharatwaj Varatharajan
Designer: Matthew Limbert

Typeset by VTeX

In the memory of Professors Christodoulos A. Floudas, M. Sam Mannan, and M. Nazmul Karim

Contents

Authors' biographies	xiii
Preface	xv
Acknowledgments	xxi

PART 1. Preliminaries

1.	Introduction to modular process intensification	3
	1.1. Introduction	3
	1.2. Definitions and principles of modular process intensification	3
	1.3. Modular process intensification technology showcases	7
	References	16
2.	Computer-aided modular process intensification: design, synthesis, and operability	19
	2.1. Conceptual synthesis and design	20
	2.2. Operability, safety, and control analysis	29
	2.3. Research challenges and key questions	35
	References	37

PART 2. Methodologies

3.	Phenomena-based synthesis representation for modular process intensification	45
	3.1. A prelude on phenomena-based PI synthesis	45
	3.2. Generalized Modular Representation Framework	47
	3.3. Driving force constraints	48
	3.4. Key features of GMF synthesis	52

viii   Contents

| 3.5. | Motivating examples | 53 |
| | References | 57 |

4. Process synthesis, optimization, and intensification — 59

4.1.	Problem statement	59
4.2.	GMF synthesis model	60
4.3.	Pseudo-capital cost estimation	68
4.4.	Solution strategy	70
4.5.	Motivating example: GMF synthesis representation and optimization of a binary distillation system	73
	Nomenclature	76
	References	77

5. Enhanced GMF for process synthesis, intensification, and heat integration — 79

5.1.	GMF synthesis model with Orthogonal Collocation	79
5.2.	GMF synthesis model with heat integration	82
5.3.	Motivating example: GMF synthesis, intensification, and heat integration of a ternary separation system	86
	References	93

6. Steady-state flexibility analysis — 95

6.1.	Basic concepts	95
6.2.	Problem definition	95
6.3.	Solution algorithms	98
6.4.	Design and synthesis of flexible processes	103
6.5.	Tutorial example: flexibility analysis of heat exchanger network	105
	References	110

Contents   ix

7. Inherent safety analysis   111

    7.1. Dow Chemical Exposure Index   111

    7.2. Dow Fire and Explosion Index   112

    7.3. Safety Weighted Hazard Index   115

    7.4. Quantitative risk assessment   120

    References   122

8. Multi-parametric model predictive control   123

    8.1. Process control basics   123

    8.2. Explicit model predictive control via multi-parametric programming   128

    8.3. The PAROC framework   135

    8.4. Case study: multi-parametric model predictive control of an extractive distillation column   139

    References   145

9. Synthesis of operable process intensification systems   147

    9.1. Problem statement   147

    9.2. A systematic framework for synthesis of operable process intensification systems   148

    9.3. Steady-state synthesis with flexibility and safety considerations   150

    9.4. Motivating example: heat exchanger network synthesis   157

    References   160

PART 3.   Case studies

10. Envelope of design solutions for intensified reaction/separation systems   163

    10.1. The Feinberg Decomposition   164

    10.2. Case study: olefin metathesis   165

    References   172

x   Contents

11. Process intensification synthesis of extractive separation systems with material selection — 173

11.1. Problem statement — 173

11.2. Case study: ethanol-water separation — 174

References — 186

12. Process intensification synthesis of dividing wall column systems — 187

12.1. Case study: methyl methacrylate purification — 188

12.2. Base case design and simulation analysis — 190

12.3. Process intensification synthesis via GMF — 193

References — 206

13. Operability and control analysis in modular process intensification systems — 207

13.1. Loss of degrees of freedom — 207

13.2. Role of process constraints — 211

13.3. Numbering up vs. scaling up — 216

13.4. Remarks — 219

References — 221

14. A framework for synthesis of operable and intensified reactive separation systems — 223

14.1. Process description — 223

14.2. Synthesis of intensified and operable MTBE production systems — 227

References — 246

15. A software prototype for synthesis of operable process intensification systems — 247

15.1. The SYNOPSIS software prototype — 247

15.2. Case study: pentene metathesis reaction — 249

References — 261

Contents xi

A. **Process modeling, synthesis, and control of reactive distillation systems** — 263

   A.1. Modeling of reactive distillation systems — 263

   A.2. Short-cut design of reactive distillation — 264

   A.3. Synthesis design of reactive distillation — 265

   A.4. Process control of reactive distillation — 266

   A.5. Software tools for modeling, simulation, and design of reactive distillation — 267

   References — 268

B. **Driving force constraints and physical and/or chemical equilibrium conditions** — 271

   B.1. Pure separation systems — 271

   B.2. Reactive separation systems — 272

   B.3. Pure reaction systems — 272

C. **Reactive distillation dynamic modeling** — 275

   C.1. Process structure — 275

   C.2. Tray modeling — 276

   C.3. Reboiler and condenser modeling — 280

   C.4. Physical properties — 280

   C.5. Initial conditions — 280

   C.6. Equipment cost correlations — 280

   References — 281

D. **Nonlinear optimization formulation of the Feinberg Decomposition approach** — 283

   References — 285

E. **Degrees of freedom analysis and controller design in modular process intensification systems** — 287

   E.1. Degrees of freedom analysis — 287

   E.2. Controller tuning for olefin metathesis case study — 291

   References — 294

xii Contents

F. MTBE reactive distillation model validation and dynamic analysis   **295**

    F.1.   MTBE reactive distillation model validation with commercial Aspen simulator   295

    F.2.   Steady-state and dynamic analyses on the selection of manipulated variable for MTBE reactive distillation   295

    References   298

Index   **299**

Authors' biographies

Dr. Yuhe Tian is an assistant professor in the Department of Chemical and Biomedical Engineering at West Virginia University. Prior to joining WVU, she received her PhD degree in Chemical Engineering from Texas A&M University under the supervision of Professor Efstratios N. Pistikopoulos (2016–2021). She holds bachelor's degrees in chemical engineering and mathematics from Tsinghua University, China (2012–2016). Her research focuses on the development and application of multi-scale systems engineering tools for modular process intensification, clean energy innovation, systems integration, and sustainable supply chain optimization.

Dr. Efstratios N. Pistikopoulos is the Director of the Texas A&M Energy Institute and the Dow Chemical Chair Professor in the Artie McFerrin Department of Chemical Engineering at Texas A&M University. He was a professor of Chemical Engineering at Imperial College London, UK (1991–2015) and the Director of its Centre for Process Systems Engineering (2002–2009). He holds a PhD degree from Carnegie Mellon University and worked with Shell Chemicals in Amsterdam before joining Imperial. He has authored or co-authored over 500 major research publications in the areas of modeling, control, and optimization of process, energy, and systems engineering applications, as well as 15 books and three patents. He is a Fellow of IChemE and AIChE, and the editor-in-chief of Computers & Chemical Engineering. In 2007, Prof. Pistikopoulos was a co-recipient of the prestigious MacRobert Award from the Royal Academy of Engineering. In 2012, he was the recipient of the Computing in Chemical Engineering Award of CAST/AIChE, while in 2020 he received the Sargent Medal from the Institution of Chemical Engineers (IChemE). He is a member of the Academy of Medicine, Engineering and Science of Texas. In 2021, he received the AIChE Sustainable Engineering Forum Research Award. He received the title of Doctor Honoris Causa in 2014 from the University Politehnica of Bucharest, and from the University of Pannonia in 2015. In 2013, he was elected Fellow of the Royal Academy of Engineering in the United Kingdom.

Preface

Today's chemical process industry is faced with pressing challenges to sustain the increasingly competitive global market with rising concerns on energy, water, food, and environment. Process intensification (PI) offers many promising opportunities to address these challenges. It aims to realize step changes in process economics, energy efficiency, and environmental impacts by developing novel process schemes and equipment. The synergistic nature between modular design and many PI technologies, which function the most effectively at small scale and feature standardized equipment, adds to the potential of modular process intensification towards flexible, agile, and efficient production systems.

Modular process intensification has gained significant impetus in the past decades featuring both successful industrial applications and burgeoning scientific research interests. However, early breakthroughs in this area mostly relied on Edisonian efforts via experimentation while lacking theory and fundamental understanding towards systematic innovation. Meanwhile, these novel technologies bring new design and operational challenges such as design complexity, safety concerns under extreme operating conditions, operation under uncertainty, unsteady-state operation, etc. Computational tools are in dire need to assess and optimize such systems at the early design stage.

In this context, computer-aided modular process intensification has become a rapidly emerging research theme in recent years. The model-based methods and tools, with expertise of the Process Systems Engineering (PSE) community, can support quantitative decision making by providing techno-economic evaluations as well as predictive capabilities on the design and operation of modular and intensified systems. This book aims to provide a unified methodology framework for the design of operable modular process intensification systems using advanced process synthesis and operability methods, as depicted in Fig. 1. Specifically, this book will cover the following topics:

Topic 1: introduction on computer-aided modular process intensification

PI can be achieved by utilizing the synergy between multi-functional phenomena, integrating multiple process steps into a single equipment, enhancing the mass, heat, and momentum transfer rate, etc. We will present an overview of the basic concepts and fundamental principles of modular process intensification from an evolutionary perspective. A number of representative intensified technologies will also be introduced, e.g. hybrid reaction/separation systems, micro-reaction systems, periodic systems. We will then discuss how computer-aided methods and tools can contribute to systematically generate innovative process design solutions with guaranteed operational performances. To this purpose, state-of-the-art methodological developments for synthesis, optimization, and control of intensified systems will be reviewed.

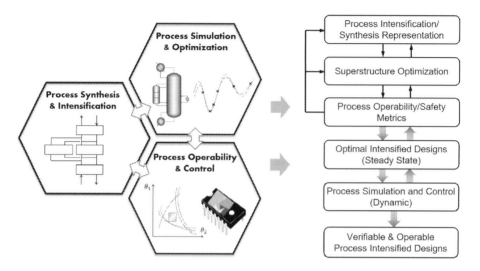

FIGURE 1 An overview of the synthesis and operability methods and a unified framework introduced in this book.

Topic 2: process intensification synthesis via a phenomena-based modular representation approach

The Generalized Modular Representation Framework (GMF), as a representative methodology for PI synthesis, will be detailed with mathematical formulation and algorithm implementation. A number of engineering application case studies will be presented to showcase GMF for the design of diverse intensified systems (reactive separation, dividing wall column, etc.). We will also explore the following research questions towards systematic process innovation: (i) how to exploit the synergy of multi-functional phenomena (e.g., reaction and separation) – without pre-postulation of tasks or equipment? (ii) how to determine the performance limits of intensified designs and benchmark with the ultimate thermodynamic or kinetics bounds? (iii) how to encapsulate intensified designs and their conventional counterparts in a unified synthesis representation and to identify when intensified designs will outperform in terms of economics, energy savings, etc.? and (iv) what is the role of functional materials (e.g., catalysts, solvents)?

Topic 3: model-based flexibility, inherent safety, and control analysis for modular process intensification systems

Advanced operability analysis, inherent safety analysis, and model-based control strategies developed in the PSE community will be introduced to assess the operational performance of chemical process systems. Particularly, we will talk about the seminal flexibility test/index approach to ensure feasible operation under uncertainty, inherent safety indices and quantitative risk analysis to evaluate process safety at the early design stage, and explicit/multi-parametric model predictive control following the PAROC (PARametric

Optimization and Control) framework to derive optimal dynamic operation strategies. Tutorial examples will be presented with step-by-step procedures to showcase the application of these methods in modular PI systems. To understand the unique operational challenges and needs resulted by modularization and intensification, we will also analyze the impact on operability of key factors including: (i) degrees of freedom, (ii) process constraints, (iii) numbering up vs. scaling up, etc.

Topic 4: a systematic framework for the synthesis of operable process intensification systems

To highlight the importance of integrating operability criteria in the design of intensified systems, we will present an integrated GMF-flexibility-safety synthesis approach which enables the automated generation of safely operable modular PI systems from phenomena level. A step-wise framework is further developed which synergizes: (i) phenomena-based process synthesis with GMF to derive novel intensified design configurations, (ii) integrated design with flexibility and inherent safety considerations, and (iii) simultaneous design and explicit model predictive control optimization to generate verifiable, operable, and optimal intensified systems. Multiple process design solutions can be delivered from the framework with the trade-offs between economic and operational performances. An integrated computer-aided software prototype will also be demonstrated to automate the distinct synthesis and operability tools as well as to implement the entire framework.

To facilitate the readers to learn and apply the techniques to their research and/or industrial problems of interest, the book is organized into three parts, respectively as *Preliminaries*, *Methodologies*, and *Case Studies*. In this way, each chapter will focus on a certain methodology, or application topic, consisting of the corresponding basic principles, model formulation, solution algorithm, and step-by-step implementation guidance on key procedures. More specifically:

Part 1: Preliminaries

Chapter 1 provides an overview of the evolution of modular process intensification definitions and fundamental principles. An overview will be given on the current status of academic research and industrial applications regarding specific modular PI technologies, including but not limited to dividing wall column, membrane-assisted separation, pressure swing adsorption, etc.

Chapter 2 highlights some recent PSE advances for modular process intensification. Process synthesis methods, particularly highlighting the use of phenomena-based representation, can systematically generate novel process solutions. Advanced operability and control strategies will also be reviewed which aim to ensure feasible process operation under uncertainty.

xviii Preface

Part 2: Methodologies

Chapter 3 introduces GMF for the representation of chemical process systems using modular phenomena building blocks, which lays the foundation for this book to drive innovation. We will discuss in detail the mass/heat exchange modular representation concepts, the key GMF synthesis features, and the driving force constraints based on total Gibbs free energy change.

Chapter 4 demonstrates GMF for process synthesis, optimization, and intensification. We will present the GMF superstructure network which can capture both conventional and novel process configurations, the mathematical model which is formulated as a mixed-integer nonlinear programming problem, and the tailored solution strategy which can efficiently screen the combinatorial design space.

Chapter 5 extends GMF as a unified approach for process intensification synthesis, heat integration, and thermal coupling. We will discuss the heat transfer feasibility constraints based on temperature gradient, which requires no pre-postulation of stream thermal properties in the process synthesis formulation. Extensions of GMF with orthogonal collocation will also be detailed to enhance intra-module representation.

Chapter 6 introduces the flexibility analysis approaches to assess if a design is feasible under expected process uncertainties. The mathematical formulation and solution algorithm of flexibility test and flexibility index will be highlighted. To synthesize flexible process systems, a multi-period design approach will be presented.

Chapter 7 discusses inherent safety analysis at the conceptual design stage. Quantitative risk analysis and inherent safety indices (e.g., Dow Fire & Explosion Index, Dow Chemical Exposure Index, Safety Weighted Hazard Index) will be introduced and demonstrated.

Chapter 8 talks about advanced model-based control strategies, with particular emphasis on explicit/multi-parametric control. We will also introduce the PAROC (PARametric Optimization and Control) framework which is an integrated framework and software platform for the design and control optimization of complex process systems.

Chapter 9 highlights integrated process intensification synthesis with operability, safety, and control considerations. We will present a holistic SYNOPSIS framework, standing for SYNthesis of Operable ProcesS Intensification Systems. Leveraging the above introduced synthesis and operability strategies, the framework can systematically and consistently address steady-state and dynamic design and operation in intensified processes.

Part 3: Case studies

Chapter 10 applies GMF with attainable region-based theory to quantitatively identify the performance limits and to develop an envelope of design solutions for reaction/separation systems, prior to establishing any specific process designs. The approach will be showcased via a case study on olefin metathesis for butene and hexene production.

Chapter 11 addresses simultaneous solvent selection and process intensification synthesis using GMF. Physical property models are explicitly incorporated in the synthesis model formulation to assess solvent performance in facilitating separation. A representative ethanol-water separation case study will be presented, considering the use of an ionic liquid solvent candidate.

Chapter 12 showcases GMF on synthesizing heterogeneous multi-component separation systems, with particular interest in exploring the use of dividing wall columns. Conventional or novel process structures, such as two-column sequences and dividing wall columns, can be systematically generated without pre-postulation of equipment design.

Chapter 13 performs rigorous model-based analyses towards a fundamental understanding of operability, safety, and control challenges in process intensification and modular designs. Comparative examples will be presented to showcase the pros and cons in intensified and modular systems versus their conventional counterparts from operational aspects.

Chapter 14 demonstrates the integrated SYNOPSIS framework to deliver verifiable, operable, and intensified systems through a methyl tert-butyl ether production case study. The approach can systematically integrate design and operability considerations at different stages (i.e., phenomena-based synthesis, steady-state design, control optimization).

Chapter 15 presents a software prototype based on the SYNOPSIS framework. The prototype comprises: (i) Process Intensification Synthesis Suite – to generate promising process configurations based on GMF, (ii) Operability and Control Suite – to ensure the actual operational performance of the resulting intensified systems, and (iii) Process Intensification Model library – with specialized steady-state and dynamic PI models.

Acknowledgments

The authors acknowledge the financial support from the Texas A&M Energy Institute, Shell, National Science Foundation PAROC Project (Grant No. 1705423), and Department of Energy RAPID Manufacturing Institute for Process Intensification SYNOPSIS Project (DE-EE0007888-09-03, Partner Organizations: Texas A&M University, Georgia Institute of Technology, Auburn University, Shell, The Dow Chemical Company, Siemens Process Systems Enterprise).

Preliminaries

1

Introduction to modular process intensification

1.1 Introduction

Facing a highly competitive global market with increasing awareness on environmental and safety issues, chemical production is making its way towards a paradigm shift to more efficient, more environmentally friendly, and more versatile. Process intensification and modular design are regarded as promising solutions to pursue this structural transformation, gaining significant recent impetus in the chemical/energy industry and the chemical engineering research community.

Process intensification (PI) aims to boost process and energy efficiency, enhance process profitability and safety, while reducing waste and emissions by utilizing the synergy between multi-functional phenomena at different time and spatial scales, as well as by enhancing process driving forces such as the mass, heat, and/or, momentum transfer rates, through the use of novel process schemes and equipment [1]. A wide range of PI technologies have been developed [2], some of which are already successfully commercialized such as reverse flow reactor, reactive distillation, and dividing wall column, to name a few.

On the other hand, modular design is a different while often concurrent concept to process intensification. It aims to dramatically reduce the size of process units to change from "the economy of scale" to small, distributed, and standardized plants with better flexibility and faster response to demand changes, especially for utilization of unconventional feedstocks and for specialty chemical production [3]. A key question for modular design is the gain or loss in cost efficiency vs. design/operation agileness when comparing "numbering up" against conventional "scaling up". For many intensified technologies (e.g., micro-reactors, membrane reactors, alternative energy sources) which inherently function the most effectively at smaller scales, the combination of PI technologies with modular design may provide an encouraging synergistic process solution [4].

In this chapter, we introduce the key concepts of modular PI, state-of-the-art research and industrial developments, and representative technology showcases.

1.2 Definitions and principles of modular process intensification

The concept of PI was first introduced into the Chemical Engineering discipline in 1983 marked by the paper of Colin Ramshaw from the ICI New Science Group, who described their studies on centrifugal fields (so-called "HiGee") in distillation processes [5]. Since

Synthesis and Operability Strategies for Computer-Aided Modular Process Intensification
https://doi.org/10.1016/B978-0-32-385587-7.00010-5
Copyright © 2022 Elsevier Inc. All rights reserved.

4 Synthesis and Operability Strategies for Computer-Aided Modular PI

then, several definitions of PI have been proposed, the differences of which mainly stem from the targeted scope in PI outcomes and the proposed strategies to achieve these outcomes. An indicative list of PI definitions is presented in Table 1.1 [11]. Interestingly, this also shows the evolution of PI principles and targets from: (i) initially emphasizing equipment size reduction and cost savings to recognizing PI with a broader impact towards more efficient, more sustainable, and safer processes, (ii) initially regarding PI as a standalone "toolbox" containing particular technology examples towards exploring PI fundamentals with respect to the role of multi-functional synergy, multi-scale driving forces, etc.

Table 1.1 Evolution of PI definitions and principles – An indicative list. (Reproduced from Tian et al. [11].)

Reference	Definition	Year
Ramshaw [5]	Devise exceedingly compact plant which reduces both the main plant item and the installation costs	1983
Stankiewicz & Moulijn [1]	Substantially decrease equipment volume, energy consumption, or waste formation; Lead to cheaper, safer, sustainable technologies	2000
Arizmendi-Sanchez & Sharratt [6]	Synergistic integration of process tasks and coupling of phenomena; Targeted intensification of transport processes	2008
Becht et al. [7]	Sustain profitability even in the presence of increasing uncertainties	2009
Van Gerven & Stankiewicz [8]	Maximize the effectiveness of intra- and inter-molecular events; Give each molecule the same processing experience; Optimize the driving forces at every scale; Maximize the synergistic effects for multitasking	2009
Lutze et al. [9]	Add/Enhance phenomena in a process through the integration of operations, functions, phenomena; Or through the targeted enhancement of phenomena	2010
Ponce-Ortega et al. [10]	Smaller equipment size; Higher throughput; Higher performance; Less usage of utility materials and feedstock	2012

To start with, Ramshaw [5] defined PI as the reduction of *"both the main plant item and the installation costs"*. A wider definition was later proposed by Stankiewicz and Moulijn [1] to recognize PI as any practice towards smaller, cheaper, safer, and/or more cost and energy efficient processes, in addition to only the reduction of unit size or costs. Becht et al. [7] further enriched the PI definition with *"... to sustain profitability even in the presence of increasing uncertainties"*, which observed PI as a more general practice and included flexibility and robustness as major PI outcomes.

There are also more holistic perspectives to define PI activities. Ponce-Ortega et al. [10], for instance, defined PI as any activity aiming at the following five outcomes: (i) smaller equipment size for a given throughput, (ii) higher throughput for a given equipment size or a given process, (iii) less holdup in equipment or less inventory in process for the same throughput, (iv) less usage of utility materials and feedstock for a given throughput, and (v) higher performance for a given unit size. This definition regarded PI as an extension of process integration activities. Based on this, they summarized the potential benefits of

PI activities to be realizing cheaper, safer, more energy efficient, and/or more environmentally friendly processes through innovation, and finally, valuing customers through just-in-time manufacturing.

Arizmendi-Sánchez and Sharratt [6] highlighted two design principles for PI: (i) synergistic integration of process tasks and coupling of phenomena, and (ii) targeted intensification of transport processes. Lutze et al. [9] extended these principles by "adding/enhancing phenomena in a process through the integration of operations, functions, phenomena or alternatively through the targeted enhancement of phenomena in an operation" to stress the importance of phenomena-based thinking in PI. Accordingly, an ongoing attempt in the Process Systems Engineering (PSE) community is to use phenomena-based synthesis tools as a means for automated generation of intensified options from a lower aggregation level without pre-postulation of plausible flowsheets, as the conventional unit operation-based synthesis strategies may hinder the exploration of out-of-the-box design alternatives due to pre-specified equipment configurations.

Van Gerven and Stankiewicz [8] suggested four principles for PI design:

- *Principle 1: Maximize the effectiveness of intra- and inter-molecular events* – improving process kinetics is a major principle for obtaining higher process performance as it is usually the underlying limiting factor for low conversion and selectivity.
- *Principle 2: Give each molecule the same processing experience which results in products with uniform properties* – uniform product distributions facilitate waste reduction, which in turn, reduce the efforts required for product separation.
- *Principle 3: Optimize the driving forces at every scale and maximize the specific surface area to which these forces apply* – thus more efficient processes can be obtained utilizing less enabling materials, which then leads to the reduction in equipment sizes.
- *Principle 4: Maximize the synergistic effects from partial processes which enable multitasking* – by combining several processing tasks together, higher process efficiencies can be achieved compared to their standalone counterparts.

As per Van Gerven and Stankiewicz [8], a completely intensified process should succeed in realizing all the above four principles via the use of one or more PI fundamental approaches, which were clustered in four domains: structure (spatial), energy (thermodynamic), functional (synergy), and temporal (time).

Based on these principles, we have summarized at least seven activities that can result in intensified processes [11]:

1. Combination of multiple process tasks or equipment into a single unit (e.g., reactive distillation, membrane reactors, sorption-enhanced reaction processes)
2. Tight process integration (i.e., material and/or energy integration)
3. Use of novel (multi)functional materials (e.g., ionic liquids, zeolites)
4. Miniaturization of process equipment (e.g., microreactors)
5. Application of enhanced driving forces (e.g., rotating reactors, ultrasonic mixing)
6. Periodic operation (e.g., pressure swing adsorption, simulated moving bed reactors)
7. Advanced operational strategies (e.g., dynamic modes via model-based control).

Over the past two decades, the scientific community has seen a rising research interest in process intensification. An increasing emphasis on the modular characteristics has also been witnessed in the recent years due to the rapid development of advanced manufacturing capabilities. A comprehensive review on the books, perspective/review articles, and technical developments can be found in Tian et al. [11]. Fig. 1.1 illustrates the growth profile based on the number of publications on the modeling and simulation of seven representative modular process intensification technologies, namely membrane reactor, simulated moving bed, dividing wall column, rotating packed bed, membrane distillation, microreactor, and reactive distillation. The statistics were collected from a recent search (updated on April 27th, 2021) of the citation database Web of Science Core Collection. From Fig. 1.1, it is clear that the number of articles published each year has experienced a steady increase, with a notable growth particularly during the past four years.

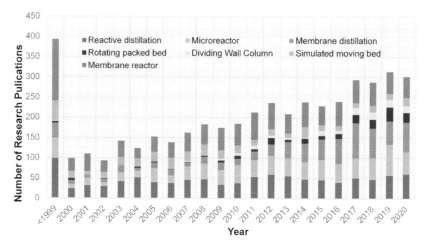

FIGURE 1.1 Statistics of modeling and simulation research articles on representative PI technologies. (Adapted from Tian et al. [11].)

In the meantime, chemical industry has also shown increasing interest in developing and deploying modular PI technologies. The 2007 European Road Mapping Analysis on process intensification [12] provided a list of 72 PI technologies with an evaluation of their potential benefits and implementation barriers. Some of the most commercialized PI technologies are: static mixers (>30 commercial models) [13], reactive distillation (>150 installations) [14], dividing wall columns (>125 installations) [15], and reverse flow reactors (>100 installations) [16]. Based on a statistics search using Google Patents (updated on March 3rd, 2018), an analysis for the number of issued patents on the above-selected seven PI technologies is shown in Fig. 1.2. The number of patents corresponds to the US, EP, and WO patents granted each year. According to Fig. 1.2, it is clear that the appeal on reactive distillation and dividing wall columns still continues and they dominate the field of PI. However, membrane-based processes and microreactors have also started to gain attention in the recent years which signifies their potential in shaping the future of (bio)chemical industry.

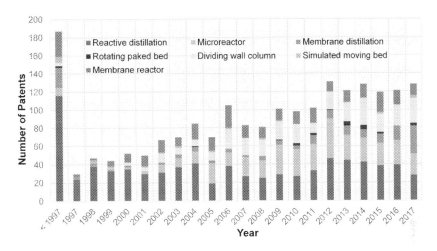

FIGURE 1.2 Statistics of patents on representative PI technologies (adapted from Tian et al. [11]).

1.3 Modular process intensification technology showcases

Reaction engineering, the heart of the many chemical production processes, can drive radical revolutions in chemical and energy industry by employing advanced reactor designs and/or cross-cutting combination with separation techniques to improve volumetric or energetic efficiency. Process intensification can be achieved at multi-scales in intensified reaction systems, namely catalyst level (e.g., with bifunctional catalysts), reaction inter-phase (e.g., monolith reactors), intra-reactor level (e.g., reverse flow reactors, reactive separations), and inter-reactor level (e.g., circulating fluid-bed reactors).

On the other hand, chemical separation operations also cast grand challenges towards a more sustainable and competitive process industry due to their high share in total energy demand. Energy used for separations constitute about half of U.S. industrial energy use, 10–15% of the nation's total energy consumption, and nearly 60% of all the energy needed by the chemical industry. Especially, distillation, a practical and widely utilized separation technology, accounts for approximately 50% of the total energy used in separation operations due to its relatively high energy intensity [17]. A diversity of routes have been explored based on modular process intensification principles to relieve the continuously growing stress on energy consumption in these processes, such as hybrid non-reactive separation (e.g., dividing wall column, membrane distillation), combined reaction/separation (e.g., reactive distillation, simulated moving bed chromatography, reactive extraction), and those assisted with external fields (e.g., HiGee, ultrasound, microwave).

8 Synthesis and Operability Strategies for Computer-Aided Modular PI

In this section, a brief overview is given on several representative modular PI technologies (dividing wall column, reactive distillation, membrane reactor, microreactor, etc.). More details on each topic can be found in the corresponding references and the literature database provided in Tian et al. [11].

1.3.1 Thermally coupled distillation and dividing wall column

In pursuit of more energy efficient distillation in multi-component separations, a promising process intensification strategy is to employ thermal coupling, which utilizes the available heat via a direct contact of material flows between separate columns instead of using separate condensers or reboilers for each column. Taking ternary separation as an example, an indicative list of the distillation-based process alternatives are depicted in Fig. 1.3, including simple column sequence (Fig. 1.3a), and partially or fully thermally coupled distillation (Fig. 1.3b-i). The fully thermally coupled distillation columns (or, Petlyuk columns, Fig. 1.3c) are known to require the least energy input, achieving approximately 30% energy savings comparing to a conventional arrangement for a given separation [18]. Most of the energy savings in partially or fully thermally coupled schemes are contributed by the prefractionator column, which is attached to the main column. The required heating or cooling for the prefractionator is satisfied through thermal coupling with the main column. It can significantly reduce the re-mixing effect occurring in the column by enabling a non-sharp split, and that in the feed tray by allowing a larger freedom to allocate the middle key component to match the column feed composition with that on one of the trays in the prefractionator. While such concepts of using one heat flux for multiple separations were introduced during the 1940s [19], the industrial use of Petlyuk column configurations did not take place until the mid-1980s [20] and the first application was based on a more compact configuration, known as dividing wall column (DWC, Fig. 1.3i).

DWCs utilize a partitioning wall in a single shell column to separate mixtures of three or more components into high-purity products. When negligible heat is transferred through the dividing wall, they are thermodynamically equivalent to Petlyuk columns while ensuring more compact design and offering reduced investment costs. DWC is one of the best examples of industry-proven process intensification technologies in distillation. More than 125 industrial applications have been established up to 2010, more than 116 of which are for ternary mixture separations [15]. The first patent which established the concept of DWC was published by Richard O. Wright from Standard Oil Development Co. in 1949 [21]. The first reported industrial application came out in 1985 as a packed DWC built by BASF [22]. Since then, the commercialization of DWCs has been catalyzed by the invention of non-welded walls by researchers in BASF SE and Montz [23], which allowed a higher flexibility in the installation process via retrofitting to avoid the considerable downtime and expenditure incurred by the use of fixed walls. The first packed, single-wall, and 4-product DWC was taken into operation in 2002 in a BASF SE plant [20].

Chapter 1 • Introduction to modular process intensification 9

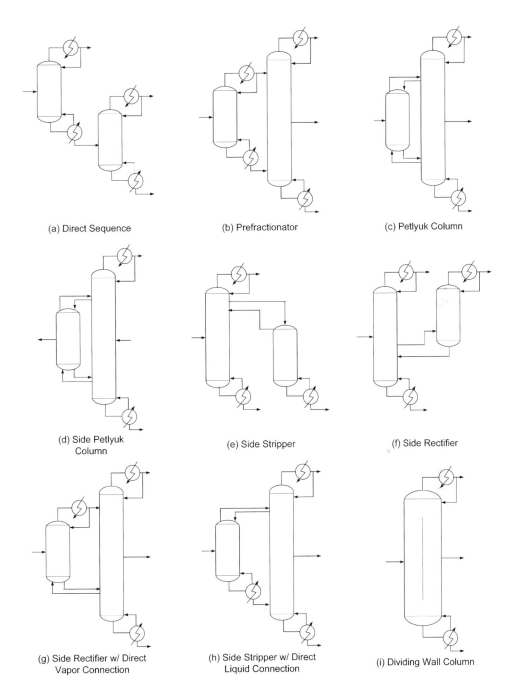

FIGURE 1.3 Distillation sequences for ternary separation.

1.3.2 Reactive distillation

A major proportion of chemical reactions are equilibrium limited which lead to low-grade products and low reactant conversions. Increasing the product purity requires higher reactor volumes, elevated operating temperatures, and/or incorporating downstream separation equipment (e.g., distillation), all resulting in increased operating and capital costs. Combining reaction and separation in a single unit is one of the major activities in PI to overcome the process bottleneck posed by such equilibrium limited reactions. The resulting reactive separation systems represent a significant class of modular PI technologies (e.g., reactive distillation, reactive adsorption, membrane reactors), most of which leverage the Le Chatelier's principle to remove reaction products in favor of further product formation.

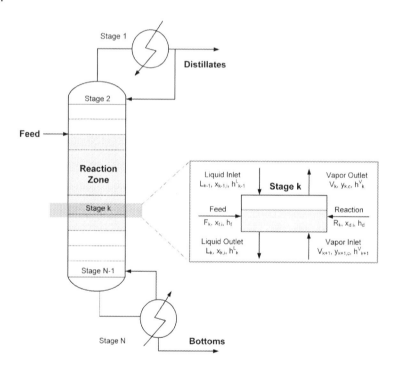

FIGURE 1.4 Reactive distillation.

Reactive distillation (RD) operations are achieved via the contact of vapor and liquid phases in a counter-current manner, while reaction takes place in the liquid phase and reaction products are simultaneously removed via vaporization (Fig. 1.4). Reactions for RD operations can be either heterogeneous/homogeneous catalysis reactions or non-catalysis reactions. In addition to the capital cost reduction due to the combination of distillation and reaction in a single unit, RD offers benefits in energy cost reduction for exothermic reactions as the reaction heat can be used to vaporize the liquid phase. Furthermore, if the catalysts are positioned above the feed location, catalyst poisoning from potential impurities in the feed can be avoided [24]. Besides the equilibrium limited reactions, RD can also be used to increase selectivity for the intermediate products in a system of consecutive reactions. For instance, in the production of ethylene glycol from ethylene oxide and water, the side reaction producing diethylene glycol can be avoided if the ethylene oxide is removed immediately after it is formed. There are many other reaction systems for which the use of RDs is reported to be advantageous, such as the production of methyl acetate, ethyl tert-butyl ether, tert-amyl methyl ether, etc. A comprehensive list of these reaction systems was summarized in Sundmacher and Kienle [25]. However, there remain some challenges – for instance, reaction temperature range needs to fall into that of vapor-liquid equilibrium of the reaction mixture which may result in a reduced process design and operating window. Additionally, catalysts should be stable enough to withstand long operation periods as the replacement of the bed is costly [24].

The first commercial RD application can be dated back to 1860s, which was used to recover ammonia in the Solvay process. The complete realization of the technology started in the early 1980s with the increased demand for methyl tert-butyl ether (MTBE). Later, the introduction and use of a task-integrated column for methyl acetate production in Eastman Chemical Company has increased the popularity of this technology [26,27]. In this task-integrated column, extractive and reactive distillation took place at separate regions of a single column which resulted in five times lower energy requirements and capital expenditure when compared with the classical process in which reaction and separation tasks took place in isolated units. There have been more than 150 successful industrial applications of reactive distillation technology for selective hydrogenation of mixed hydrocarbons, selective desulfurization of mid catalytic naphtha, isomerization of n-olefins to iso-olefins, and etc. [14] Currently, there exist more than 1000 patents on reactive distillation which also indicate the acceptance and interest toward RDs in the chemical industry. An extensive patent review on reactive distillation can be found in Lutze et al. [28].

1.3.3 Membrane-assisted separation

Various membrane technologies have been developed to yield better process performances in terms of product quality, environmental impact, and energy use [29]: (i) molecular separations (e.g., reverse osmosis), (ii) chemical transformations (e.g., membrane reactors, catalytic membranes), and (iii) enhanced mass and energy transfer between different phases by using hybrid separation techniques in a single unit (e.g., membrane distillation, membrane crystallizer). In what follows, we briefly discuss two membrane-assisted separation technologies. For more information in this area, Buonomenna [30] provided a comprehensive review on a number of membrane technologies with applications to water treatment, gas separation, (petro)chemical purification and recovery, regenerative medicine, etc.

Reverse osmosis
For water desalination, membrane-based technologies (e.g., reverse osmosis, nanofiltration, electrodialysis) are reported to be 10 times more energetically efficient than thermal

the original batch or semi-batch production. 44% of the analyzed reactions were benefited from continuous operation with microreactors.

However, the incapability of microreactors in handling solid components, as well as their higher CAPEX than conventional batch plants, hinders the practical implementation of this technology. Despite their advantages in numbering up instead of scaling up, the capital cost for microreactors scales linearly with the capacity when compared to a power of 0.65 in conventional reactors, which poses challenges over their cost-competitiveness for large-scale applications [16].

1.3.5 Periodic operation

From the PI view point, the periodically operated systems offer the advantage to integrate multiple tasks into a single unit at different time scales. A number of periodic systems have been developed and employed in chemical process industry. For example, cyclic distillation can be retrofitted from traditional distillation columns by changing the internals and operation mode but to offer higher tray efficiencies and lower energy consumption as already been demonstrated in industrial applications [50]. Other technologies, such as simulated moving bed and pressure swing adsorption, also provide the benefits of standardization and parallelization by using modular process equipment [51]. However, operational challenges result from: (i) the lack of a true steady state but only attaining a pseudo (or cyclic) steady state in these systems, (ii) the design and coordination of switching strategies, and (iii) the requirement of simultaneous design, operability, and control due to the intrinsic interlink of these aspects in periodic systems. In this section, we provide an introductory overview on several representative periodic reaction/separation systems.

Pressure swing adsorption

Pressure swing adsorption (PSA, Fig. 1.6a) provides an efficient method for gas separation or purification based on the differences in species' affinity towards an adsorbent material. A PSA process normally consists of four major steps repeated in each operating cycle (i.e., pressurization, high-pressure adsorption, depressurization, and low pressure desorption). This technology is characterized with high technology availability (>99%), flexible operation (e.g., typical operation range can be varied from 25% to 100%), high recovery rates (up to 90%), and fully automated operation under uncertainty [52]. PSA is a well-established industrial technology with its first commercial inception in the 1960s (i.e., Skarstrom cycle and Guerin-Domine cycle) and has undergone rapid development since then (Yang, 1988). The early commercial developments before 1985 can be found in Yang [53]. Over the years, PSA has been used in a number of areas, such as carbon dioxide sequestration [54], hydrogen purification [55], methane purification [56], air separation [57], etc. Ongoing efforts are being made to address the selection or synthesis of appropriate adsorbents, the optimization of design and operating variables, and the control issues posed by the absence of a steady-state operating point and the highly nonlinear process dynamics.

Chapter 1 • Introduction to modular process intensification 15

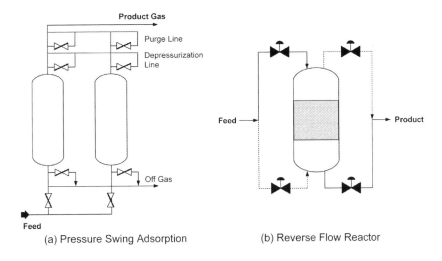

FIGURE 1.6 Periodic systems.

Sorption enhanced reaction processes
Another method of separation that can be combined with reaction is the adsorption. Chromatographic separators can be combined with reaction by inserting catalytic functionality to the adsorbents to yield chromatographic reactors. Chromatographic reactors are particularly useful and constitute an alternative to reactive distillation when the chemicals to be treated are sensitive to temperature and/or non-volatile, which are usually the cases for some fine chemical or pharmaceutical applications [58]. On the contrary of usual batch operation of chromatographic separators, periodic operation as used in simulated moving bed reactors or pressure swing adsorptive reactor (PSAR) yields intensified processes, which are generally termed as Sorption Enhanced Reaction Processes (SERP).

Taking the PSAR technology as an example, it combines the PSA process with reaction in a single unit while the overall operation is facilitated in a cyclic manner. Regeneration of the adsorbent might be realized via temperature swing as well and a second column can be utilized. SERP processes offer significant reduction in operating temperatures and, still, yields high conversions. Major challenge for combining adsorption and reaction, however, is the development of multi-functional materials which have desired selectivity for separation, high activity for reaction at the same temperature, and also stability to withstand the periodic nature of the process. This idea was first presented by Vaporciyan and Kadlec [59] and patented by Sircar et al. [60] from Air Products and Chemicals for methane steam reforming to produce hydrogen. After that, SERP systems have been investigated for oxidation dehydrogenation of ethane to ethylene, dehydrogenation of methylcyclohexane to toluene, propene metathesis, ethanol steam reforming, water-gas shift reaction, etc. [61].

Reverse flow reactors

An early review on the benefits of dynamic operation in reactors via periodic changes in feed concentration, flow rate, flow direction, and/or temperature was provided by Stankiewicz and Kuczynski [62]. Reverse flow reactor (RFR, Fig. 1.6b) utilizes the flow direction reversal in a periodic manner. The feed stream is first fed from one side while switched to the other side after some reaction time, where now the inlet becomes the outlet and vice versa. This procedure is periodically repeated and provides benefits in terms of suppressing the hot spot formation and, hence, increasing catalyst stability, and also reducing the utility required for feed preheating and/or product cooling. Furthermore, the cycles can be designed as either symmetric (same reaction in both directions) or asymmetric (reaction changes with direction). While symmetric cycles are used in mildly exothermic reactions as in autothermal reforming, asymmetric cycles can be used in execution of sequential endothermic and exothermic reactions in different cycles.

RFR is an industrially accepted technology, which was reported to be utilized in Russia before 1995 [62]. In the review presented by Harmsen [16], RFRs were reported to be used in Shell for catalytic combustion of waste gas streams from ethylene oxide plants. It was speculated therein that there were more than 100 industrial applications. Several recent patent applications exist for RFRs. For instance, Sankaranarayanan et al. [63] proposed a patent to control the combustion, thus to improve the thermal efficiency of the bed used in reverse flow. Hershkowitz et al. [64] disclosed a reactor design for RFRs with minimal dead volume between the valve and the reactor.

References

[1] A.I. Stankiewicz, J.A. Moulijn, Process intensification: transforming chemical engineering, Chemical Engineering Progress 96 (2000) 22–34.

[2] F.J. Keil, Process intensification, Reviews in Chemical Engineering 34 (2018) 135–200.

[3] M. Baldea, T.F. Edgar, B.L. Stanley, A.A. Kiss, Modularization in chemical processing, Chemical Engineering Progress 114 (2018) 46–54.

[4] J. Bielenberg, I. Palou-Rivera, The RAPID Manufacturing Institute – reenergizing US efforts in process intensification and modular chemical processing, Chemical Engineering and Processing: Process Intensification (2019).

[5] C. Ramshaw, HIGEE distillation – an example of process intensification, Chemical Engineering (1983) 13–14.

[6] J.A. Arizmendi-Sánchez, P. Sharratt, Phenomena-based modularisation of chemical process models to approach intensive options, Chemical Engineering Journal 135 (2008) 83–94.

[7] S. Becht, R. Franke, A. Geißelmann, H. Hahn, An industrial view of process intensification, Chemical Engineering and Processing: Process Intensification 48 (2009) 329–332.

[8] T. Van Gerven, A. Stankiewicz, Structure, energy, synergy, time – the fundamentals of process intensification, Industrial & Engineering Chemistry Research 48 (2009) 2465–2474.

[9] P. Lutze, R. Gani, J.M. Woodley, Process intensification: a perspective on process synthesis, Chemical Engineering and Processing: Process Intensification 49 (2010) 547–558.

[10] J.M. Ponce-Ortega, M.M. Al-Thubaiti, M.M. El-Halwagi, Process intensification: new understanding and systematic approach, Chemical Engineering and Processing: Process Intensification 53 (2012) 63–75.

[11] Y. Tian, S.E. Demirel, M.F. Hasan, E.N. Pistikopoulos, An overview of process systems engineering approaches for process intensification: state of the art, Chemical Engineering and Processing: Process Intensification (2018).

Chapter 1 • Introduction to modular process intensification   17

[12] Creative Energy, European roadmap for process intensification, http://efce.info/efce_media/-p-531-EGOTEC-phbgr90gkpjgr8lqp8nesd0ic1.pdf?rewrite_engine=id, 2017. (Accessed 10 May 2021).

[13] G. Stephanopoulos, Synthesis of process flowsheets: an adventure in heuristic design or a utopia of mathematical programming?, Foundations of Computer-Aided Chemical Process Design 2 (1981) 439.

[14] G.J. Harmsen, Reactive distillation: the front-runner of industrial process intensification: a full review of commercial applications, research, scale-up, design and operation, Chemical Engineering and Processing: Process Intensification 46 (2007) 774–780.

[15] Ö. Yildirim, A.A. Kiss, E.Y. Kenig, Dividing wall columns in chemical process industry: a review on current activities, Separation and Purification Technology 80 (2011) 403–417.

[16] J. Harmsen, Process intensification in the petrochemicals industry: drivers and hurdles for commercial implementation, Chemical Engineering and Processing: Process Intensification 49 (2010) 70–73.

[17] D.S. Sholl, R.P. Lively, Seven chemical separations to change the world, Nature News 532 (2016) 435.

[18] F.B. Petlyuk, Thermodynamically optimal method for separating multicomponent mixtures, International Chemical Engineering 5 (1965) 555–561.

[19] B.A. Johan, Process and device for fractional distillation of liquid mixtures, more particularly petrolcum, US Patent 2,295,256, 1942.

[20] I. Dejanović, L. Matijašević, Ž. Olujić, Dividing wall column – a breakthrough towards sustainable distilling, Chemical Engineering and Processing: Process Intensification 49 (2010) 559–580.

[21] R.O. Wright, Fractionation apparatus, US Patent 2,471,134, 1949.

[22] G. Parkinson, Dividing-wall columns find greater appeal, Chemical Engineering Progress 103 (2007) 8–11.

[23] G. Kaibel, M. Stroezel, U. Rheude, Dividing wall column for continuous fractionation of multicomponent mixtures by distillation, US Patent 5,914,012, 1999.

[24] A. Tuchlenski, A. Beckmann, D. Reusch, R. Düssel, U. Weidlich, R. Janowsky, Reactive distillation – industrial applications, process design & scale-up, Chemical Engineering Science 56 (2001) 387–394.

[25] K. Sundmacher, A. Kienle, Reactive Distillation: Status and Future Directions, John Wiley & Sons, 2006.

[26] V.H. Agreda, L.R. Partin, Reactive distillation process for the production of methyl acetate, US Patent 4,435,595, 1984.

[27] V. Agreda, W. Heise, High-purity methyl acetate via reactive distillation, Chemical Engineering Progress 86 (1990) 40–46.

[28] P. Lutze, E.A. Dada, R. Gani, J.M. Woodley, Heterogeneous catalytic distillation – a patent review, Recent Patents on Chemical Engineering 3 (2010) 208–229.

[29] E. Drioli, A.I. Stankiewicz, F. Macedonio, Membrane engineering in process intensification – an overview, Journal of Membrane Science 380 (2011) 1–8.

[30] M.G. Buonomenna, Membrane processes for a sustainable industrial growth, RSC Advances 3 (2013) 5694–5740.

[31] P. Bernardo, E. Drioli, G. Golemme, Membrane gas separation: a review/state of the art, Industrial & Engineering Chemistry Research 48 (2009) 4638–4663.

[32] L. Sidney, S. Srinivasa, High flow porous membranes for separating water from saline solutions, US Patent 3,133,132, 1964.

[33] C. Fritzmann, J. Löwenberg, T. Wintgens, T. Melin, State-of-the-art of reverse osmosis desalination, Desalination 216 (2007) 1–76.

[34] P. Lutze, A. Gorak, Reactive and membrane-assisted distillation: recent developments and perspective, Chemical Engineering Research and Design 91 (2013) 1978–1997.

[35] F. Lipnizki, R.W. Field, P.-K. Ten, Pervaporation-based hybrid process: a review of process design, applications and economics, Journal of Membrane Science 153 (1999) 183–210.

[36] Y.K. Ong, G.M. Shi, N.L. Le, Y.P. Tang, J. Zuo, S.P. Nunes, T.-S. Chung, Recent membrane development for pervaporation processes, Progress in Polymer Science 57 (2016) 1–31.

[37] J. Holtbrügge, A.K. Kunze, A. Niesbach, P. Schmidt, R. Schulz, D. Sudhoff, M. Skiborowski, Reactive and Membrane-Assisted Separations, Walter de Gruyter GmbH & Co KG, 2016.

[38] M.V. Kothare, Dynamics and control of integrated microchemical systems with application to microscale fuel processing, Computers & Chemical Engineering 30 (2006) 1725–1734.

[39] J.-C. Charpentier, In the frame of globalization and sustainability, process intensification, a path to the future of chemical and process engineering (molecules into money), Chemical Engineering Journal 134 (2007) 84–92.

[40] K.F. Jensen, Microreaction engineering – is small better?, Chemical Engineering Science 56 (2001) 293–303.

[41] S.J. Haswell, R.J. Middleton, B. O'Sullivan, V. Skelton, P. Watts, P. Styring, The application of micro reactors to synthetic chemistry, Chemical Communications (2001) 391–398.

[42] D.M. Roberge, L. Ducry, N. Bieler, P. Cretton, B. Zimmermann, Microreactor technology: a revolution for the fine chemical and pharmaceutical industries?, Chemical Engineering & Technology: Industrial Chemistry-Plant Equipment-Process Engineering-Biotechnology 28 (2005) 318–323.

[43] M. Doble, Green reactors, Chemical Engineering Progress 104 (2008) 33–42.

[44] M.S. Mettler, G.D. Stefanidis, D.G. Vlachos, Scale-out of microreactor stacks for portable and distributed processing: coupling of exothermic and endothermic processes for syngas production, Industrial & Engineering Chemistry Research 49 (2010) 10942–10955.

[45] A.S. Patil, T.G. Dubois, N. Sifer, E. Bostic, K. Gardner, M. Quah, C. Bolton, Portable fuel cell systems for America's army: technology transition to the field, Journal of Power Sources 136 (2004) 220–225.

[46] M. Winter, R.J. Brodd, What are batteries, fuel cells, and supercapacitors?, Chemical Reviews 104 (2004) 4245–4270.

[47] M.W. Ellis, M.R. Von Spakovsky, D.J. Nelson, Fuel cell systems: efficient, flexible energy conversion for the 21st century, Proceedings of the IEEE 89 (2001) 1808–1818.

[48] A. Kirubakaran, S. Jain, R. Nema, A review on fuel cell technologies and power electronic interface, Renewable & Sustainable Energy Reviews 13 (2009) 2430–2440.

[49] J.D. Holladay, Y. Wang, E. Jones, Review of developments in portable hydrogen production using microreactor technology, Chemical Reviews 104 (2004) 4767–4790.

[50] C.S. Bîldea, C. Pătruţ, S.B. Jørgensen, J. Abildskov, A.A. Kiss, Cyclic distillation technology – a mini-review, Journal of Chemical Technology and Biotechnology 91 (2016) 1215–1223.

[51] M. Baldea, T.F. Edgar, Dynamic process intensification, Current Opinion in Chemical Engineering 22 (2018) 48–53.

[52] C. Voss, Applications of pressure swing adsorption technology, Adsorption 11 (2005) 527–529.

[53] R.T. Yang, Gas Separation by Adsorption Processes, Vol. 1, World Scientific, 1997.

[54] D. Ko, R. Siriwardane, L.T. Biegler, Optimization of pressure swing adsorption and fractionated vacuum pressure swing adsorption processes for CO_2 capture, Industrial & Engineering Chemistry Research 44 (2005) 8084–8094.

[55] H. Khajuria, E.N. Pistikopoulos, Optimization and control of pressure swing adsorption processes under uncertainty, AIChE Journal 59 (2013) 120–131.

[56] A. Olajossy, A. Gawdzik, Z. Budner, J. Dula, Methane separation from coal mine methane gas by vacuum pressure swing adsorption, Chemical Engineering Research and Design 81 (2003) 474–482.

[57] M. Hassan, D. Ruthven, N. Raghavan, Air separation by pressure swing adsorption on a carbon molecular sieve, Chemical Engineering Science 41 (1986) 1333–1343.

[58] F. Lode, M. Houmard, C. Migliorini, M. Mazzotti, M. Morbidelli, Continuous reactive chromatography, Chemical Engineering Science 56 (2001) 269–291.

[59] G.G. Vaporciyan, R.H. Kadlec, Equilibrium-limited periodic separating reactors, AIChE Journal 33 (1987) 1334–1343.

[60] S. Sircar, J.R. Hufton, S. Nataraj, Process and apparatus for the production of hydrogen by steam reforming of hydrocarbon, US Patent 6,103,143, 2000.

[61] A.E. Rodrigues, C.S. Pereira, J.C. Santos, Chromatographic reactors, Chemical Engineering & Technology 35 (2012) 1171–1183.

[62] A. Stankiewicz, M. Kuczynski, An industrial view on the dynamic operation of chemical converters, Chemical Engineering and Processing: Process Intensification 34 (1995) 367–377.

[63] K. Sankaranarayanan, F. Hershkowitz, J.W. Frederick, R. Agnihotri, Controlled combustion for regenerative reactors with mixer/flow distributor, US Patent 7,815,873, 2010.

[64] F. Hershkowitz, R.J. Basile, J.W. Frederick, J.W. Fulton, P.F. Keusenkothen, B.A. Patel, A.R. Szafran, Reactor with reactor head and integrated valve, US Patent 8,524,159, 2013.

2

Computer-aided modular process intensification: design, synthesis, and operability

Despite the technology advances introduced in Chapter 1, modular process intensification has mostly been regarded as an Edisonian effort driven by breakthrough engineering thinking and experimentation discoveries, while lacking theoretical developments on the fundamentals. Therefore, efforts have been made from the Process Systems Engineering (PSE) perspective to provide more systematic approaches for the design and operation of modular and intensified process systems. A number of recent perspective and review papers are summarized in Table 2.1, which have highlighted the challenges and opportunities for model-based computer-aided synthesis, design, optimization, and operational analysis of modular PI systems.

Table 2.1 Recent review/perspective articles on computer-aided process intensification and modular design – an indicative list (adapted from Pistikopoulos et al. [95]).

Reference	Highlight	Year
Weinfeld et al. [1]	Experimentation, modeling, and dynamic control of reactive dividing wall columns	2018
Baldea et al. [2]	Status, challenges, and opportunities for modular manufacturing processes	2018
Tian et al. [3]	PSE approaches for PI design, synthesis, and operability analysis	2018
Baldea and Edgar [4]	Process intensification via dynamic/periodic operation	2018
Skiborowski [5]	PSE approaches for design and synthesis of intensified reaction and/or separation systems	2018
Daoutidis et al. [6]	Distributed optimization, control, and monitoring of large-scale PI systems	2019
Dias and Ierapetritou [7]	Optimal operation, control, and scheduling of PI processes	2019
Sitter et al. [8]	Process synthesis and optimization approaches for PI design and retrofit	2019
Masuku and Biegler [9]	Optimization-based tools for PI in gas-to-liquids processes	2019
Tian and Pistikopoulos [10]	Challenges and opportunities for synthesis of operable PI systems	2019
Demirel et al. [11]	Systematic approaches and procedures to generate and analyze PI solutions	2019
Aglave et al. [12]	Role of simulation and digitalization to assist modular process intensification	2019
Jiang and Agrawal [13]	Thermally coupled distillation processes for multi-component separation	2019
Gazzaneo et al. [14]	Model-based operability theory, applications, and tools for modular PI systems	2019
Tula et al. [15]	Perspectives on computer-aided process intensification	2020

20 Synthesis and Operability Strategies for Computer-Aided Modular PI

In this chapter, we introduce some recent advances for modular process intensification through PSE approaches, particularly addressing the following topics:

i. **Process intensification synthesis and design** – i.e. how to systematically represent, generate, and evaluate modular and intensified process configurations? How to explore the combinatorial design space in a computationally efficient way?
ii. **Operability and control in PI and modular design** – i.e. how to ensure the actual operational performance of modular and intensified systems by developing model-based control, operability, and safety analysis approaches?

2.1 Conceptual synthesis and design

Process synthesis strives for selecting the optimal equipment among many alternatives and also for determining the interconnections between them while providing optimum operating conditions [16–18]. Approaches for synthesizing intensified flowsheets can be clustered into three main categories: (i) knowledge-based methods, (ii) optimization-based methods, and (iii) hybrid methods.

2.1.1 Knowledge-based methods

A knowledge-based approach features a trial and error hierarchical procedure using heuristic rules and analyses to generate better and intensified process alternatives. Rigorous optimization methods or tools are not required here. Design decisions are made iteratively to overcome process bottlenecks and/or to reduce the number of unit operations. An intensified process solution will be obtained when PI objectives are met or no further equipment reductions are possible. However, this approach depends on available analysis methods and heuristic rules which may not result in the optimal intensified design solution.

Siirola [19] proposed a task-oriented approach by which a conceptual design was obtained via the identification of tasks and the means-ends analysis. This approach was applied to develop a reactive-extractive distillation process for methyl acetate production, which combined multiple tasks within a single equipment [20]. As illustrated in Fig. 2.1a, an existing process was essential to initiate the means-ends analysis. In this case, the conventional flowsheet consisted of a reactor to produce methyl acetate and water from methanol and acetic acid, followed by a train of distillation columns to separate the methyl acetate product from the azeotropic quaternary mixture.

The first improvement trials were made to introduce extractive distillation tasks after the distillation task (Fig. 2.1b). The top product from the distillation task gave a ternary mixture, comprising methyl acetate and methanol which were at saddle points. Based on the residue curve map, extractive distillation could help to separate methyl acetate towards the top of the column while to separate methanol towards the bottom. The bottom product was then sent to another extractive distillation task to concentrate acetic acid. However, it was further noticed that the separation of acetic acid from water (co-product) was costly.

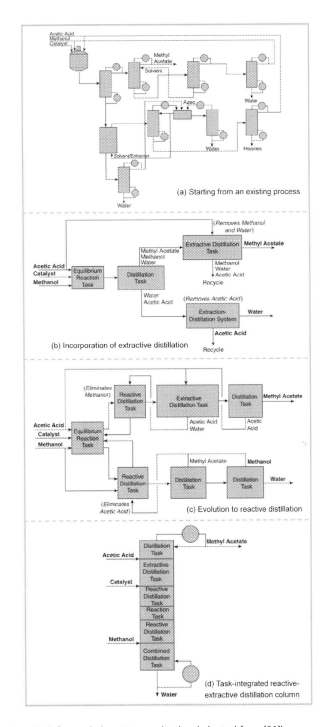

FIGURE 2.1 Means-ends analysis for methyl acetate production (adapted from [20]).

22   Synthesis and Operability Strategies for Computer-Aided Modular PI

To this purpose, a reactive distillation task was used right after the reactor instead of a distillation task. In this way, acetic acid could be completely consumed and the separation of acetic acid-water was no longer needed (Fig. 2.1c). Similarly, another reactive distillation task could be added to consume all the remaining methanol in the reactor effluent via reaction. As can be noted, the resulting process consisted of seven tasks, six of which utilized distillation-based operation and were closely interconnected via recycle streams. Thus, all the tasks could be integrated into a single column. In this way, only one reboiler and one condenser were required as shown in Fig. 2.1d. The new task-integrated and intensified column design could save around 80% capital cost and energy consumption comparing to the original conventional process.

Another example of knowledge-based methods towards process intensification is proposed by Commenge and Falk [21]. A 3-step sequential technique was proposed which featured: (i) Step 1 – to identify the process limitation among a list of 16 possibilities that covered a large spectrum of cases, (ii) Step 2 – the identified limitations were related to a set of intensification strategies, (iii) Step 3 – the identified strategies were related to the technologies that utilized these strategies. The proposed method supplied a short list of appropriate solutions to be considered in technical design and economic evaluations together with a list of innovation strategies.

2.1.2 Optimization-based methods

Optimization-based methods utilize mathematical programming for overall flowsheet synthesis. First, a superstructure containing all the plausible flowsheet alternatives is proposed. The postulated superstructure is then translated into an optimization problem with an objective function – which can be defined with respect to process economics, operability, safety, environmental impacts, etc. This generally leads to a Mixed Integer Nonlinear Programming (MINLP) type of mathematical model formulation, in which discrete decisions are used to decide between different processing routes. Optimization-based process synthesis methods are advantageous in terms of their comprehensive outlook over flowsheet synthesis. However, the superstructure representation utilized in constructing the optimization problem is crucial as it determines the extent of the solution space included in the problem formulation, the solution strategies, and the efficiency of the overall method in terms of its ability to be translated into computer algorithms.

There exist several superstructure representation methods developed for process synthesis. A review of the state-of-the-art superstructure-based process synthesis approaches is provided by Mencarelli et al. [22]. In what follows, we briefly introduce several representative approaches to illustrate the evolution of this concept, with particular emphasis on process intensification synthesis. Yeomans and Grossmann [23] proposed two different representations based on three chemical process components (i.e., states, tasks, and equipment): State-Task-Network (STN) and State-Equipment-Network (SEN) as depicted in Fig. 2.2. In these representations, *states* were regarded as physicochemical properties defining the streams, *tasks* were transformations between adjacent states (i.e., streams), and *equipment* were where tasks were carried out. In STN representation, task nodes

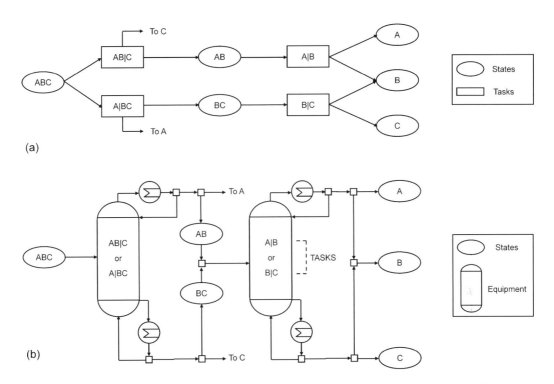

FIGURE 2.2 Superstructure representation for synthesis of ternary distillation sequence with sharp split: (a) State-Task-Network, (b) State-Equipment-Network.

were used to connect two different states while the equipment assignment was handled by the mathematical model. In SEN, on the other hand, states and equipment were defined while the assignment of tasks to equipment was determined via the model. Wu et al. [26] proposed the unit-port-conditioning stream (UPCS) framework for superstructure representation and developed connectivity rules to generate fully connected minimal superstructures with all feasible routes (Fig. 2.3a). Friedler et al. [24] proposed Process graph (P-graph) as a representation method which used material and operation nodes connected with arcs. Based on P-graph, a polynomial algorithm was devised for generating the maximal superstructure containing all the possible structures for a given process synthesis problem. All of the feasible solutions embedded in the maximal superstructure could also be identified via the P-graph method. Farkas and Lelkes [25] proposed R-graph in which nodes constituted the input and output ports of the possible units. While the output port nodes were regarded as stream splitters, input nodes were used as stream mixers and connections from output to input nodes represented the process streams (Fig. 2.3b). The authors also investigated the structural multiplicity and redundancy in process network superstructures.

Design and synthesis of a certain class of intensified equipment are made possible by using the above superstructure-based optimization approaches. As an illustrative exam-

24 Synthesis and Operability Strategies for Computer-Aided Modular PI

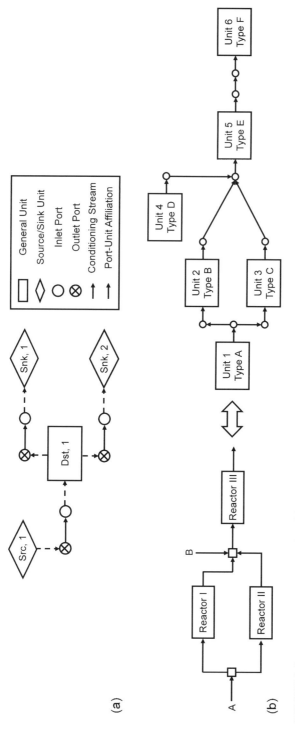

FIGURE 2.3 Superstructure representation: (a) Distillation column representation using UPCS network, (b) Reaction network representation using R graph. (Reproduced from [26] and [25])

ple, we provide an overview on the synthesis of reactive distillation systems in Appendix A. A more extensive list of synthesis studies for modular PI can be found in Table 19 of Tian et al. [3]. Yet, a major challenge towards a generic optimization-based process intensification method is how to propose a general superstructure representation method that can enable the identification of novel intensified processes without pre-specifying process alternatives. The pre-specification of plausible unit operations or process configurations, required by conventional process synthesis representation approaches, may restrict the design search space to be the known equipment types while hindering the consideration of novel and intensified process alternatives at the early design stage and leaves the creativity to the expertise of the design team involved in the decision making.

In this regard, many promising methods for systematic process intensification have been developed at the phenomena level utilizing optimization-based methods. These works start from a lower aggregation level than the equipment-based approaches and regard chemical processes as the combination of several phenomena such as mixing, splitting, phase contact, phase change, reaction, heating, cooling, etc. As design activities become less focused on conventional unit operations, but go deeper into task and phenomena levels, thinking becomes more fundamental and more intensification opportunities can be revealed.

Several methods have emerged for modular chemical process intensification to define and represent the combinatorial search space with phenomenological representations. Papalexandri and Pistikopoulos [27] developed a Generalized Modular Representation Framework (GMF) for the modular process intensification synthesis of chemical processes. Built on aggregated multifunctional mass/heat exchange modules and pure heat exchange modules (Fig. 2.4), GMF disclosed intensification possibilities by optimizing mass and heat transfer performances. Conventional or even unconventional process flowsheets were generated via the solution of an optimization-based superstructure problem with MINLP formulation. Herein, the synthesis alternatives were not pre-postulated as process units, but explored as mass/heat exchange feasibilities determined by Gibbs free energy-based driving force constraints [28,29]. The proposed synthesis framework has been demonstrated for combined separation/reaction systems, material selection, thermally coupled distillation systems, etc. In Chapters 3–5 of this book, we will discuss in details the synthesis representation and mathematical formulation of GMF, which lays out the foundation towards a systematic framework for the synthesis of operable PI systems to be introduced in Chapter 9. A book by Georgiadis and Pistikopoulos [30] is also available on GMF applications in energy and heat integration systems.

Manousiouthakis and co-workers proposed the Infinite DimEnsionAl State Space (IDEAS) approach [31] as a representation and synthesis strategy for process intensification. IDEAS approach was built on a state-space representation of processing routes where a set of process operators (OP), i.e. unit operations, were connected to each other with a distribution network (DN), i.e. flow operations, to consider all inlet-outlet possibilities (please see Fig. 2.5 as an example on reactive flash separator network for reactive distillation synthesis). The IDEAS approach has been applied to the synthesis of heat/power integrated distillation processes, azeotropic distillation processes, nonideal reactor network synthesis, reactive distillation processes, and energetic intensification of a hydrogen production process [32].

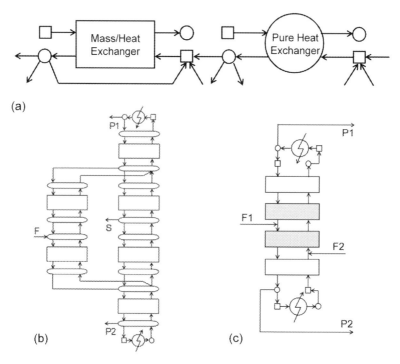

FIGURE 2.4 Generalized Modular Representation Framework: (a) Modular building blocks, (b) A Petlyuk column representation, (c) A reactive distillation representation.

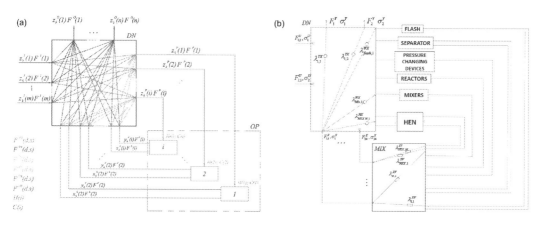

FIGURE 2.5 Infinite DimEnsionAl State Space representation: (a) A reactive flash separator network (adapted from [33]), (b) A chemical process flowsheet (adapted from [32]).

Chapter 2 • Computer-aided modular PI: design, synthesis, and operability   27

Hasan and co-workers [34] proposed a building block superstructure for systematic representation and synthesis of intensified chemical processes. Building blocks were defined as abstract modules that were used to represent different chemical phenomena, tasks, and materials. These building blocks were positioned on a 2-D grid superstructure where each block was connected to each other via inter-block streams (Fig. 2.6). In order to expedite the search for intensified flowsheets, the proposed superstructure was modeled as a MINLP problem containing thermodynamics and material/energy balance constraints. The presented method has been shown to automatically generate intensified flowsheets, with simultaneously mass and/or heat integration considerations, using membrane reactors, reactive absorption, and reactive distillation units [35]. A computer-aided framework, SPICE (Systematic Process Intensification of Chemical Enterprises), was developed which (i) contained an interface for collecting the model input data on feeds, products, and materials, (ii) generated optimal block configurations by solving the MINLP model to optimality using a commercial solver without requiring any superstructure *a priori*, and (iii) converted the block configurations into classical process flow diagrams [36].

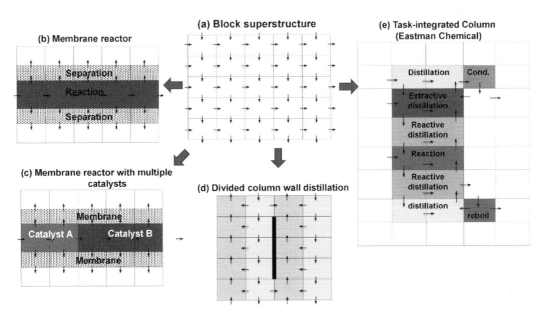

FIGURE 2.6 Building block-based representation of chemical processes and new process intensification superstructure (adapted from [34]).

2.1.3 Hybrid methods

A hybrid approach decomposes the complex mathematical problem employed in optimization-based approaches to smaller subproblems as per a knowledge-based hierarchical order. A hybrid approach was proposed by Gani and co-workers [37] for sustainable process synthesis-intensification using the phenomena building blocks (PBBs). The proposed method comprises three stages. It started from the synthesis stage (Stage-1), to find optimal processing route based on well-known technologies to convert raw materials to desired products. Stage-2 was for the design of selected processing route from Stage-1. This stage aimed to identify process bottlenecks and to improve by performing detailed process simulation and analyses. Stage-3 was for innovation, to find more sustainable and intensified process alternatives using the phenomena building block approach (Fig. 2.7). Nine major classes of PBBs were selected to represent the basic structures for process alternatives: Reaction (R), Mixing (M), 2-phase mixing (2phM), phase contact (PC), phase transition (PT), phase separation (PS), heating (H), cooling (C), and dividing (D). This methodology has been demonstrated in a number of intensified systems as summarized in Tula et al. [38], e.g. isopropyl acetate production using membrane-based processes with sustainability considerations [39].

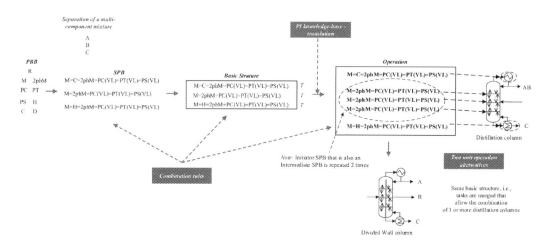

FIGURE 2.7 Phenomena Building Block representation of process flowsheet (adapted from [38]).

2.1.4 Software tools for process synthesis

Since the definition of the process synthesis problem, there have been extensive efforts on the development of computer-aided software tools to enable automated process synthesis. These attempts have led to a number of academic-based software platforms for process integration which were extensively reviewed by Lam et al. [40]. AIDES [41] was the first software platform for flowsheet synthesis which relied on heuristics for flowsheet generation. MIPSYN [42] was one of the pioneers to introduce synthesis capabilities into an algebraic modeling environment with custom modeling – note that so far there is no available commercial software supporting process synthesis activities. P-Graph Studio (http://www.p-graph.com), another successful software package for process flowsheet synthesis, was based on the P-graph representation of process networks [24]. These software tools,

Chapter 2 • Computer-aided modular PI: design, synthesis, and operability   29

however, mainly aim for process flowsheet synthesis based on pre-specified equipment types, and, hence, their applicability is limited in terms of modular process intensification.

More recently, attempts have been initiated from academia to develop software prototypes for computer-aided modular process intensification. ProCAFD [38], developed by Gani and co-workers as part of the ICAS tool set, supported flowsheet synthesis and design using process-groups to represent unit operations. It follows a hierarchical procedure to generate optimal process alternatives with considerations of process safety, sustainability, etc. More recently, the authors have published another toolkit, ProCACD [43], which enabled controller design and simultaneous design-control. As part of the Department of Energy IDAES project [44], Pyosyn presented a newer synthesis framework using Generalized Disjunctive Programming model based in Python/Pyomo algebraic modeling platform [45]. Another ongoing software prototype development, as part of the RAPID SYNOPSIS project [46], aimed to fully integrate the synthesis, design, operation, and control of intensified process configurations. This will be the topic of Chapter 15 in this book.

2.2 Operability, safety, and control analysis

Practical implementation of any process cannot be possible if it is not operable and safe. Compared with conventional unit operations, intensified processes typically feature higher degree of integration, less degrees of freedom, narrower operating windows, and faster dynamics. In this context, the consideration of control, operability, and safety aspects at conceptual design stage becomes even more critical for intensified process systems under process disturbances and parameter uncertainty. In this section, we discuss the PSE approaches to assess/improve the operability and control of chemical processes, in particular highlighting the ones that have been applied to intensified processes and modular systems.

2.2.1 Process flexibility analysis

Flexibility/Feasibility refers to the capability of a process to satisfy all relevant constraints under varying operating conditions, e.g. variations in product demand, quality targets, material properties, etc. Flexibility analysis features sophisticated mathematical formulations to characterize the feasible region of operation as depicted in Fig. 2.8. The flexibility analysis strategies (e.g., flexibility test problem, flexibility index problem) pioneered by Grossmann and co-workers have been proved to be a powerful tool to determine the maximum parameter range that can be achieved during steady-state operation [47–51]. In Chapter 6, we will give an introductory review on the mathematical basis of flexibility analysis approaches, the solution algorithms, and the applications on analyzing and/or synthesizing process networks with motivating examples. Ostrovsky et al. [52] further suggested a bi-level optimization approach using the branch-and-bound method to determine the upper and lower bounds for the afore-introduced feasibility test approach. For the quantification of process feasibility on convex regions, a detailed review can be found in Ierapetritou et al. [53]. To deal with nonconvex regions with concave or quasi-convex

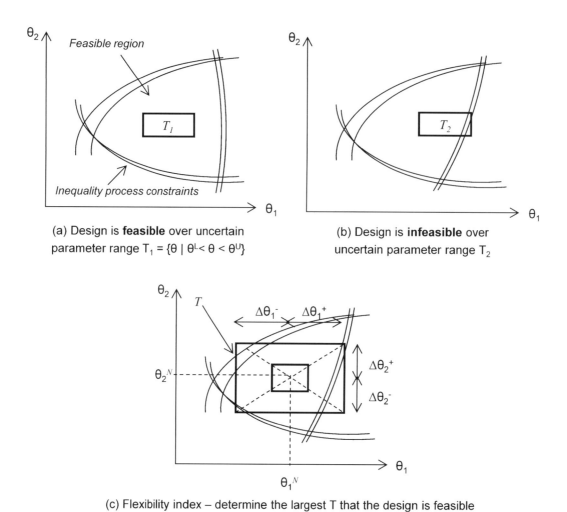

FIGURE 2.8 Illustration of flexibility analysis concepts.
Note – θ denotes uncertain parameters, T defines the range of uncertain parameters.

constraints, Goyal and Ierapetritou [54] identified the operating envelope by systematically evaluating the infeasibility areas with outer approximation, together with simplicial approximation for expanded feasible space constructed with the exclusion of nonconvex constraints. A later work from this group [55] proposed a more generalized method applicable to convex, nonconvex, and disjoint problems by considering the feasible region as an object and applying surface reconstruction ideas to define its shape. To take a further step towards the global solution of the discontinuous and nonconvex flexibility programming problem, Floudas et al. [56] introduced a deterministic global optimization-based algorithm which relied on the convexification/relaxation of the feasible region coupled with convex underestimation schemes within a branch-and-bound framework. Surrogate models, motivated by the unknown analytical form of the feasibility function and the need for reducing computational complexity of the underlying simulations, have also been applied for flexibility analysis. This topic has been extensively discussed in a recent review by Bhosekar and Ierapetritou [57].

Dimitriadis and Pistikopoulos [58] extended the steady-state approach by Swaney and Grossmann [48] to deal with dynamic feasibility and flexibility analysis. It was shown that when the variation of the uncertain parameters over time was known or when the critical points that limited feasibility were vertices of the time-varying uncertainty space, the problems could be transformed into dynamic optimization. However, in other cases, they proposed an explicit discretization scheme for the differential equations and combined it with an active constraint strategy to transform the problems into MINLP problems. Bansal et al. [59] proposed a stochastic flexibility index to evaluate the performance of linear dynamic process systems in the presence of uncertain parameters, described by Gaussian probability distribution functions. Dynamic flexibility consideration can also be integrated into simultaneous design and control optimization. Mohideen et al. [60] developed a unified framework featuring explicit flexibility, controller design, and stability criteria at the conceptual design stage of dynamic systems under time-varying disturbances and parameter uncertainty. With the use of orthogonal collocation on finite elements to discretize the state and control variables, this mixed-integer dynamic optimization (MIDO) problem was converted to a MINLP problem. A different solution algorithm for this approach was presented in Bansal et al. [61] based on variant-2 of the Generalized Benders Decomposition (v2-GBD). More recently, a two-level strategy which integrated design, control, and flexibility was introduced by Malcolm et al. [62].

Operability-based analysis is a similar yet not identical concept compared to flexibility analysis: while flexibility analysis aims to determine the maximum disturbance set that can be handled at the nominal operating point, operability-based approaches synthesize an operable process design by linear/nonlinear mapping among available input set (AIS), achievable output set (AOS), desired output set (DOS), desired input set (DIS), and expected disturbance set (EDS) [63,64]. The pioneering works for operability-based approaches come from Georgakis and co-workers [65–67]. More recently, this approach has been applied to generate modular process intensification systems by Lima and co-workers [68,69]. They proposed a bilevel/parallel programming-based approach to simultaneously address process design constraints and modular process intensification targets using high-dimensional nonlinear process models. Motivated by the advent of the shale gas revolution, their proposed strategy was demonstrated on the design of a catalytic membrane reactor for direct methane aromatization, indicating potential footprint reduction and efficiency maximization via modular plant design.

2.2.2 Process inherent safety analysis

Inherently safer design [70,71] has received increasing attention from the chemical process industry and the research community when developing new processes, as it is believed to be the best way to prevent accidents by avoiding, or at least minimizing, hazards in the first place rather than by placing control layers or protective add-ons later on. The principles of inherently safer design can be summarized as:

- *Minimize* – to use smaller quantities of hazardous materials or to reduce the size of equipment.
- *Substitute* – to use the materials, chemistry, or processes which are less hazardous.
- *Moderate* – to reduce hazards by dilution, refrigeration, or utilization of process alternatives that operate at less hazardous conditions (e.g., lower temperature or pressure).
- *Simplify* – to eliminate unnecessary complexity and design user friendly plants to reduce opportunities for error.

Process intensification, with one of its motivations to create substantially smaller process, offers a promising opportunity to significantly reduce the inventory of dangerous substances and the consequences of a hazardous process failure. However, despite the statement "small is safer", there have been long-lasting concerns on the safety performance of intensified equipment which hurdle the commercialization of many PI technologies. Hence, it is critical to identify the pros and cons resulted by process intensification in terms of process safety [72,73], an indicative list of which is presented below:

Pros

- Reduced number of process operations/equipment, leading to reduced chances of leaks, etc.
- Reduced vessel volume to stand the maximum pressure of any credible explosion
- Reduced inventory for continuous versus batch operations
- May provide better control of exothermic reactions to avoid runaway by using microchannel devices
- May reduce the number of incidents due to transient operations in batch systems – such as start-up, shutdown, etc.

Cons

- Potential for greater energy release if operated at higher temperature/pressure
- Significantly increased energy/product generation rates due to enhanced reaction rates
- May introduce new, or magnify the present, hazards or risks due to design integration or innovation
- Increased difficulty and complexity for control due to multi-functional equipment and faster process dynamics
- Potential need to adapt the safety and risk management approaches specifically for modular PI systems (e.g., Layers of Protection Analysis)

To include inherent safety considerations as part of early design, several key questions need to be addressed: (i) how to develop a standardized assessment approach or protocol to quantify safety performance with a wide range of existing chemicals, operating conditions, and process equipment – while taking modular and intensified process systems into consideration? (ii) how to define a model-based metric and integrate it with process mod-

Chapter 2 • Computer-aided modular PI: design, synthesis, and operability   33

els to properly monitor the process inherent safety performance under varying operating conditions? (iii) how to systematically suggest grassroots or retrofit design solutions leading to enhanced safety?

Classical safety evaluation methods (e.g., hazard and operability study – HAZOP) are often based on semi-heuristics and are employed as posterior evaluation tool after the implementation of detailed plant design. However, during the conceptual design stage, only very limited equipment/plant design/operating information is available. Evaluation based on initial data (e.g., chemicals, chemistry routes, operating conditions, process equipment) will provide a more useful and practical basis for the estimation of process risk and safety. Inherent safety index and quantitative risk analysis are two representative classes of approaches which can be applied for this purpose. In Chapter 7, we will provide a tutorial on how to apply such model-based metrics for inherent safety analysis as well as their integration with design optimization. In what follows, we present a brief literature review on process inherent safety studies from process systems point of view.

A comprehensive review was provided by Roy et al. [74] on available safety indices in chemical or biochemical process design, where twenty-five representative indices were summarized according to their application level, required input information, and output result type. Some of these indices, such as Process Route Index (PRI) [75] and Process Stream Index (PSI) [76], have been integrated with process simulators (e.g., Aspen HYSYS). On the efforts to apply these safety metrics in the conceptual design stage, Ortiz-Espinoza et al. [77] introduced a methodology for the selection of technology alternatives by considering safety, economics, and environment factors, where the inherent safety assessment of design alternatives was based on PRI and PSI. This procedure was applied to the production of ethylene and methanol; whereafter process modifications for enhanced safety performance were suggested by the highest index values of PRI and PSI. New metrics have also been proposed by extending the traditional economic performance metric (e.g., return on investment) to include safety (and sustainability)-relevant performance criteria [78]. The integrated metric thus outputted a financial-based evaluation with weighted impact on safety (and sustainability). However, these important attempts mainly address the need for quantifying inherent safety performances and screening among given process routes. Further efforts are expected to incorporate such assessment approaches in the optimization environment to enable automated process synthesis and design with inherent safety considerations.

Quantitative risk analysis (QRA), accounting for equipment failure frequency and consequence severity, is a more complex approach for risk assessment. Several studies have used this methodology in distillation-based processes. Medina-Herrera et al. [79] proposed to perform pre-screening of solvents according to safety properties (e.g., flammability, toxicity) and then to conduct a consequence analysis calculating an average distance likely to cause death as the major safety indicator. This method gave rise to a multi-objective optimization problem using genetic algorithms to investigate the trade-off between the conflicting economic and safety factors. It was demonstrated with an extractive distillation case study. In addition, Nemet et al. [80] employed QRA analysis for heat exchanger

34 Synthesis and Operability Strategies for Computer-Aided Modular PI

network synthesis for thermal intensification, considering the risks introduced by substance property on toxicity, flammability, and explosiveness. Herein, the risk tolerance was incorporated as a constraint in the synthesis model and systematically addressed in the superstructure-based optimization process.

2.2.3 Process control

The inherent process physics and dynamics in PI and modular systems can be very different from those in well-established conventional processes, normally posing more demanding requirements on operation and control. Extensive literature reviews have been provided on the control of specific PI and modular processes (e.g., reactive distillation, dividing wall column, pressure swing adsorption) [81,3].

Major research efforts in this area are oriented towards advanced model-based and optimization-based control (particularly model predictive control, MPC), while conventional PID control is still the most widely applied in industrial practice. MPC has been proved to provide superior control performance than proportional–integral–derivative (PID) control in a number of intensified/modular systems by accurately capturing their highly nonlinear behavior, complex dynamic characteristics, and strong variable interactions. However, the main technical challenges for MPC implementation in conventional and intensified chemical processes include [82]: (i) development and validation of rigorous dynamic models, (ii) computational complexity which hinders online MPC application and real-time optimization, (iii) need for proper model reduction scheme, and (iv) realization of nonlinear MPC. Pistikopoulos et al. [83] further highlighted the need for simultaneous design and control optimization, since intensified systems can be more vulnerable to constraint violation under uncertainties and/or control actions thus resulting in a notable mismatch between the dynamic and steady-state feasible regions of operation.

An alternative to online MPC is the explicit/multi-parametric MPC which can reduce the online computational load by reformulating and solving the MPC problem offline via multi-parametric programming (Fig. 2.9) [84,85]. Based on this, Pistikopoulos and co-workers proposed the PAROC (PARametric Optimization and Control) framework which is an integrated framework and software platform for the optimization and advanced model-based control of real-world process systems [86]. The proposed framework can also handle operational uncertainties on different time scales, i.e. control, design, scheduling [87]. The details will be discussed in Chapter 8.

In addition to regulatory control, dynamic operation can also be driven by demand response such as fluctuating energy pricing, product demand, raw material availability, daily/seasonal disturbances, etc. – particularly in modular systems which provide improved operational flexibility and agileness. A promising approach to tackle this problem is via the integrated scheduling and control leveraging dynamic optimization formulations. A representative application is the air separation units (ASU) under real-time electricity pricing and customer demands. For example, Dias et al. [88] proposed a framework for the integration of scheduling and MPC control, which consisted of: (i) an outer loop to integrate state-space process model, offline-MPC, and scheduling problem, and

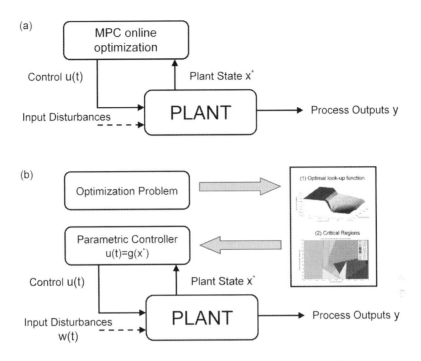

FIGURE 2.9 Illustration of model predictive control strategies: (a) Online MPC, (b) Explicit/Multi-parametric MPC.

(ii) an inner loop to solve online MPC problems. Results of the study showed cost reductions up to 1.4% for the ASU operating under real-time energy pricing changes. As the extension to intensified process systems, Otashu and Baldea [89] developed a demand response-oriented dynamic modeling and scheduling approach with application to a power-intensive membrane-based chlor-alkali process in response to electricity price changes in the short-term electricity market (i.e., fifteen minute). A 7.3% saving in energy cost has been reported by fluctuating current electrolysis density following electricity price signals as well as by depleting stored product to sustain demand during peak electricity price periods.

2.3 Research challenges and key questions

Although the phenomena-based process synthesis representation methods clearly offer the potential to derive novel intensified systems, major research challenges remain open.

The focus on the representation and integration of multiple tasks, central to most PI methods, often results in a very large search space with possibly excessive computational efforts. To alleviate this disadvantage, the definition and mathematical description of phenomenological building blocks need to be physically and computationally compact. Employing (multi-scale) driving force concepts [27,90] or bounding strategies to tighten the search space (such as the attainable region-based method for reactor network synthesis [91]) can be beneficial.

Most synthesis frameworks for modular PI applications have typically focused on steady-state reaction/separation process systems (e.g., dividing wall columns, reactive separation, and membrane-assisted systems). Expanding the scope to other important classes of PI designs, such as those characterized by the significant enhancement of mass/heat transfer phenomena, is essential (e.g., microreactors, rotating packed bed reactors) – especially if the degree of modularization is desired. Another open question lies on how to formally incorporate the temporal domain within the main building blocks of phenomena-based representation methods. The analogy of discretization of partial differential algebraic equations to both spatial and temporal domain [92] may offer a feasible and promising direction if computational issues can be equally addressed.

Operability concerns result from the violation of inequality process constraints during actual operation (under uncertainty and disturbances). Theoretical developments are necessitated to fully understand the unique operability characteristics in modular and intensified systems due to task-integration, equipment size reduction, etc. For example, their impacts on degrees of freedom (DOF), process dynamics, operating window. Key open questions include but not limited to: (i) How does the loss of DOFs affect the operation of an intensified task-integrated process compared to its conventional process counterpart? (ii) For the role of process constraints, what is the difference between intensified vs. conventional processes? (iii) What is the trade-off between PI design numbering up vs. scaling up considering the gain or loss in cost efficiency vs. design/operation agileness? (v) Inspired by the periodic processes, is it possible to operate the process in a dynamic manner following an optimal trajectory (such as to minimize energy consumption)? etc.

In tandem with the theory development, model-based analysis approaches discussed in Section 2.2 should be tailored to address these operational challenges. Moreover, operability metrics which can simultaneously assess multiple operational aspects of such systems are very limited (e.g., flexibility, operability, safety, controllability, etc.).

The integration of operability and control metrics at an early PI design stage still poses a formidable challenge. The match of information between design approaches and operational analysis methods is important to integrate these decisions as early as possible and to provide a consistent estimation of operational performance throughout the multi-scale design procedure. In this context, operability analysis for the phenomena-based synthesis intensification is kind of challenging since the synthesis is done using abstract phenomena building blocks (e.g., reaction, separation) with no available equipment information (e.g., reactor, membrane). More suitable criteria for operability at this stage may include elements of flexibility [93] and structural controllability [94]. On the other hand, most existing process safety analysis approaches [79,74,77] require very detailed equipment or flowsheet information – which may not be readily available at the PI synthesis level. Therefore, a hierarchical step-wise framework to decompose the problem can be beneficial to integrate operability, safety, and dynamic control at different design stages.

From software tools point of view, to the best of our knowledge, currently there is no widely applied commercial software which can: (i) generate novel conceptual process designs without pre-postulation of equipment or flowsheet configurations while accounting for modular process intensification opportunities, (ii) provide tailored operability and control analysis approaches or metrics to modular PI systems operation, (iii) integrate operability analysis with design and synthesis tasks to deliver optimal process solutions with guaranteed operability performance, and (iv) supply readily developed PI model library consisting of process models and physical property models on a consistent basis.

To address some of the above challenges for computer-aided design and operation of modular PI systems, in the following chapters of this book we introduce:

- **A process intensification synthesis approach** with compact physicochemical building block representation, which can efficiently screen the large design space and systematically generate a wide range of intensified designs
- **The operability, safety, and control metrics** developed in the PSE community and the extensions to assess operational characteristics in intensified and modular systems
- **A systematic framework** which can integrate PI synthesis, operability, safety, and control to ensure the feasibility and optimality of the derived intensified structures during actual operation
- **A software prototype** to support systematic modular process intensification with operability, safety, and control

References

[1] J.A. Weinfeld, S.A. Owens, R.B. Eldridge, Reactive dividing wall columns: a comprehensive review, Chemical Engineering and Processing: Process Intensification 123 (2018) 20–33.

[2] M. Baldea, T.F. Edgar, B.L. Stanley, A.A. Kiss, Modularization in chemical processing, Chemical Engineering Progress 114 (2018) 46–54.

[3] Y. Tian, S.E. Demirel, M.F. Hasan, E.N. Pistikopoulos, An overview of process systems engineering approaches for process intensification: state of the art, Chemical Engineering and Processing: Process Intensification (2018).

[4] M. Baldea, T.F. Edgar, Dynamic process intensification, Current Opinion in Chemical Engineering 22 (2018) 48–53.

[5] M. Skiborowski, Process synthesis and design methods for process intensification, Current Opinion in Chemical Engineering 22 (2018) 216–225.

[6] P. Daoutidis, A. Allman, S. Khatib, M.A. Moharir, M.J. Palys, D.B. Pourkargar, W. Tang, Distributed decision making for intensified process systems, Current Opinion in Chemical Engineering 25 (2019) 75–81.

[7] L.S. Dias, M.G. Ierapetritou, Optimal operation and control of intensified processes – challenges and opportunities, Current Opinion in Chemical Engineering 25 (2019) 82–86.

[8] S. Sitter, Q. Chen, I.E. Grossmann, An overview of process intensification methods, Current Opinion in Chemical Engineering 25 (2019) 87–94.

[9] C.M. Masuku, L.T. Biegler, Recent advances in gas-to-liquids process intensification with emphasis on reactive distillation, Current Opinion in Chemical Engineering 25 (2019) 95–100.

[10] Y. Tian, E.N. Pistikopoulos, Synthesis of operable process intensification systems: advances and challenges, Current Opinion in Chemical Engineering 25 (2019) 101–107.

[11] S.E. Demirel, J. Li, M.F. Hasan, Systematic process intensification, Current Opinion in Chemical Engineering 25 (2019) 108–113.

[12] R. Aglave, J. Lusty, J. Nixon, Using simulation and digitalization for modular process intensification, Chemical Engineering Progress 115 (2019) 45–49.

[13] Z. Jiang, R. Agrawal, Process intensification in multicomponent distillation: a review of recent advancements, Chemical Engineering Research and Design 147 (2019) 122–145.

[14] V. Gazzaneo, J.C. Carrasco, D.R. Vinson, F.V. Lima, Process operability algorithms: past, present, and future developments, Industrial & Engineering Chemistry Research 59 (2019) 2457–2470.

[15] A.K. Tula, M.R. Eden, R. Gani, Computer-aided process intensification: challenges, trends and opportunities, AIChE Journal 66 (2020) e16819.

[16] N. Nishida, G. Stephanopoulos, A.W. Westerberg, A review of process synthesis, AIChE Journal 27 (1981) 321–351.

[17] I.E. Grossmann, G. Guillén-Gosálbez, Scope for the application of mathematical programming techniques in the synthesis and planning of sustainable processes, Computers & Chemical Engineering 34 (2010) 1365–1376.

[18] Q. Chen, I. Grossmann, Recent developments and challenges in optimization-based process synthesis, Annual Review of Chemical and Biomolecular Engineering 8 (2017) 249–283.

[19] J.J. Siirola, Industrial applications of chemical process synthesis, in: Advances in Chemical Engineering, Vol. 23, Elsevier, 1996, pp. 1–62.

[20] S.D. Barnicki, J.J. Siirola, Separations process synthesis, Kirk-Othmer Encyclopedia of Chemical Technology (2000).

[21] J.-M. Commenge, L. Falk, Methodological framework for choice of intensified equipment and development of innovative technologies, Chemical Engineering and Processing: Process Intensification 84 (2014) 109–127.

[22] L. Mencarelli, Q. Chen, A. Pagot, I.E. Grossmann, A review on superstructure optimization approaches in process system engineering, Computers & Chemical Engineering 136 (2020) 106808.

[23] H. Yeomans, I.E. Grossmann, A systematic modeling framework of superstructure optimization in process synthesis, Computers & Chemical Engineering 23 (1999) 709–731.

[24] F. Friedler, K. Tarjan, Y. Huang, L. Fan, Graph-theoretic approach to process synthesis: axioms and theorems, Chemical Engineering Science 47 (1992) 1973–1988.

[25] T. Farkas, Z. Lelkes, Process flowsheet superstructures: structural multiplicity and redundancy: part I: basic GDP and MINLP representations, Computers & Chemical Engineering 29 (2005) 2180–2197.

[26] W. Wu, C.A. Henao, C.T. Maravelias, A superstructure representation, generation, and modeling framework for chemical process synthesis, AIChE Journal 62 (2016) 3199–3214.

[27] K.P. Papalexandri, E.N. Pistikopoulos, Generalized modular representation framework for process synthesis, AIChE Journal 42 (1996) 1010–1032.

[28] S.R. Ismail, P. Proios, E.N. Pistikopoulos, Modular synthesis framework for combined separation/reaction systems, AIChE Journal 47 (2001) 629–649.

[29] Y. Tian, E.N. Pistikopoulos, Synthesis of operable process intensification systems – steady-state design with safety and operability considerations, Industrial & Engineering Chemistry Research 58 (2018) 6049–6068.

[30] M.C. Georgiadis, E.N. Pistikopoulos, Energy and Process Integration, Begell House, 2006.

[31] S. Wilson, V. Manousiouthakis, IDEAS approach to process network synthesis: application to multicomponent MEN, AIChE Journal 46 (2000) 2408–2416.

[32] P. Pichardo, V.I. Manousiouthakis, Infinite dimensional state-space as a systematic process intensification tool: energetic intensification of hydrogen production, Chemical Engineering Research and Design 120 (2017) 372–395.

[33] F.E. da Cruz, V.I. Manousiouthakis, Process intensification of reactive separator networks through the IDEAS conceptual framework, Computers & Chemical Engineering 105 (2017) 39–55.

[34] S.E. Demirel, J. Li, M.F. Hasan, Systematic process intensification using building blocks, Computers & Chemical Engineering 105 (2017) 2–38.

[35] J. Li, S.E. Demirel, M.F. Hasan, Process synthesis using block superstructure with automated flowsheet generation and optimization, AIChE Journal 64 (2018) 3082–3100.

[36] M.S. Monjur, S.E. Demirel, J. Li, M.F. Hasan, SPICE_MARS: a process synthesis framework for membrane-assisted reactive separations, Industrial & Engineering Chemistry Research (2021).

[37] P. Lutze, R. Gani, J.M. Woodley, Process intensification: a perspective on process synthesis, Chemical Engineering and Processing: Process Intensification 49 (2010) 547–558.

[38] A.K. Tula, D.K. Babi, J. Bottlaender, M.R. Eden, R. Gani, A computer-aided software-tool for sustainable process synthesis-intensification, Computers & Chemical Engineering 105 (2017) 74–95.

[39] D.K. Babi, J. Holtbruegge, P. Lutze, A. Gorak, J.M. Woodley, R. Gani, Sustainable process synthesis–intensification, Computers & Chemical Engineering 81 (2015) 218–244.

[40] H.L. Lam, J.J. Klemeš, Z. Kravanja, P.S. Varbanov, Software tools overview: process integration, modelling and optimisation for energy saving and pollution reduction, Asia-Pacific Journal of Chemical Engineering 6 (2011) 696–712.

[41] J.J. Siirola, D.F. Rudd, Computer-aided synthesis of chemical process designs. From reaction path data to the process task network, Industrial & Engineering Chemistry Fundamentals 10 (1971) 353–362.

[42] Z. Kravanja, I.E. Grossmann, PROSYN – an MINLP process synthesizer, Computers & Chemical Engineering 14 (1990) 1363–1378.

[43] A.K. Tula, J. Wang, X. Chen, S.S. Mansouri, R. Gani, ProCACD: a computer-aided versatile tool for process control, Computers & Chemical Engineering 136 (2020) 106771.

[44] A. Lee, J.H. Ghouse, J.C. Eslick, C.D. Laird, J.D. Siirola, M.A. Zamarripa, D. Gunter, J.H. Shinn, A.W. Dowling, D. Bhattacharyya, et al., The idaes process modeling framework and model library – flexibility for process simulation and optimization, Journal of Advanced Manufacturing and Processing (2021) e10095.

[45] Q. Chen, S. Kale, J. Bates, R. Valentin, D.E. Bernal, M. Bynum, J.D. Siirola, I.E. Grossmann, Pyosyn: a collaborative ecosystem for process design advancement, in: 2019 AIChE Annual Meeting, 2019.

[46] E.N. Pistikopoulos, M.M.F. Hasan, J.S. Kwon, M.J. Realff, F. Boukouvala, M.R. Eden, S. Cremaschi, B.J. Tatarchuk, J.B. Powell, L. Spanu, R. Bindlish, S. Leyland, SYNOPSIS – Synthesis of Operable Process Intensification Systems. RAPID Institute Project 9.3, DE-EE0007888-09-03, 2020.

[47] K.P. Halemane, I.E. Grossmann, Optimal process design under uncertainty, AIChE Journal 29 (1983) 425–433.

[48] R.E. Swaney, I.E. Grossmann, An index for operational flexibility in chemical process design. Part I: formulation and theory, AIChE Journal 31 (1985) 621–630.

[49] I.E. Grossmann, C.A. Floudas, Active constraint strategy for flexibility analysis in chemical processes, Computers & Chemical Engineering 11 (1987) 675–693.

[50] E. Pistikopoulos, T. Mazzuchi, A novel flexibility analysis approach for processes with stochastic parameters, Computers & Chemical Engineering 14 (1990) 991–1000.

[51] D.A. Straub, I.E. Grossmann, Design optimization of stochastic flexibility, Computers & Chemical Engineering 17 (1993) 339–354.

[52] G. Ostrovsky, Y.M. Volin, E. Barit, M. Senyavin, Flexibility analysis and optimization of chemical plants with uncertain parameters, Computers & Chemical Engineering 18 (1994) 755–767.

[53] M.G. Ierapetritou, C. Floudas, P. Pardalos, Bilevel optimization: feasibility test and flexibility index, 2009.

[54] V. Goyal, M.G. Ierapetritou, Framework for evaluating the feasibility/operability of nonconvex processes, AIChE Journal 49 (2003) 1233–1240.

[55] I. Banerjee, M.G. Ierapetritou, Feasibility evaluation of nonconvex systems using shape reconstruction techniques, Industrial & Engineering Chemistry Research 44 (2005) 3638–3647.

[56] C.A. Floudas, Z.H. Gümüş, M.G. Ierapetritou, Global optimization in design under uncertainty: feasibility test and flexibility index problems, Industrial & Engineering Chemistry Research 40 (2001) 4267–4282.

[57] A. Bhosekar, M. Ierapetritou, Advances in surrogate based modeling, feasibility analysis and optimization: a review, Computers & Chemical Engineering 108 (2017) 250–267.

[58] V.D. Dimitriadis, E.N. Pistikopoulos, Flexibility analysis of dynamic systems, Industrial & Engineering Chemistry Research 34 (1995) 4451–4462.

[59] V. Bansal, J. Perkins, E. Pistikopoulos, Flexibility analysis and design of dynamic processes with stochastic parameters, Computers & Chemical Engineering 22 (1998) S817–S820.

[60] M.J. Mohideen, J.D. Perkins, E.N. Pistikopoulos, Optimal design of dynamic systems under uncertainty, AIChE Journal 42 (1996) 2251–2272.

[61] V. Bansal, J. Perkins, E. Pistikopoulos, R. Ross, J. Van Schijndel, Simultaneous design and control optimisation under uncertainty, Computers & Chemical Engineering 24 (2000) 261–266.

[62] A. Malcolm, J. Polan, L. Zhang, B.A. Ogunnaike, A.A. Linninger, Integrating systems design and control using dynamic flexibility analysis, AIChE Journal 53 (2007) 2048–2061.

[63] C. Georgakis, D. Uztürk, S. Subramanian, D.R. Vinson, On the operability of continuous processes, Control Engineering Practice 11 (2003) 859–869.

[64] F.V. Lima, Z. Jia, M. Ierapetritou, C. Georgakis, Similarities and differences between the concepts of operability and flexibility: the steady-state case, AIChE Journal 56 (2010) 702–716.

[65] D.R. Vinson, C. Georgakis, A new measure of process output controllability, Journal of Process Control 10 (2000) 185–194.

[66] S. Subramanian, C. Georgakis, Steady-state operability characteristics of idealized reactors, Chemical Engineering Science 56 (2001) 5111–5130.

[67] D. Uztürk, C. Georgakis, Inherent dynamic operability of processes: general definitions and analysis of SISO cases, Industrial & Engineering Chemistry Research 41 (2002) 421–432.

[68] J.C. Carrasco, F.V. Lima, An optimization-based operability framework for process design and intensification of modular natural gas utilization systems, Computers & Chemical Engineering 105 (2017) 246–258.

[69] B.A. Bishop, F.V. Lima, Novel module-based membrane reactor design approach for improved operability performance, Membranes 11 (2021) 157.

[70] T.A. Kletz, Inherently safer plants, Plant/Operations Progress 4 (1985) 164–167.

[71] M.S. Mannan, O. Reyes-Valdes, P. Jain, N. Tamim, M. Ahammad, The evolution of process safety: current status and future direction, Annual Review of Chemical and Biomolecular Engineering 7 (2016) 135–162.

[72] J. Etchells, Process intensification: safety pros and cons, Process Safety and Environmental Protection 83 (2005) 85–89.

[73] P. Yelvington, W. Grieco, A. Gokhale, L. Nara, The link between process safety and process intensification, https://www.aiche.org/academy/webinars/link-between-process-safety-and-process-intensification, 2019. (Accessed 9 July 2021).

[74] N. Roy, F. Eljack, A. Jiménez-Gutiérrez, B. Zhang, P. Thiruvenkataswamy, M. El-Halwagi, M.S. Mannan, A review of safety indices for process design, Current Opinion in Chemical Engineering 14 (2016) 42–48.

[75] C.T. Leong, A.M. Shariff, Process route index (PRI) to assess level of explosiveness for inherent safety quantification, Journal of Loss Prevention in the Process Industries 22 (2009) 216–221.

[76] A.M. Shariff, C.T. Leong, D. Zaini, Using process stream index (PSI) to assess inherent safety level during preliminary design stage, Safety Science 50 (2012) 1098–1103.

[77] A.P. Ortiz-Espinoza, A. Jiménez-Gutiérrez, M.M. El-Halwagi, Including inherent safety in the design of chemical processes, Industrial & Engineering Chemistry Research 56 (2017) 14507–14517.

[78] K. Guillen-Cuevas, A.P. Ortiz-Espinoza, E. Ozinan, A. Jimenez-Gutierrezez, N.K. Kazantzis, M.M. El-Halwagi, Incorporation of safety and sustainability in conceptual design via a return on investment metric, ACS Sustainable Chemistry & Engineering 6 (2018) 1411–1416.

[79] N. Medina-Herrera, I.E. Grossmann, M.S. Mannan, A. Jiménez-Gutiérrez, An approach for solvent selection in extractive distillation systems including safety considerations, Industrial & Engineering Chemistry Research 53 (2014) 12023–12031.

[80] A. Nemet, J.J. Klemeš, I. Moon, Z. Kravanja, Safety analysis embedded in heat exchanger network synthesis, Computers & Chemical Engineering 107 (2017) 357–380.

[81] A.A. Kiss, C.S. Bildea, A control perspective on process intensification in dividing-wall columns, Chemical Engineering and Processing: Process Intensification 50 (2011) 281–292.

[82] N.M. Nikačević, A.E. Huesman, P.M. Van den Hof, A.I. Stankiewicz, Opportunities and challenges for process control in process intensification, Chemical Engineering and Processing: Process Intensification 52 (2012) 1–15.

[83] E.N. Pistikopoulos, Y. Tian, R. Bindlish, Operability and control in process intensification and modular design: challenges and opportunities, AIChE Journal (2021) e17204.

[84] A. Bemporad, M. Morari, V. Dua, E.N. Pistikopoulos, The explicit linear quadratic regulator for constrained systems, Automatica 38 (2002) 3–20.

[85] E.N. Pistikopoulos, N.A. Diangelakis, R. Oberdieck, Multi-Parametric Optimization and Control, John Wiley & Sons, 2020.

[86] E.N. Pistikopoulos, N.A. Diangelakis, R. Oberdieck, M.M. Papathanasiou, I. Nascu, M. Sun, PAROC – an integrated framework and software platform for the optimisation and advanced model-based control of process systems, Chemical Engineering Science 136 (2015) 115–138.

[87] B. Burnak, N.A. Diangelakis, E.N. Pistikopoulos, Integrated process design and operational optimization via multiparametric programming, Synthesis Lectures on Engineering, Science, and Technology 2 (2020) 1–258.

[88] L.S. Dias, R.C. Pattison, C. Tsay, M. Baldea, M.G. Ierapetritou, A simulation-based optimization framework for integrating scheduling and model predictive control, and its application to air separation units, Computers & Chemical Engineering 113 (2018) 139–151.

[89] J.I. Otashu, M. Baldea, Demand response-oriented dynamic modeling and operational optimization of membrane-based chlor-alkali plants, Computers & Chemical Engineering 121 (2019) 396–408.

[90] T. Lopez-Arenas, M. Sales-Cruz, R. Gani, E.S. Pérez-Cisneros, Thermodynamic analysis of the driving force approach: reactive systems, Computers & Chemical Engineering 129 (2019) 106509.

[91] M. Feinberg, P. Ellison, General kinetic bounds on productivity and selectivity in reactor-separator systems of arbitrary design: principles, Industrial & Engineering Chemistry Research 40 (2001) 3181–3194.

[92] A. Agarwal, L.T. Biegler, S.E. Zitney, A superstructure-based optimal synthesis of psa cycles for post-combustion CO2 capture, AIChE Journal 56 (2010) 1813–1828.

[93] C.-T. Chang, V.S.K. Adi, Deterministic Flexibility Analysis: Theory, Design, and Applications, CRC Press, 2017.

[94] C.-T. Lin, Structural controllability, IEEE Transactions on Automatic Control 19 (1974) 201–208.

[95] E.N. Pistikopoulos, Y. Tian, R. Bindlish, Operability and control in process intensification and modular design: Challenges and opportunities, AIChE Journal 67 (2021) e17204.

PART
2

Methodologies

3

Phenomena-based synthesis representation for modular process intensification

As discussed in Section 2.1, process synthesis can provide systematic approaches for the discovery of novel intensified designs. However, key research questions remain on: (i) how to define the driving forces to intensify process schemes, given the diverse modular process intensification technology options? (ii) how to systematically derive process solutions with significantly improved process metrics (e.g., energy, cost, sustainability)? and (iii) how to explore the combinatorial design space in a computationally efficient way?

To address these challenges, we introduce a process intensification (PI) synthesis approach using the Generalized Modular Representation Framework (GMF) towards systematic innovation of chemical process designs. In this chapter, we focus on the GMF representation capabilities which enable to capture both conventional and intensified process systems in a unified bottom-up approach. In Chapter 4, we will extend the discussion to GMF synthesis leveraging superstructure optimization to generate optimal and intensified process solutions.

3.1 A prelude on phenomena-based PI synthesis

Before we introduce GMF in details, we would like to first clarify several questions regarding phenomena-based synthesis strategies for modular process intensification:

- **What is the "phenomena" in phenomena-based synthesis?**
  In general, "phenomena" are defined at a lower aggregated level in contrast to the well-established concept of "unit operation". As depicted in Fig. 3.1, they stand for the fundamental chemical process functions (e.g., mass transfer, heat transfer) which repetitively take place and the assemble of which makes a certain functional task (e.g., separation, heating, cooling), and then forms unit operations or flowsheets (e.g., reactor, distillation, membrane). Different sets of phenomena and tasks have been used to define building blocks in different phenomena-based synthesis approaches (e.g., [1], [2], [3], [4]). Although lacking a consensus yet, most building blocks include mass transfer, reaction, phase separation, selective separation, mixing, splitting, heating/cooling, etc. *The difference between the sets of building blocks mainly lies in their generality in definition* – for example, a mass transfer building block can encapsulate phase separation, selective separation, or even reaction if well defined.

Synthesis and Operability Strategies for Computer-Aided Modular Process Intensification
https://doi.org/10.1016/B978-0-32-385587-7.00013-0
Copyright © 2022 Elsevier Inc. All rights reserved

45

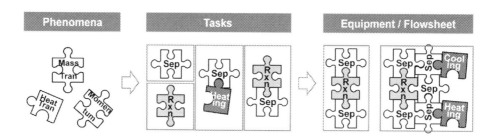

FIGURE 3.1 Phenomena-based representation of a chemical process.

- **Why "phenomena-based synthesis"?**
  Conventional unit operation-based synthesis has shown tremendous power in energy systems design and operational optimization to improve process cost, energy, and sustainability performances [5]. However, phenomena-based synthesis brings in the opportunity to *re-invent unit operations, or even flowsheets, without any pre-postulation of process alternatives which may hinder the discovery of non-intuitive design solutions*. In this way, it drives innovation in a systematic manner rather than merely relying on breakthroughs via engineering expertise or trial-and-error.

- **What "results" to obtain from phenomena-based synthesis?**
  At least two types of information can be extracted from the phenomena-based synthesis results: (i) *Performance targets* – with the conceptual phenomenological building blocks which can be flexibly and synergistically integrated or combined, a performance target can be identified, e.g. how much process improvements can be achieved; and (ii) *Instructive process alternatives* – the structural and operational information obtained from the resulting phenomena-based process solutions can provide insights (e.g., de-bottlenecking, heat/mass integration, task combination) to the optimal unit operations or flowsheet alternatives. However, phenomena-based synthesis cannot directly generate process units or flowsheets with all the design details. Instead, it contributes to rapidly and systematically screen the extended design space with potentially "out-of-the-box" designs. Then the promising process solutions should be translated to unit operations and validated via rigorous process design.

- **What are the "challenges" of phenomena-based synthesis?**
  A major challenge is the computational complexity. Let us consider a flowsheet with one plug flow reactor (PFR) and two succeeding distillation columns. Assume that the PFR can be represented with five reaction building blocks to account for axial distribution. Every distillation column consists of fifteen column trays, each of which corresponds to a separation building block. The superstructure formulation expands from the originally three unit blocks (1 reactor + 2 distillation columns) to thirty-five building blocks (5 reaction + 15×2 separation), which results in an explosion of binary variables for structural combinations and of nonlinearities for repetitively describing the same phenomena. If more detailed phenomena classification is used, such as phase separa-

tion and selective separation, the scale of the resulting synthesis problem grows even larger. *To address this challenge, the definition of phenomena building blocks should be highly compact in terms of both physical representation and computational modeling.*

- **Can "every" modular PI technology be captured by phenomena-based synthesis?**
  Theoretically we would hope this answer to be "yes". But currently "not yet". To the best of our knowledge, phenomena-based synthesis has been mostly applied to conventional scale reactors, advanced distillation systems (e.g., dividing wall columns, reactive distillation columns,), and membrane-assisted systems. So far there is a lack of contributions on periodic systems, micro-scale systems, rotating systems, and other alternative energy assisted systems (e.g., microwave).

3.2 Generalized Modular Representation Framework

A process operation can be generally characterized by a set of mass- and heat-transfer phenomena, concerning mainly the mass transfer of one component from one phase to another (e.g., distillation) or from one substance to another (e.g., reaction) due to the difference in their chemical potentials [1]. In this context, two types of building blocks are utilized in GMF to represent chemical processes as shown in Fig. 3.2, namely:

- *Mass/Heat exchange module* – where mass and heat transfer take place between two contacting streams in the form of reaction, separation, or reactive separation. The mass transfer feasibility within this type of module is ensured by imposing the driving force constraints derived based on the change of total Gibbs free energy.

- *Pure heat exchange module* – where the participating streams do not come into mass active contact but only perform heat exchange in the form of utility heating, utility cooling, or process stream integration. The heat exchange feasibility is governed by temperature gradient.

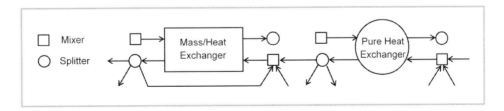

FIGURE 3.2 Mass/Heat exchange module and pure heat exchange module.

Using these modular building blocks with higher level of abstraction, GMF follows a bottom-up representation hierarchy as shown in Fig. 3.3. It starts from phenomena building blocks (i.e. mass and/or heat transfer) to identify diverse process tasks (e.g., heating, cooling, reaction and/or separation), and finally leads to diverse equipment/flowsheet designs with the synergistic combination of multi-functional tasks (e.g., reactive distillation).

In this way, GMF provides a generic strategy to overcome process bottlenecks by directly improving mass/heat transfer and/or shifting reaction equilibrium, thus readily to discover novel intensified pathways without pre-postulation of plausible units/flowsheets.

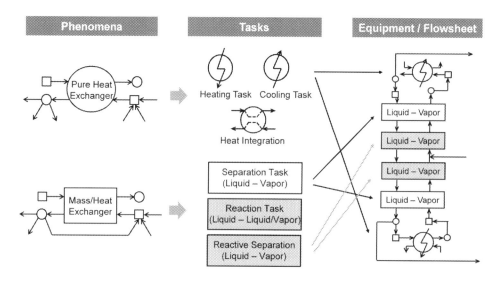

FIGURE 3.3 GMF phenomena-task-process representation hierarchy.

More examples on GMF representation of conventional and intensified chemical processes are illustrated in Fig. 3.4. A single mass/heat exchange module can represent a reactor, a flash separator, or a reactive absorption column – determined by the type of task and stream phase conditions in the module. With an aggregation of the mass/heat exchange modules and pure heat exchange modules, distillation-based processes can be captured. A number of intensified systems have been synthesized using GMF including reactive distillation [6], reactive absorption [7], homogeneous azeotropic separation [8,9], heat-integrated distillation [10], (reactive) Petlyuk column and dividing wall column [11,12] etc. We will demonstrate GMF on more engineering applications in Part 3 of this book.

3.3 Driving force constraints

Based on Figs. 3.3 and 3.4, a key question is how to identify different tasks from a single modular building block? For example, how to determine whether a separation, reaction or reactive separation task is taking place in a mass/heat exchange module? Moreover, what is the resulting benefit for such representation considerations? To answer these questions, we introduce the driving force constraints employed in GMF, which provide a unified Gibbs free energy-based formulation to systematically identify necessary reaction or separation tasks. The task identification capabilities of a pure heat exchange module will be discussed in Chapter 5.

Chapter 3 • Phenomena-based synthesis representation for modular PI 49

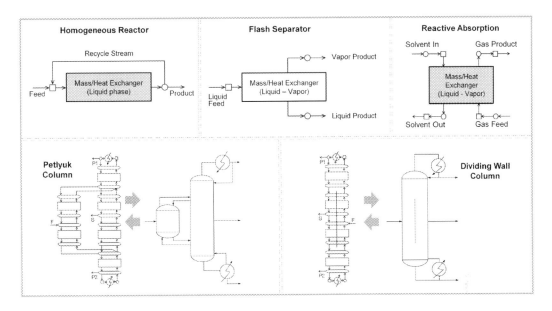

FIGURE 3.4 GMF modular representation examples.

FIGURE 3.5 A mass/heat exchange module and associated stream variables.
Superscripts – LI: Liquid inlet stream, LO: Liquid outlet stream. VI: Vapor inlet stream, VO: Vapor outlet stream.
Variables – f: Flowrate, x: Composition, T: Temperature, i: Component.

Consider a most generic mass/heat exchange module containing multicomponent liquid-vapor mixture as shown in Fig. 3.5. If constant temperature and pressure are assumed, mass transfer between inlet streams can occur when the total Gibbs free energy $(nG)^{tot}$ is decreasing in this module, namely:

$$d(nG)^{tot}_{T,P} \leq 0 \tag{3.1}$$

To assist the expression of this decreasing trend, dn_i^L is introduced to describe the molar amount changes of component i in the liquid inlet and outlet streams as a result of the reaction and/or separation task taking place in the module:

$$dn_i^L = f^{LO} x_i^{LO} - f^{LI} x_i^{LI} \tag{3.2}$$

Thus, Eq. (3.1) is consistent with the following "driving force constraints":

$$G1_i \times G2_i \leq 0, \quad \forall i = 1, ..., NC \tag{3.3}$$

50 Synthesis and Operability Strategies for Computer-Aided Modular PI

where

$$G1_i = dn_i^L = f^{LO}x_i^{LO} - f^{LI}x_i^{LI} \tag{3.4}$$

$$G2_i = \left[\frac{\partial(nG)^{tot}}{\partial(n_i^L)}\right]_{T,P} \tag{3.5}$$

Before proceeding with the expansion of the $G2_i$ term, several remarks can be made here to clarify the physical meanings of the driving force constraints:

- The mass transfer feasibility in a GMF mass/heat exchange module is constrained by $d(nG)_{T,P}^{tot} \leq 0$ for a decreasing total Gibbs free energy, instead of $d(nG)_{T,P}^{tot} = 0$ for the minimization of total Gibbs free energy. Therefore, the mass transfer in the module is not enforced to reach the physical equilibrium.

- $G1_i$ actually defines the net mass transfer direction for each component i. If $G1_i \leq 0$, then component i is transferred from the liquid phase to the vapor phase or is consumed by reaction in the liquid phase. If $G1_i \geq 0$, then component i is transferred from vapor to liquid or is generated by reaction in the liquid phase.

- $G2_i$ can be used to indicate how far the component is from its physical and/or chemical equilibrium state. A detailed derivation of the relationship between $G2_i = 0$ with chemical/physical equilibrium conditions is presented in Appendix B.

To describe the numerator in $G2_i$, let us look at the general form of total Gibbs free energy for a system with T, P as independent variables:

$$d(nG)^{tot} = -(nS)dT + (nV)dP + \sum_i \mu_i dn_i \tag{3.6}$$

where n gives molar amount, S defines entropy, T denotes temperature, V is volume, P stands for pressure, and μ_i is chemical potential for each component i. With an assumption of constant T and P,

$$d(nG)_{T,P}^{tot} = \sum_i \mu_i dn_i = \sum_i (\mu_i^L dn_i^L + \mu_i^V dn_i^V) \tag{3.7}$$

Consider the component-based mass conservation in this module which contains a multicomponent liquid-vapor mixture with k reactions taking place:

$$n_i^L + n_i^V = n_i^0 + \sum_k v_{ik}\epsilon_k \tag{3.8}$$

where n_i^0 denotes the initial number of moles for each component i, v_{ik} is the stoichiometric coefficient for each component i in reaction k, and ϵ is the extent of reaction k

Chapter 3 • Phenomena-based synthesis representation for modular PI   51

determined by reaction rate, catalyst mass, etc. Differentiation of Eq. (3.8) gives:

$$dn_i^L + dn_i^V = \sum_k v_{ik} d\epsilon_k$$

$$dn_i^V = -dn_i^L + \sum_k v_{ik} d\epsilon_k$$

(3.9)

Eq. (3.7) can be reformulated as:

$$d(nG)_{T,P}^{tot} = \sum_i \mu_i^L dn_i^L + \sum_i \mu_i^V \left(-dn_i^L + \sum_k v_{ik} d\epsilon_k\right)$$

$$= \sum_i (\mu_i^L - \mu_i^V) dn_i^L + \sum_i \sum_k v_{ik} \mu_i^V d\epsilon_k$$

(3.10)

Thus, if approximating $\partial \mu_i / \partial n_i^L$ to be 0,

$$G2_i = \left[\frac{\partial (nG)^{tot}}{\partial (n_i^L)}\right]_{T,P} = \mu_i^L - \mu_i^V + \sum_i \sum_k v_{ik} \mu_i^V \frac{\partial \epsilon_k}{\partial n_i^L}$$

(3.11)

Note that $\frac{\partial \epsilon_k}{\partial n_i^L}$ needs to be analytically derived for each reaction (see Ismail et al. [6] for illustration examples). Substituting with the following expressions for chemical potential:

$$\mu_i^L = \Delta G_i^f + RT \ln(\gamma_i^L x_i^L P_i^{sat,L})$$

$$\mu_i^V = \Delta G_i^f + RT \ln(\phi_i^V x_i^V P)$$

(3.12)

where ΔG_i^f is the standard Gibbs function of formation for each component i from its elements at T and 1 atm, γ_i^L denotes liquid activity coefficient for each component i, ϕ_i^V gives vapor fugacity coefficient for each component i, and $P^{sat,L}$ is saturated vapor pressure for each component i. $G2_i$ is then expanded as:

$$G2_i = RT[\ln(\gamma_i^L x_i^L P_i^{sat,L}) - \ln(\phi_i^V x_i^V P)] + \sum_i \sum_k v_{ik} \left[\Delta G_i^f + RT \ln(\phi_i^V x_i^V P)\right] \frac{\partial \epsilon_k}{\partial n_i^L}$$

(3.13)

Divided by RT which will not affect the sign of this term, the expression used for Eq. (3.3), i.e. the driving force constraints, can be obtained:

$$G2_i = \ln\left[\frac{\gamma_i^L x_i^L P_i^{sat,L}}{\phi_i^V x_i^V P}\right] + \sum_i \sum_k \left[\frac{v_{ik} \Delta G_i^f}{RT} + v_{ik} \ln(\phi_i^V x_i^V P)\right] \frac{\partial \epsilon_k}{\partial n_i^L}$$

(3.14)

The next question is how to define the participating Liquid (L) and Vapor (V) streams for $G2_i$ in Eq. (3.14). The following two modeling considerations can be adapted which also render different representation capabilities for GMF:

52 Synthesis and Operability Strategies for Computer-Aided Modular PI

- Definition 1 – $G2_i$ is defined between the module outlet streams LO and VO. In this case, a mass/heat exchange module is equivalent to a single distillation tray where the liquid and vapor outlet streams can be in liquid-vapor phase equilibrium or non-equilibrium.

- Definition 2 – $G2_i$ is defined at the two ends of the module between LI and VO as well as between LO and VI. *The resulting driving force constraints are adapted in this work unless otherwise explicitly stated.* This formulation enables a larger design space to be captured by a GMF mass/heat exchange module (in analogy to a column section compared to a single distillation tray) and allows for more degrees of freedom for stream variables – e.g., the LO and VO streams can have different temperatures while satisfying the overall energy balance.

3.4 Key features of GMF synthesis

Based on the GMF driving force constraints formulation, we highlight the following key features of GMF representation which enable the generation of optimal and intensified process systems while overcoming potential computational challenges:

(i) Systematic identification of reaction and/or separation tasks from heat and/or mass transfer phenomena

As can be noted, $G2_i$ comprises two components:

- Separation: $\ln\left[\frac{\gamma_i^L x_i^L P_i^{sat,L}}{\phi_i^V x_i^V P_{tot}}\right]$

- Reaction: $\sum_i \sum_k \left[\frac{v_{ik}\Delta G_i^f}{RT} + v_{ik}\ln(\phi_i^V x_i^V P_{tot})\right]\frac{\partial \epsilon_k}{\partial n_i^L}$

If a pure separation or pure reaction process is of interest, the corresponding $G2_i$ separation or reaction component can be used without losing generality. However, for a combined reaction/separation system, a key question is how to dictate the identity of a mass/heat exchange module to perform pure separation task, pure reaction task, or reactive separation task. To this purpose, we introduce two auxiliary sets of binary variables to denote the existence of separation and reaction phenomena in each mass/heat exchange module, i.e. y_{sep} and y_{rxn}. If $y_{sep} = 1$, separation takes place in the module; if $y_{rxn} = 1$, reaction takes place in the module (otherwise the binary variables take the value of 0). If both $y_{sep} = 1$ and $y_{rxn} = 1$, separation and reaction take place simultaneously.

(ii) Automated identification of equilibrium/non-equilibrium tasks and intensification towards ultimate thermodynamic limits

As mentioned earlier, no physical or chemical equilibrium is enforced in GMF representation. Thus, equilibrium-limited tasks ($G1_i \times G2_i = 0$) or non-equilibrium tasks ($G1_i \times G2_i < 0$) will be identified as per the process inherent characteristics or the optimal process physicochemical status. Also benefited from the Gibbs free energy-based driving force constraints, GMF enables to improve process performances (with respect to energy,

Chapter 3 • Phenomena-based synthesis representation for modular PI   53

cost, sustainability, etc.) by exploiting the full thermodynamic space. More discussions will be offered on this topic in Chapter 10.

(iii) Enabling selection of functional materials within process tasks

Material selection can be achieved in GMF by: (i) utilizing rigorous thermodynamic models (e.g., NRTL, Redlich-Kwong-Soave equation) to calculate phase equilibrium parameters (e.g., liquid activity coefficient γ_i, vapor fugacity coefficient ϕ_i) which are then used to describe the nonideal mixture properties, (ii) employing rigorous reaction kinetics (e.g., reaction rate r_k) to capture the impact of catalysts. The driving force constraints (Eq. 3.14), as an explicit function of these physical property parameters, also enable to directly investigate the impact of materials on process driving forces. This will be demonstrated later in Chapter 11 using an extractive separation case study with solvent selection.

(iv) Compact/Aggregated representation to avoid combinatorial explosion

Another key question is what dictates the required number of mass/heat exchange modules to represent a process in Fig. 3.4. Actually, each module is characterized by a certain mass transfer pattern dictated by $G1_i$. In a reactive absorption example by Algusane et al. [7], GMF is applied, coupled with orthogonal collocation techniques, to synthesize a 70-stage absorption column with two mass/heat exchange modules. Thus, due to this aggregated representation capability, the GMF synthesis optimization problem is in a more compact size avoiding combinatorial explosion.

3.5 Motivating examples

In this section, we apply GMF for the representation and optimization of simple reaction and/or separation systems. Via the two motivating examples, we aim to demonstrate: (i) how GMF represents reaction/separation systems using mass/heat exchange modules, (ii) how GMF can identify more promising process improvements than equipment-based modeling using the driving force constraints, and (iii) the conjunctive thermodynamic basis of GMF with other physical/chemical equilibrium-based equipment models.

3.5.1 Four-tray simulation in distillation column

In what follows, we consider a set of four distillation trays/modules from a pentene (PEN), butene (BUT), and hexene (HEX) separation process (Fig. 3.6). The liquid and vapor inlet stream conditions are given at the two ends of the column section in a counter-current manner. We apply GMF for: (i) equilibrium-based simulation which is identical to tray-by-tray simulation (i.e., no degrees of freedom for optimization), and (ii) modular optimization to maximize hexene molar fraction in the liquid outlet stream using the driving force constraints. The resulting liquid molar fraction profiles from the above two types of GMF modeling setups are summarized in Table 3.1.

As can be noted from the equilibrium-based simulation results, from the first separation module to the last separation module (numbered in a descending order), pentene and butene consistently transfer from the liquid phase to the vapor phase, while hexene from

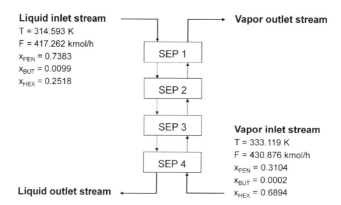

FIGURE 3.6 Four-tray simulation in a ternary distillation column.

Table 3.1 Results for equilibrium-based simulation vs. modular optimization.

Module Number	x_{PEN}, mol/mol		x_{BUT}, mol/mol		x_{HEX}, mol/mol	
	Equilibrium	Modular	Equilibrium	Modular	Equilibrium	Modular
1	0.6038	0.7383	0.0027	0.0069	0.3935	0.2548
2	0.4219	0.4226	0.0006	0.0014	0.5775	0.5760
3	0.2457	0.4217	0.0001	0.0002	0.7542	0.5781
4	0.1278	0.0946	2.27E-05	1.0E-5	0.8722	0.9054

vapor to liquid. This monotonic trend is also correctly captured by the GMF modular optimization. Moreover, the modular optimization is able to identify a higher (or optimal) hexene composition in the liquid outlet stream compared to the equilibrium-based simulation. This provides the opportunity to evaluate the maximum possible performance improvements from an enriched design space within the thermodynamic limits, with both known and unknown process solutions (beyond the four-tray configuration). The promising process solutions identified by GMF can be further validated with equipment-based simulation. It can also be observed that, the composition profile obtained from modular optimization is not as smooth as that from the equilibrium-based simulation. This indicates that the GMF modules are not always optimized at their equilibrium conditions, which is also approved by the non-zero $G2_i$ values in these modules.

3.5.2 Reactor-separator simulation

We consider a combined reaction/separation process for pentene metathesis to produce butene and hexene (Eq. 3.15), which is an equilibrium-limited reaction (Eq. 3.15) taking place in liquid phase at 1 atm. The phase behavior can be described using ideal vapor-liquid equilibrium [13]. The reaction kinetics are detailed in Eq. (3.16) [14]. The raw material is a saturated liquid stream of 100 kmol/h pure pentene, and the design objective is to maximize butene production rate. Two process configurations are of interest, i.e. an integrated CSTR and flash column process and a reactive flash column process (Fig. 3.7).

Chapter 3 • Phenomena-based synthesis representation for modular PI   55

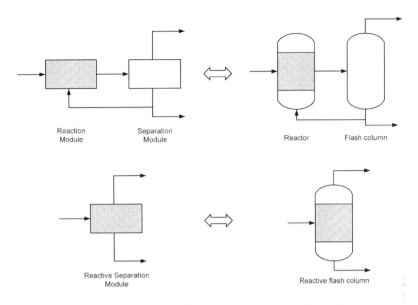

FIGURE 3.7 Combined reaction/separation systems. (Left: GMF representation, Right: Equipment-based modeling.)

$$2C_5H_{10} \rightleftharpoons C_4H_8 + C_6H_{12} \tag{3.15}$$

$$r = k_f(x_{C_5H_{10}}^2 - \frac{x_{C_4H_8}x_{C_6H_{12}}}{K_{eq}}) \quad (h^{-1})$$

$$k_f = 1.0661 \times 10^5 e^{(-3321.2/T(K))} \quad (h^{-1}) \tag{3.16}$$

$$K_{eq} = 0.25$$

The following models are utilized in this analysis:

i. **GMF representation** – With the driving force constraints introduced in Section 3.3, a reaction module, a separation module, and a reactive separation module can be used to describe reactor, separator, and reactive separator, respectively. Since only 1-stage reaction or separation is considered in this example, the driving force constraints are set between the liquid and vapor outlet streams of the mass/heat exchange module, instead of constraining the mass transfer feasibility at the two ends of the module.

ii. **Rigorous equipment-based models** – A CSTR reactor model with reaction equilibrium expressions and a flash column with liquid-vapor phase equilibrium calculations are used to describe the integrated reactor and flash column process. A reactive flash column model is used which assumes chemical and physical equilibrium.

iii. **Attainable region calculation** – The kinetic attainable region-based approach (details can be found in Chapter 10) is applied to identify the maximum butene production potential and to benchmark the solutions obtained from the above equipment-based modeling and GMF representation.

The maximum butene production rates, pertaining to different reaction volumes, are summarized in Fig. 3.8a. As can be noted, the maximum butene production rates from the sequential reaction and separation process are identical for equipment-based modeling and GMF representation. In other words, the GMF reaction module and separation module in this case are respectively at chemical and physical equilibrium. This is because the maximum production rate in a single liquid-phase reactor is reached at the reaction equilibrium. The same is with the 1-stage separator to achieve the best butene separation performance at vapor-liquid equilibrium conditions.

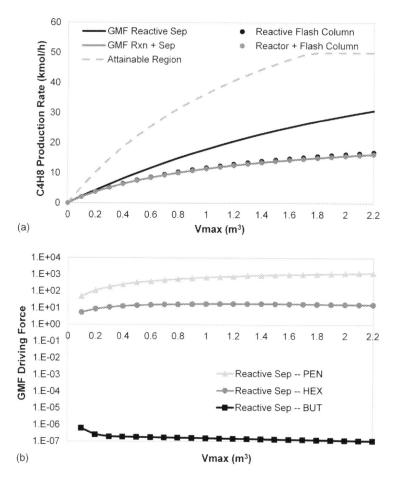

FIGURE 3.8 Combined reaction/separation systems comparison.
(a) Butene (C4H8) production rates, (b) GMF driving forces in reactive separation module.

If reaction and separation are integrated into a single unit, the production rates can be shifted to higher than the pure reactor equilibrium condition. The equipment-based reactive flash column offers a slightly better butene production rate by vaporizing the butene instantaneously to the vapor phase. By allowing for additional design degrees of freedom to intensify the module mass transfer capability (e.g., different vapor and liquid temperatures), GMF identifies a higher butene production rate than the reactive flash column, but well within the attainable region. As shown in Fig. 3.8b, only the desired product component, i.e. butene, is pushed to reach a driving force ($G1_i \times G2_i$) near 0 in the reactive separation module. The pentene and butene components are at non-equilibrium conditions. As will be observed from the following chapters, for a GMF reactive separation module, the optimal process design may not always be obtained at $G1_i \times G2_i = 0$. A detailed derivation of the relationship between GMF zero driving forces with chemical/physical equilibrium conditions is presented in Appendix B.

References

[1] K.P. Papalexandri, E.N. Pistikopoulos, Generalized modular representation framework for process synthesis, AIChE Journal 42 (1996) 1010–1032.

[2] A.K. Tula, D.K. Babi, J. Bottlaender, M.R. Eden, R. Gani, A computer-aided software-tool for sustainable process synthesis-intensification, Computers & Chemical Engineering 105 (2017) 74–95.

[3] F.E. da Cruz, V.I. Manousiouthakis, Process intensification of reactive separator networks through the ideas conceptual framework, Computers & Chemical Engineering 105 (2017) 39–55.

[4] S.E. Demirel, J. Li, M.F. Hasan, Systematic process intensification using building blocks, Computers & Chemical Engineering 105 (2017) 2–38.

[5] Q. Chen, I. Grossmann, Recent developments and challenges in optimization-based process synthesis, Annual Review of Chemical and Biomolecular Engineering 8 (2017) 249–283.

[6] S.R. Ismail, P. Proios, E.N. Pistikopoulos, Modular synthesis framework for combined separation/reaction systems, AIChE Journal 47 (2001) 629–649.

[7] T.Y. Algusane, P. Proios, M.C. Georgiadis, E.N. Pistikopoulos, A framework for the synthesis of reactive absorption columns, Chemical Engineering and Processing: Process Intensification 45 (2006) 276–290.

[8] S.R. Ismail, E.N. Pistikopoulos, K.P. Papalexandri, Modular representation synthesis framework for homogeneous azeotropic separation, AIChE Journal 45 (1999) 1701–1720.

[9] Y. Tian, E.N. Pistikopoulos, A process intensification synthesis framework for the design of extractive separation systems with material selection, Journal of Advanced Manufacturing and Processing (2021) e10097.

[10] P. Proios, N.F. Goula, E.N. Pistikopoulos, Generalized modular framework for the synthesis of heat integrated distillation column sequences, Chemical Engineering Science 60 (2005) 4678–4701.

[11] Y. Tian, E.N. Pistikopoulos, Synthesis of operable process intensification systems – steady-state design with safety and operability considerations, Industrial & Engineering Chemistry Research 58 (2018) 6049–6068.

[12] P. Proios, E.N. Pistikopoulos, Hybrid generalized modular/collocation framework for distillation column synthesis, AIChE Journal 52 (2006) 1038–1056.

[13] R.C. Reid, J.M. Prausnitz, B.E. Poling, The properties of gases and liquids, 1987.

[14] M.J. Okasinski, M.F. Doherty, Design method for kinetically controlled, staged reactive distillation columns, Industrial & Engineering Chemistry Research 37 (1998) 2821–2834.

4

Process synthesis, optimization, and intensification

In Chapter 3, we have discussed the modular mass/heat exchange building block concepts and the Gibbs free energy-based driving force constraints, which supports the representation validity and versatility of chemical process systems using the Generalized Modular Representation Framework (GMF). In this chapter, we further elucidate the process synthesis and optimization capabilities of GMF to systematically generate optimal, and potentially intensified, process design solutions.

4.1 Problem statement

The generalized synthesis problem addressed in this chapter is defined as follows:

Given:
- A set of process streams to be used as raw materials with given compositions (their flowrates and supply temperatures can be either given or incorporated as optimization variables);
- A set of desired products and specifications on their flowrates, temperatures, and/or purities;
- A set of available heating/cooling utilities such as steam and cooling water with their availability, supply temperatures, and compositions;
- A set of available mass utilities such as mass separating agents (e.g., solvent, adsorbent) and reaction catalysts;
- All reaction schemes and kinetics data;
- All physical property models;
- Cost data of raw materials, mass and heat utilities, etc.

Objective:

To synthesize optimal process design(s), consisting of conventional or intensified unit operations, which can satisfy the afore-defined product specifications. The optimality of the solution can be evaluated with respect to economic performances (e.g., operating cost or total annualized cost). Other process performance criterion can also be readily incorporated into this problem formulation, taking the minimization of energy usage or carbon footprint as examples.

Synthesis and Operability Strategies for Computer-Aided Modular Process Intensification
https://doi.org/10.1016/B978-0-32-385587-7.00014-2
Copyright © 2022 Elsevier Inc. All rights reserved.

4.2 GMF synthesis model

As discussed in Chapter 3, GMF utilizes mass/heat exchange modules and pure heat exchange modules to represent chemical processes. For simplification purposes in this chapter, each mass/heat exchange module (M/H) is explicitly connected to two utility heat exchange modules as shown in Fig. 4.1. Thus, no heat integration is considered. Heat integration via GMF will be showcased in Chapter 5. To allow for all possible interconnections between the modular building blocks, auxiliary mixers and splitters are assigned to these modules at their inlet and outlet streams, respectively. An assumption is made that no phase change is allowed at mixers or splitters. An illustration of the GMF superstructure network is given in Fig. 4.2.

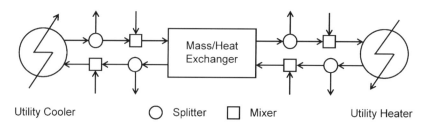

FIGURE 4.1 GMF modular building blocks.

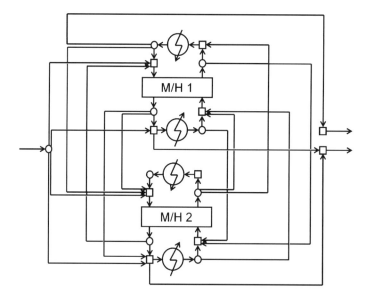

FIGURE 4.2 GMF superstructure network.

Chapter 4 • Process synthesis, optimization, and intensification   61

To formulate the GMF synthesis problem, the following sets are defined. Note that each GMF process stream is defined by its temperature, pressure, compositions, and phase (i.e., liquid, vapor), while no lean or rich stream properties are pre-postulated for mass change and no hot or cold steam properties are pre-postulated for heat exchange.

$$
\begin{aligned}
C &= \{c \mid \text{components}\} \\
I &= \{n \mid \text{available feeds}\} \\
P &= \{p \mid \text{product streams}\} \\
E &= \{e \mid \text{available mass/heat exchange modules}\} \\
LI_e &= \{s \mid \text{liquid inlet stream in module } e\} \\
LO_e &= \{s \mid \text{liquid outlet stream in module } e\} \\
VI_e &= \{s \mid \text{vapor inlet stream in module } e\} \\
VO_e &= \{s \mid \text{vapor outlet stream in module } e\} \\
HI_e &= \{s \mid \text{inlet stream in utility heater of module } e\} \\
HO_e &= \{s \mid \text{outlet stream in utility heater of module } e\} \\
CI_e &= \{s \mid \text{inlet stream in utility cooler of module } e\} \\
CO_e &= \{s \mid \text{outlet stream in utility cooler of module } e\} \\
LL_{e,e'} &= \{s \mid \text{interconnecting stream from liquid outlet splitter of } e \text{ to liquid inlet mixer of } e'\} \\
LH_{e,e'} &= \{s \mid \text{interconnecting stream from liquid outlet splitter of } e \text{ to utility heater of } e'\} \\
LC_{e,e'} &= \{s \mid \text{interconnecting stream from liquid outlet splitter of } e \text{ to utility cooler of } e'\} \\
LP_{e,p} &= \{s \mid \text{interconnecting stream from liquid outlet splitter of } e \text{ to final mixer of product } p\} \\
VV_{e,e'} &= \{s \mid \text{interconnecting stream from vapor outlet splitter of } e \text{ to vapor inlet mixer of } e'\} \\
VC_{e,e'} &= \{s \mid \text{interconnecting stream from vapor outlet splitter of } e \text{ to utility cooler of } e'\} \\
VP_{e,p} &= \{s \mid \text{interconnecting stream from vapor outlet splitter of } e \text{ to final mixer of product } p\} \\
IL_{n,e} &= \{s \mid \text{interconnecting stream from initial stream } n \text{ to liquid inlet mixer of } e\} \\
IV_{n,e} &= \{s \mid \text{interconnecting stream from initial stream } n \text{ to vapor inlet mixer of } e\} \\
IH_{n,e} &= \{s \mid \text{interconnecting stream from initial stream } n \text{ to utility heater of } e\} \\
IC_{n,e} &= \{s \mid \text{interconnecting stream from initial stream } n \text{ to utility cooler of } e\} \\
IP_{n,p} &= \{s \mid \text{interconnecting stream from initial stream } n \text{ to final mixer of product } p\} \\
HL_{e,e'} &= \{s \mid \text{interconnecting stream from utility heater of } e \text{ to liquid inlet mixer of } e'\} \\
HV_{e,e'} &= \{s \mid \text{interconnecting stream from utility heater of } e \text{ to vapor inlet mixer of } e'\} \\
HP_{e,p} &= \{s \mid \text{interconnecting stream from utility heater of } e \text{ to final mixer of product } p\} \\
CL_{e,e'} &= \{s \mid \text{interconnecting stream from utility cooler of } e \text{ to liquid inlet mixer of } e'\} \\
CP_{e,p} &= \{s \mid \text{interconnecting stream from utility cooler of } e \text{ to final mixer of product } p\}
\end{aligned}
$$

Available feed streams and product streams can be further classified as per their phase states (i.e., liquid or vapor):

$$
I = N^{\text{liq}} \cup N^{\text{vap}}
$$
$$
P = P^{\text{liq}} \cup P^{\text{vap}}
$$

62　Synthesis and Operability Strategies for Computer-Aided Modular PI

Binary variables are introduced to denote the existence (or not) of each module (y_e), the existence (or not) of reaction or separation task in each module (y_{rxn}, y_{sep}), and the existence (or not) of process streams (Table 4.1). However, the binary variables for process streams are not required, but will facilitate the formulation of structural model and simplify the solution of this synthesis problem.

Table 4.1　GMF binary variables for process streams.

Variable	To define the existence of
yh_e	utility heater of module e
yc_e	utility cooler of module e
$yll_{ee'}$	interconnecting stream from liquid outlet splitter of e to liquid inlet mixer of e'
$ylh_{ee'}$	interconnecting stream from liquid outlet splitter of e to utility heater of e'
$ylc_{ee'}$	interconnecting stream from liquid outlet splitter of e to utility cooler of e'
ylp_{ep}	interconnecting stream from liquid outlet splitter of e to final mixer of product p
$yvv_{ee'}$	interconnecting stream from vapor outlet splitter of e to vapor inlet mixer of e'
$yvc_{ee'}$	interconnecting stream from vapor outlet splitter of e to utility cooler of e'
yvp_{ep}	interconnecting stream from vapor outlet splitter of e to final mixer of product p
yIl_{ne}	interconnecting stream from initial stream n to liquid inlet mixer of e
yIv_{ne}	interconnecting stream from initial stream n to vapor inlet mixer of e
yIh_{ne}	interconnecting stream from initial stream n to utility heater of e
yIc_{ne}	interconnecting stream from initial stream n to utility cooler of e
yIp_{np}	interconnecting stream from initial stream n to final mixer of product p
$yhl_{ee'}$	interconnecting stream from utility heater of e to liquid inlet mixer of e'
$yhv_{ee'}$	interconnecting stream from utility heater of e to vapor inlet mixer of e'
yhp_{ep}	interconnecting stream from utility heater of e to final mixer of product p
$ycl_{ee'}$	interconnecting stream from utility cooler of e to liquid inlet mixer of e'
ycp_{ep}	interconnecting stream from utility cooler of e to final mixer of product p

The resulting GMF synthesis model comprises: (i) a physical model to represent the physical phenomena taking place in each module (with mass and energy balances, mass transfer driving force constraints, phase defining constraints, etc.), (ii) a structural model to generate structural alternatives (with logical constraints), and (iii) an objective function to evaluate process optimality. In what follows, we present the detailed mathematical model formulations for each sub-model. The list of nomenclature for the variables can be found at the end of this chapter.

I. Physical model

A. Mass balances for total stream flows at

- initial stream splitters

$$f_n^I - \sum_e f_{ne}^{IL} - \sum_e f_{ne}^{IH} - \sum_e f_{ne}^{IC} - \sum_{p \in P^{liq}} f_{np}^{IP} = 0 \qquad \forall n \in N^{liq} \qquad (4.1)$$

$$f_n^I - \sum_e f_{ne}^{IV} - \sum_e f_{ne}^{IC} - \sum_{p \in P^{vap}} f_{np}^{IP} = 0 \qquad \forall n \in N^{vap} \qquad (4.2)$$

Chapter 4 • Process synthesis, optimization, and intensification 63

- splitters at the outlets of each side of module e

$$f_e^{LO} - \sum_{e'}(f_{ee'}^{LL} + f_{ee'}^{LH} + f_{ee'}^{LC}) - \sum_{p \in P^{liq}} f_{ep}^{LP} = 0 \tag{4.3}$$

$$f_e^{VO} - \sum_{e'}(f_{ee'}^{VV} + f_{ee'}^{VC}) - \sum_{p \in P^{vap}} f_{ep}^{VP} = 0 \tag{4.4}$$

$$f_e^{H} - \sum_{e'}(f_{ee'}^{HV} + f_{ee'}^{HL}) - \sum_{p} f_{ep}^{HP} = 0 \tag{4.5}$$

$$f_e^{C} - \sum_{e'} f_{ee'}^{CL} - \sum_{p \in P^{liq}} f_{ep}^{CP} = 0 \tag{4.6}$$

- mixers at the inlets of each side of module e

$$f_e^{LI} - \sum_{n \in N^{liq}} f_{ne}^{IL} - \sum_{e'}(f_{e'e}^{LL} + f_{e'e}^{CL}) - \sum_{e'} f_{e'e}^{HL} = 0 \tag{4.7}$$

$$f_e^{VI} - \sum_{n \in N^{vap}} f_{ne}^{IV} - \sum_{e'}(f_{e'e}^{VV} + f_{e'e}^{HV}) = 0 \tag{4.8}$$

$$f_e^{C} - \sum_{n} f_{ne}^{IC} - \sum_{e'}(f_{e'e}^{VC} + f_{e'e}^{LC}) = 0 \tag{4.9}$$

$$f_e^{H} - \sum_{n \in N^{liq}} f_{ne}^{IH} - \sum_{e'} f_{e'e}^{LH} = 0 \tag{4.10}$$

- final product mixers

$$f_p^{P} - \sum_{n \in N^{liq}} f_{np}^{IP} - \sum_{e}(f_{ep}^{LP} + f_{ep}^{CP} + f_{ep}^{HP}) = 0 \qquad \forall p \in P^{liq} \tag{4.11}$$

$$f_p^{P} - \sum_{n \in N^{vap}} f_{np}^{IP} - \sum_{e}(f_{ep}^{VP} + f_{ep}^{HP}) = 0 \qquad \forall p \in P^{vap} \tag{4.12}$$

B. Mass balances for each component at

- mixers prior to liquid and vapor sides and utility exchangers of module e

$$f_e^{LI} x_{ei}^{LI} - \sum_{n \in N^{liq}} f_{ne}^{IL} x_{ni}^{I} - \sum_{e'}(f_{e'e}^{LL} x_{e'i}^{LO} + f_{e'e}^{CL} x_{e'i}^{C}) - \sum_{e'} f_{e'e}^{HL} x_{e'i}^{H} = 0 \tag{4.13}$$

$$f_e^{VI} x_{ei}^{VI} - \sum_{n \in N^{vap}} f_{ne}^{IV} x_{ni}^{I} - \sum_{e'}(f_{e'e}^{VV} x_{e'i}^{VO} + f_{e'e}^{HV} x_{e'i}^{H}) = 0 \tag{4.14}$$

$$f_e^{C} x_{ei}^{C} - \sum_{n} f_{ne}^{IC} x_{ni}^{I} - \sum_{e'}(f_{e'e}^{VC} x_{e'i}^{VO} + f_{e'e}^{LC} x_{e'i}^{LO}) = 0 \tag{4.15}$$

$$f_e^{H} x_{ei}^{H} - \sum_{n \in N^{liq}} f_{ne}^{IH} x_{ni}^{I} - \sum_{e'} f_{e'e}^{LH} x_{e'i}^{LO} = 0 \tag{4.16}$$

64 Synthesis and Operability Strategies for Computer-Aided Modular PI

- final mixer of each product stream

$$f_p^P - \sum_{n \in N^{liq}} f_{np}^{IP} x_{ni}^I - \sum_e (f_{ep}^{LP} x_{ei}^{LO} + f_{ep}^{CP} x_{ei}^C + f_{ep}^{HP} x_{ei}^H) = 0 \qquad \forall p \in P^{liq} \qquad (4.17)$$

$$f_p^P x_i^P - \sum_{n \in N^{vap}} f_{np}^{IP} x_{ni}^I - \sum_e (f_{ep}^{VP} x_{ei}^{VO} + f_{ep}^{HP} x_{ei}^H) = 0 \qquad \forall p \in P^{vap} \qquad (4.18)$$

- around each mass/heat exchange module

$$f_e^{LI} x_{ei}^{LI} + f_e^{VI} x_{ei}^{VI} - f_e^{LO} x_{ei}^{LO} - f_e^{VO} x_{ei}^{VO} + \sum_k v_{ik} r_{ek} Mcat = 0 \qquad (4.19)$$

C. Energy balances at

- mixers prior to liquid and vapor sides and utility exchangers of module e

$$f_e^{LI} h_e^{LI} - \sum_{n \in N^{liq}} f_{ne}^{IL} h_n^I - \sum_{e'} (f_{e'e}^{LL} h_{e'}^{LO} + f_{e'e}^{CL} h_{e'}^{CO}) - \sum_{e'} f_{e'e}^{HL} h_{e'}^{HO} = 0 \qquad (4.20)$$

$$f_e^{VI} h_e^{VI} - \sum_{n \in N^{vap}} f_{ne}^{IV} h_n^I - \sum_{e'} (f_{e'e}^{VV} h_{e'}^{VO} + f_{e'e}^{HV} h_{e'}^{HO}) = 0 \qquad (4.21)$$

$$f_e^C h_e^{CI} - \sum_n f_{ne}^{IC} h_n^I - \sum_{e'} (f_{e'e}^{VC} h_{e'}^{VO} + f_{e'e}^{LC} h_{e'}^{LO}) = 0 \qquad (4.22)$$

$$f_e^H h_e^{HI} - \sum_{n \in N^{liq}} f_{ne}^{IH} h_n^I - \sum_{e'} f_{e'e}^{LH} h_{e'}^{LO} = 0 \qquad (4.23)$$

- final mixer of each product stream

$$f_p^P h^P - \sum_{n \in N^{liq}} f_{np}^{IP} h_n^I - \sum_e (f_{ep}^{LP} h_e^{LO} + f_{ep}^{CP} h_e^{CO} + f_{ep}^{HP} h_e^{HO}) = 0 \qquad \forall p \in P^{liq}$$

$$(4.24)$$

$$f_p^P h^P - \sum_{n \in N^{vap}} f_{np}^{IP} h_n^I - \sum_e (f_{ep}^{VP} h_e^{VO} + f_{ep}^{HP} h_e^{HO}) = 0 \qquad \forall p \in P^{vap} \qquad (4.25)$$

- around each mass/heat exchange module

$$f_e^{LI} h_e^{LI} + f_e^{VI} h_e^{VI} - f_e^{LO} h_e^{LO} - f_e^{VO} h_e^{VO} - \sum_k r_{ek} Mcat \Delta H_{reac,k} = 0 \qquad (4.26)$$

- around utility exchangers of each module e

$$Qh_e - f_e^H (h_e^{HO} - h_e^{HI}) = 0 \qquad (4.27)$$

$$Qc_e - f_e^C (h_e^{CI} - h_e^{CO}) = 0 \qquad (4.28)$$

D. Summation of molar fractions

- for streams $s = LI, LO, VI, VO, C, H, P$

$$\sum_i x_{ei}^s - 1 = 0 \tag{4.29}$$

E. Phase defining constraints

- for liquid streams

$$\sum_i (\gamma_{ei}^s P_{ei}^{sat,s} x_{ei}^s)/(\phi_{ei}^s P_e) \leq 1 \tag{4.30}$$

- for vapor streams

$$\sum_i (x_{ei}^s \phi_{ei}^s P_{tot})/(\gamma_{ei}^s P_{ei}^{sat,s}) \leq 1 \tag{4.31}$$

F. Mass transfer driving force constraints for each component

$$G1_{ei} \times G2_{ei} \leq 0 \tag{4.32}$$

$$G1_{ei} = f_e^{LO} x_{ei}^{LO} - f_e^{LI} x_{ei}^{LI} \tag{4.33}$$

$$G2_{ei} = \ln\left[\frac{\gamma_{ei}^L x_{ei}^L P_{ei}^{sat,L}}{\phi_{ei}^{VI} x_{ei}^V P_{tot}}\right] + \sum_i \sum_k \left[\frac{\nu_{ik} \Delta G_i^f}{RT_e^L} + \nu_{ik} \ln(\phi_{ei}^V x_{ei}^V P_e)\right] \frac{\partial \epsilon_k}{\partial n_{ei}^L} \tag{4.34}$$

G. Thermodynamic property calculation

- reaction rate

$$r_{ek} = f(x_{ei}^{LO}, T_e^{LO}) \tag{4.35}$$

- density

$$\rho_e^{LO} = f(x_{ei}^{LO}, T_e^{LO}) \tag{4.36}$$

- saturated vapor pressure, enthalpy, activity coefficient, and fugacity coefficient for $s = LI, LO, VI, VO, HI, HO, CI, CO, P$

$$p_{ei}^{sat,s} = f(T_e^s) \tag{4.37}$$

$$h^s = f(x_{ei}^s, T_e^s) \tag{4.38}$$

$$\gamma_{ei}^s = f(x_{ei}^s, T_e^s, P_e) \tag{4.39}$$

$$\phi_{ei}^s = f(x_{ei}^s, T_e^s, P_e) \tag{4.40}$$

II. Structural model

A. To define the existence of process streams – i.e. no stream flows exist if the module does not exist

$$[f_e^{LI} + f_e^{LO} + f_e^{VI} + f_e^{VO}] - y_e F^{max} \leq 0 \tag{4.41}$$

where F^{max} is the upper bound for stream flowrates.

66 Synthesis and Operability Strategies for Computer-Aided Modular PI

B. To define the existence of interconnecting streams

$$f^s_{ee'} - y^s_{ee'} F^{max} \leq 0 \qquad (4.42)$$

C. To define the existence of utility exchangers

$$yh_e - y_e \leq 0 \qquad (4.43)$$

$$yc_e - y_e \leq 0 \qquad (4.44)$$

$$f^C_e - yc_e F^{max} \leq 0 \qquad (4.45)$$

$$f^H_e - yh_e F^{max} \leq 0 \qquad (4.46)$$

$$Qh_e - yh_e Q^{max} \leq 0 \qquad (4.47)$$

$$Qc_e - yc_e Q^{max} \leq 0 \qquad (4.48)$$

where Q^{max} is the upper bound for heat transfer loads.

D. To define that if a module exists, there will be an inlet flow

$$y_e - [\sum_{n \in N^{liq}} yIl_{ne} + \sum_{e'} yll_{e'e} + \sum_{e'} ycl_{e'e} + \sum_{e'} yhl_{e'e}] \leq 0 \qquad (4.49)$$

$$y_e - [\sum_{n \in N^{vap}} yIv_{ne} + \sum_{e'} yvv_{e'e} + \sum_{e'} yhv_{e'e}] \leq 0 \qquad (4.50)$$

$$yh_e - [\sum_{n \in N^{liq}} yIh_{ne} + \sum_{e'} ylh_{e'e}] \leq 0 \qquad (4.51)$$

$$yc_e - [\sum_{n} yIc_{ne} + \sum_{e'} yvc_{e'e} + \sum_{e'} ylc_{e'e}] \leq 0 \qquad (4.52)$$

E. To define that if a module exists, there will be an outlet flow

$$y_e - [\sum_{e'} yll_{ee'} + \sum_{e'} ylh_{ee'} + \sum_{e'} ylc_{ee'} + \sum_{p \in P^{liq}} ylp_{ep}] \leq 0 \qquad (4.53)$$

$$y_e - [\sum_{e'} yvv_{ee'} + \sum_{e'} yvc_{ee'} + \sum_{p \in P^{vap}} yvp_{ep}] \leq 0 \qquad (4.54)$$

$$yh_e - [\sum_{e'} yhv_{ee'} + \sum_{e'} yhl_{ee'} + \sum_{p} yhp_{ep}] \leq 0 \qquad (4.55)$$

$$yc_e - [\sum_{e'} ycl_{ee'} + \sum_{p} ycp_{ep}] \leq 0 \qquad (4.56)$$

F. To ensure that at least one feed point per initial stream

$$1 - [\sum_{e} yIl_{ne} + \sum_{e} yIh_{ne} + \sum_{e} yIc_{ne} + \sum_{p \in P^{liq}} yIp_{np}] \leq 0 \qquad n \in N^{liq} \qquad (4.57)$$

$$1 - [\sum_{e} yIv_{ne} + \sum_{e} yIc_{ne} + \sum_{p \in P^{vap}} yIp_{np}] \leq 0 \qquad n \in N^{vap} \qquad (4.58)$$

Chapter 4 • Process synthesis, optimization, and intensification   67

G. To ensure that at least one source per product

$$1 - [\sum_{n \in N^{liq}} yIp_{np} + \sum_e ylp_{ep} + \sum_e ycp_{ep} + \sum_e yhp_{ep}] \leq 0 \qquad p \in P^{liq} \qquad (4.59)$$

$$1 - [\sum_{n \in N^{vap}} yIp_{np} + \sum_e yvp_{ep} + \sum_e yhp_{ep}] \leq 0 \qquad p \in P^{vap} \qquad (4.60)$$

H. To constrain the number of streams allowed at existing mixers

$$[\sum_{n \in N^{liq}} yIl_{ne} + \sum_{e'} yll_{e'e} + \sum_{e'} ycl_{e'e} + \sum_{e'} yhl_{e'e}] - y_e Nmix^{max} \leq 0 \qquad (4.61)$$

$$[\sum_{n \in N^{vap}} yIv_{ne} + \sum_{e'} yvv_{e'e} + \sum_{e'} yhv_{e'e}] - y_e Nmix^{max} \leq 0 \qquad (4.62)$$

$$[\sum_{n \in N^{liq}} yIh_{ne} + \sum_{e'} ylh_{e'e}] - yh_e Nmix^{max} \leq 0 \qquad (4.63)$$

$$[\sum_n yIc_{ne} + \sum_{e'} yvc_{e'e} + \sum_{e'} ylc_{e'e} - yc_e Nmix^{max} \leq 0 \qquad (4.64)$$

I. To constrain the number of splits allowed at existing splitters

$$[\sum_{e'} yll_{ee'} + \sum_{e'} ylh_{ee'} + \sum_{e'} ylc_{ee'} + \sum_{p \in P^{liq}} ylp_{ep}] - y_e Nsplit^{max} \leq 0 \qquad (4.65)$$

$$[\sum_{e'} yvv_{ee'} + \sum_{e'} yvc_{ee'} + \sum_{p \in P^{vap}} yvp_{ep}] - y_e Nsplit^{max} \leq 0 \qquad (4.66)$$

$$[\sum_{e'} yhl_{ee'} + \sum_p yhp_{ep}] - yh_e Nsplit^{max} \leq 0 \qquad (4.67)$$

$$[\sum_{e'} ycl_{ee'} + \sum_p ycp_{ep}] - yc_e Nsplit^{max} \leq 0 \qquad (4.68)$$

J. To constrain the number of splits allowed at initial splitters

$$[\sum_e yIl_{ne} + \sum_e yIh_{ne} + \sum_e yIc_{ne} + \sum_{p \in P^{liq}} yIp_{np}] - Nsplit^{max} \leq 0 \qquad n \in N^{liq} \quad (4.69)$$

$$[\sum_e yIv_{ne} + \sum_e yIc_{ne} + \sum_{p \in P^{vap}} yIp_{np}] - Nsplit^{max} \leq 0 \qquad n \in N^{vap} \quad (4.70)$$

K. To constrain the number of product sources

$$[\sum_{n \in N^{liq}} yIp_{np} + \sum_e ylp_{ep} + \sum_e ycp_{ep} + \sum_e yhp_{ep}] - Nmix^{max} \leq 0 \qquad p \in P^{liq} \quad (4.71)$$

$$[\sum_{n \in N^{vap}} yIp_{np} + \sum_e yvp_{ep} + \sum_e yhp_{ep}] - Nmix^{max} \leq 0 \qquad p \in P^{vap} \quad (4.72)$$

68   Synthesis and Operability Strategies for Computer-Aided Modular PI

where $Nmix^{max}$ and $Nsplit^{max}$ give the maximum number of streams allowed to be mixed or split respectively at mixers or splitters. Initially, $Nmix^{max} = Nsplit^{max} = 3$ is used and is relaxed if the constraint is active. These bounds are introduced to aid the solution of the resulting optimization problem.

L. Non-redundancy constraints

$$y_{e+1} - y_e \leq 0 \tag{4.73}$$

III. Objective function

The objective function (Eq. 4.74) can be formulated to optimize cost performances, considering: (i) cost of raw materials, (ii) cost of heating and cooling utilities, and (iii) cost of GMF modules.

$$Obj = \sum_n C_n f_I^n + \sum_e C_{cw} Q_{c_e} + \sum_e C_{st} Q_{h_e} + \sum_e Cost_e y_e \tag{4.74}$$

While operating costs can be calculated via the heat transfer loads in utility heat exchange modules, capital costing correlations cannot be applied as no equipment information is available at this stage. A pseudo-capital module cost function as shown in Eq. (4.75) was developed to estimate the module cost as a function of operating conditions (e.g., stream flows, module holdup). Despite that a truly economically optimal process option cannot be obtained, this approximation can provide an initial idea on equipment sizing, with the assumption that more GMF modules correspond to larger equipment units. The detailed derivation of the pseudo-capital module cost for reaction and/or separation systems is provided in the next section.

$$Cost_e(\$) = \begin{cases} 99.507(\frac{MW_e^{VI}}{\rho_e^{VI}})^{0.25}\sqrt{f_e^{VI}}H^{0.802}(2.18 + F_c) \\ \qquad\qquad \text{if (reactive) separation module} \\ 583.295V_e^{0.623}(2.18 + F_c) \\ \qquad\qquad \text{if reaction module} \end{cases} \tag{4.75}$$

4.3 Pseudo-capital cost estimation

In what follows, we introduce how to estimate the pseudo-capital cost for GMF mass/heat exchange modules, as developed by Ismail et al. [1]. The cost estimation is applicable to reaction and/or separation systems.

Let us approximate a GMF mass/heat exchange module as a pressure vessel and start with the Guthrie's costing correlation:

$$\text{Installed Cost}, \$ = \frac{M\&S}{280} 101.9 D^{1.066} H^{0.802} (2.18 + F_c) \tag{4.76}$$

where D is diameter (ft), H is height (ft). F_c takes the value of 1 for a carbon steel vessel with a maximum pressure of 3.4 atm, or takes the value of 1.5 for pressure up to 13.6 atm.

Chapter 4 • Process synthesis, optimization, and intensification   69

4.3.1 Separation and reactive separation modules

A mass/heat exchange module, in which liquid-vapor (reactive) separation takes place, can be estimated as a section of distillation column as discussed in Chapter 3. Thus, the pseudo-capital cost of this module is majorly determined by its: (i) inlet/outlet vapor flowrate, (ii) liquid molar holdup, and (iii) extent of mass exchange.

The diameter of the module can be obtained via Eq. (4.77):

$$A = \frac{\pi D^2}{4} = \frac{\text{Vapor volumetric flowrate}}{\text{Vapor velocity}} \tag{4.77}$$

where A is the cross-sectional area of the module. The vapor volumetric flowrate can be calculated by dividing the vapor molar flowrate (i.e., f^V in the GMF synthesis model) by the vapor molar density (e.g., estimated as ideal gas). The vapor velocity can be assumed as 80% of the flooding velocity, the latter of which can be calculated via Eq. (4.78) [2]:

$$\frac{v_{flood}}{\rho_m^V} = 1.5 \tag{4.78}$$

where v_{flood} is flooding velocity (ft/s), ρ_m^V is vapor mass density (lb/ft^3).

Based on Eq. (4.78) and further considering a 12% of the cross-sectional area to be occupied by internals, Eq. (4.77) can be reformulated as:

$$D = 0.2734(\frac{MW^V}{\rho^V})^{0.25}\sqrt{f^V} \tag{4.79}$$

where f^V is vapor molar flowrate (kmol/h), MW^V is the vapor mixture molecular weight (g/mol), and ρ^V is the vapor molar density (mol/m^3).

The height of the module (H) is a function of the liquid holdup and the extent of mass transfer in the module. H can be estimated via Eq. (4.80):

$$H = 2 + \frac{4V}{\pi D^2} \tag{4.80}$$

Based on Eq. (4.76), take $M\&S$ as 1000 and approximate 1.066 as 1, then the pseudo-capital cost of a liquid-vapor module can be estimated by:

$$Cost_{L-V}(\$) = 99.507(\frac{MW^V}{\rho^V})^{0.25}\sqrt{f^V} H^{0.802}(2.18 + F_c) \tag{4.81}$$

where Eqs. (4.79) and (4.80) should be used to calculate H (ft).

4.3.2 Reaction modules

A mass/heat exchange module, in which liquid phase reaction takes place, can be approximated as a reactor. Thus, the pseudo-capital cost of the module is determined by the liquid

70   Synthesis and Operability Strategies for Computer-Aided Modular PI

holdup volume and/or the catalyst volume if applicable. Assuming the module is a cylindrical vessel, its liquid holdup volume is calculated via:

$$V = \frac{\pi D^2}{4} H = \frac{6\pi D^3}{4} \tag{4.82}$$

The diameter of the module can then be obtained by:

$$D = (\frac{2V}{3\pi})^{1/3} \tag{4.83}$$

Considering a height to diameter ratio of 6:

$$D^{1.066} H^{0.802} = 6^{0.802} D^{1.868} = 6^{0.802} (\frac{2V}{3\pi})^{0.623} = 1.602 V^{0.623} \tag{4.84}$$

Again take the $M\&S$ value as 1000, Eq. (4.76) can be re-arranged as:

$$Cost_{L-L}(\$) = 583.295 V^{0.623} (2.18 + F_c) \tag{4.85}$$

In sum, the pseudo-capital cost for a GMF mass/heat exchange module e can be estimated as per Eq. (4.86). Although an accurate cost calculation cannot be performed at this stage, an approximation of the cost for GMF modules can be obtained with corresponding sizing parameters.

$$Cost\ (\$) = \begin{cases} 99.507(\frac{MW^V}{\rho^V})^{0.25}\sqrt{f^{VI}}H^{0.802}(2.18 + F_c) \\ \qquad\qquad \text{if (reactive) separation module} \\ 583.295 V^{0.623}(2.18 + F_c) \\ \qquad\qquad \text{if reaction module} \end{cases} \tag{4.86}$$

4.4 Solution strategy

Several methods can be used to solve the resulting mixed-integer nonlinear programming (MINLP) problem [3], such as Branch and Bound [4], Generalized Benders Decomposition (GBD) [5], Outer Approximation [6], etc. Herein, the GBD method is adapted because of the control it provides during the solution procedure, which is essential for this type of highly nonconvex MINLP problems to avoid the large infeasible portion of the design space. Due to the highly nonconvex and nonlinear nature of the model, the GMF synthesis problem cannot be guaranteed to solve to global optimality. However, arguably even intermediate solutions can provide significant insights to the process designs. In what follows, we provide a brief review on the GBD algorithm with application to the GMF/MINLP problem.

The GMF synthesis model, presented in detail in the previous sections, can be recast in the following compact mathematical form:

$$\begin{aligned}
\min \quad & z = f(x, y) && \text{objective function} \\
\text{s.t.} \quad & h(x) = 0 && \text{mass and energy balances,} \\
& && \text{physical/kinetic properties, etc.} \\
& g_1(x, y) \leq 0 && \text{existence of modules, streams, etc.} \\
& g_2(y) \leq 0 && \text{structural constraints}
\end{aligned} \quad (4.87)$$

where x is the vector of continuous variables (e.g., flowrate, temperature, composition), and y is the vector of structural binary variables.

The GBD algorithm decomposes a large-scale MINLP problem (Eq. 4.87) to:

- A Primal Problem, or nonlinear programming (NLP) subproblem: by fixing the binary variables y, it reduces to a NLP problem to solve for the continuous variables x. The solution of this primal problem provides an upper bound to the overall solution of Eq. (4.87);
- A Master Problem, or mixed-integer programming (MIP) subproblem: the binary variables y are determined based on the information obtained from solving the Primal Problem (e.g., Lagrange multipliers). The solution of this master problem provides a lower bound to the overall solution of Eq. (4.87).

The iteration steps and convergence criteria of the GBD algorithm are depicted in Fig. 4.3 and detailed below:

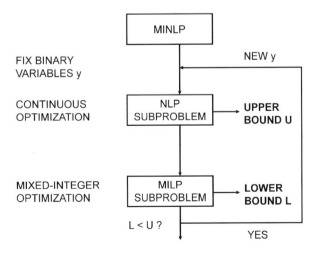

FIGURE 4.3 GBD algorithm workflow.

72 Synthesis and Operability Strategies for Computer-Aided Modular PI

Step 1. Initialization – Set $k = 1$, $z^u = +\infty$, select y^1

Step 2. Primal Problem in the kth iteration

$$z(y^k) = \min \quad f(x, y^k)$$
$$\text{s.t.} \quad h(x) = 0 \qquad\qquad (4.88)$$
$$g_1(x, y^k) \leq 0$$

With y^k fixed by initial guesses or by solving the Master Problem in the $(k-1)^{th}$ iteration, the solution of the kth Primal Problem provides: (i) the optimal values for continuous variables x in the kth iteration, i.e. x^k, (ii) the vector of Lagrange multipliers for mixed-integer model constraints (not including pure integer constraints) λ^k.

If $z(y^k) < z^U$, then assign $z^U = z(y^k)$, and update candidate MINLP solutions $x^* = x^k$, $y^* = y^k$.

Step 3. Master Problem in the kth iteration

$$z_\mu^k = \min \quad \mu$$

$$\text{s.t.} \quad \mu \geq f(x^k, y) + \sum_{i=1}^{t} \lambda_i^k g_{1,i}(x^k, y_i) \quad k = 1, ..., K$$

$$g_2(y) \leq 0 \qquad\qquad (4.89)$$

$$y \in \{0, 1\}^m$$

$$\sum_{i \in B^k} y_i - \sum_{i \in N^k} y_i \leq |B^k| - 1 \qquad k = 1, ..., K$$

The last constraint in Eq. (4.89) defines the integer cut to exclude the y combinations which have been examined in the previous iterations. B^k is the set to store binary variables assigned as 1 in each iteration, while N^k is the set to store binary variables assigned to be 0. Infeasibility cut constraints (Eq. 4.90) can also be added to Eq. (4.89) in the case of infeasible Primal Problem in any iteration k:

$$\sum_{i=1}^{t} \lambda_i^k g_{1,i}(x^k, y_i) \leq 0 \qquad\qquad (4.90)$$

With x^k and μ^k fixed, the binary variables y are determined via the solution of this MIP problem. In the meantime, z_μ^k provides the lower bound to the overall solution.

If $z_\mu^k \geq z^U$, stop (as shown in Fig. 4.4). The solution of the original MINLP problem (Eq. 4.87) is given by (x^*, y^*, z^U). Otherwise, set $k = k + 1$, return to step 2 with new y^{k+1} values.

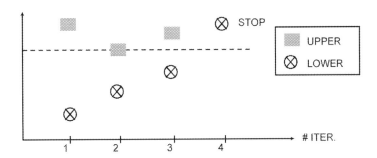

FIGURE 4.4 GBD convergence criteria.

Several notes regarding GBD for the solution of GMF problems:

- The structure of GMF model formulation (i.e., a physical model with continuous variables and a structural model with mixed-integer variables) fits very well with the GBD decomposition strategy. This will help to decompose the computational complexities resulted respectively by model nonlinearities and by binary variables during the solution procedure. The infeasible space – which can be quite large for these phenomena-based synthesis approaches – can also be efficiently ruled out with the infeasibility cut constraints (Eq. 4.90), after which the previous feasible solutions can be recovered to re-initialize the iteration.
- The iterative characteristic of GBD enables the record of any intermediate feasible solutions, which can be used as intermediate design solutions, generated throughout the solution procedure. The integer cut constraint can also be used to rule out the optimal GBD solution, and to generate the second or third-best solutions with different design structures determined by the different sets of binary variables.
- GBD also allows for flexible manipulation of binary variable combinations – in the case of retrofit design, certain binary variables can be fixed to maintain the original structure while optimizing other ones for minor design changes. All these characteristics will be highlighted via the case studies in the following chapters.

4.5 Motivating example: GMF synthesis representation and optimization of a binary distillation system

In this example by Georgiadis and Pistikopoulos [7], GMF is applied for the optimization of a binary distillation system. For illustration purpose, we fix the binary variables of the GMF structural model to represent a specific column configuration. The physical model is utilized to describe the liquid-vapor separation process and is validated against tray-by-tray distillation optimization. GMF capabilities on generating alternative structures (i.e. without pre-postulations on binary variables) will be showcased via the case studies in the Part 3 of this book.

4.5.1 Process description

A binary mixture of benzene and toluene is to be separated using a simple distillation column to obtain high purity benzene product. The process data are based on the EX5FEED test problem provided by Viswanathan and Grossmann [8].

The feed stream consists of 0.7 mol/mol benzene and 0.3 mol/mol toluene at a flowrate of 100 mol/s. The feed temperature and pressure are respectively at 359.98 K and 1 atm (i.e., at bubble point). The mixture phase behavior can be accurately described using ideal liquid-vapor equilibrium. The parameters for other physical properties (e.g., heat capacity, enthalpy of vaporization, saturated vapor pressure) are adapted from Reid et al. [9].

The distillation column is operated at a fixed pressure at 1 am. The process specification is to obtain a benzene product from the distillate, with a purity higher than 0.99 mol/mol. The optimization objective function is to maximize $(P1 - 50r)$ for energy saving considerations (where $P1$ is distillate rate, r is reflux ratio).

4.5.2 GMF problem formulation

Fig. 4.5 illustrates the GMF representation of the simple distillation column for benzene-toluene separation. The black solid lines indicate the activated streams and modules (i.e., with the associated binary variable as 1), while the gray dashed lines show the deactivated streams and modules based on the full superstructure connections shown in Fig. 4.2.

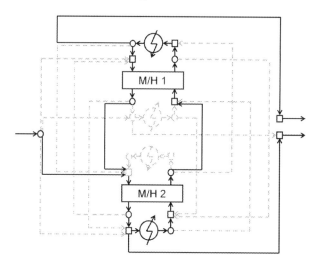

FIGURE 4.5 GMF structural variables for a simple distillation column representation.

The GMF structural sets and binary variables need to be specified according to the unique design structure. For example, two mass/heat exchange modules (M/H) are utilized to respectively represent the stripping section and the rectifying section, i.e. $E = \{M/H1, M/H2\}$ numbered in a descending order. Two pure heat exchange modules are used respectively to represent the reboiler and the condenser to the column. The pure heat exchange module for cooling task is associated with $M/H1$ and that for heating task is with $M/H2$, i.e. $yh_2 = 1$ and $yc_1 = 1$. The other non-zero structural sets and binary variables are listed in Table 4.2.

Table 4.2 GMF structural sets and variables for a simple distillation column representation.

	Structural Sets	Binary Variables
Modules	$E = \{M/H1,\ M/H2\}$	$y_1 = y_2 = 1$
		$yh_2 = 1,\ yc_1 = 1$
Feed stream	$I = \{F\}$	$yIl_{F,2} = 1$
Product streams	$P = \{P1,\ P2\}$	$ylp_{2,P2} = 1,\ ycp_{1,P1} = 1$
Interconnecting streams		$yll_{1,2} = 1,\ yvv_{2,1} = 1$
		$ycl_{1,1} = 1,\ yvc_{1,1} = 1$
		$yhv_{2,2} = 1,\ ylh_{2,2} = 1$

Then the GMF physical model is applied to describe the mass exchange and heat exchange phenomena taking place in each module. Additional modeling equations are added for this specific case to: (i) fix the module pressure at 1 atm, and (ii) explicitly calculate the column reflux ratio as $r = f_{1,1}^{CL}/f_{1,P1}^{CP}$.

4.5.3 Results and discussions

By fixing the binary variables, the mathematical model is reduced to a NLP problem which is solved using CONOPT. The optimal GMF solution provides the module design and operating parameters. To validate the optimality and accuracy of the GMF solution, we compare it with a rigorous tray-by-tray model of Viswanathan and Grossmann [10] for distillation column optimization. The optimization objective for the tray-by-tray model, using MESH equations, is consistent with that for GMF synthesis. Based on the EX5FEED test example, the design of the distillation column is fixed with 25 equilibrium stages. The feed stream enters the column at the 11th stage from the bottom.

The comparisons of GMF versus tray-by-tray model on energy consumption and objective function are presented in Table 4.3. It can be noted that the results are in quantitative agreement, although GMF tends to further intensify the process towards lower energy consumption with the abstract modular representation. Fig. 4.6 compares the molar faction profile and temperature profile between GMF representation and tray-by-tray model. As can be noted, GMF captures well the major trend of separation taking place in the binary column, although tray-wise information within each column section is missing. However, this is beyond the purpose of the GMF, which aims to provide rapid screening capability at the synthesis level instead of performing rigorous simulation studies.

Moreover, from modeling statistics point of view, GMF provides a more compact representation compared to the tray-by-tray model. As shown in Fig. 4.6, GMF only evaluates 7 points out of the full binary distillation column against the 25 points examined by the tray-by-tray model, while it can still provide a valid physical representation and identify the optimal operating conditions. The detailed model statistics of the resulting NLP problems are given in Table 4.4, which further demonstrate the compactness in the GMF representation. It is also worth noting that, with the aggregation nature of mass/heat exchange modules, the GMF model size expands only if the number of column sections increases but not with the number of column trays – which is a notable difference with the tray-by-tray model.

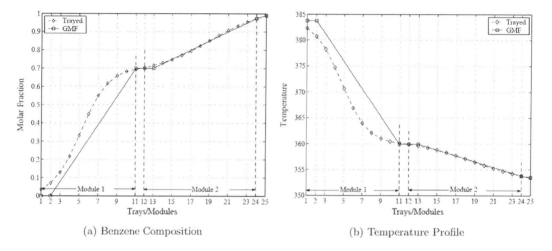

(a) Benzene Composition   (b) Temperature Profile

FIGURE 4.6 Validation of GMF for binary distillation.
(a) Molar fraction profile, (b) Temperature profile (reproduced from [7]).

Table 4.3 Variable validation of GMF.

	GMF	Trayed Model
Reboiler duty (kW)	4194	4401
Condenser duty (kW)	4149.5	4398.6
$P1$ (mol/s)	70.7	69.7
r	0.91	0.97

Table 4.4 Optimization statistics.

	GMF	Trayed	% Difference
Continuous variables	134	283	53%
Continuous equality equations	127	282	55%

Nomenclature

a	activity
D	module diameter
f	molar flowrate

Chapter 4 • Process synthesis, optimization, and intensification   77

F^{max}	upper bound of molar flowrate
G	Gibbs free energy
H	module height
h	enthalpy
k	rate constant
Ka	reaction equilibrium constant
$Level$	liquid level
M	big M constraint
$Mcat$	catalyst mass
MW	molecular weight
P^{sat}	saturated vapor pressure
P_{tot}	pressure
Q	heat load
Q^{max}	upper bound of heat load
r	reaction rate
T	temperature
V	module volume
x	molar fraction
y	binary variable

Subscripts

c	component
e, e'	module
n	feed stream
p	product stream

Greek letters

ϵ	reaction extent
γ	activity coefficient
μ	chemical potential
ν	stoichiometric coefficient
ϕ	fugacity coefficient
ρ	density

References

[1] S.R. Ismail, P. Proios, E.N. Pistikopoulos, Modular synthesis framework for combined separation/reaction systems, AIChE Journal 47 (2001) 629–649.

[2] J. Douglas, A hierarchical decision procedure for process synthesis, AIChE Journal 31 (1985) 353–362.

[3] C.A. Floudas, Nonlinear and Mixed-Integer Optimization: Fundamentals and Applications, Oxford University Press, 1995.

[4] E.M.L. Beale, Integer programming, in: K. Schittkowski (Ed.), Computational Mathematical Programming, Springer Berlin Heidelberg, Berlin, Heidelberg, 1985, pp. 1–24.

[5] A.M. Geoffrion, Generalized benders decomposition, Journal of Optimization Theory and Applications 10 (1972) 237–260.

[6] M.A. Duran, I.E. Grossmann, An outer-approximation algorithm for a class of mixed-integer nonlinear programs, Mathematical Programming 36 (1986) 307–339.

[7] M.C. Georgiadis, E.N. Pistikopoulos, Energy and Process Integration, Begell House, 2006.

[8] J. Viswanathan, I.E. Grossmann, A combined penalty function and outer-approximation method for minlp optimization, Computers & Chemical Engineering 14 (1990) 769–782.

[9] R.C. Reid, J.M. Prausnitz, B.E. Poling, The properties of gases and liquids, 1987.

[10] J. Viswanathan, I.E. Grossmann, Optimal feed locations and number of trays for distillation columns with multiple feeds, Industrial & Engineering Chemistry Research 32 (1993) 2942–2949.

5

Enhanced GMF for process synthesis, intensification, and heat integration

In Chapter 4, we have discussed the process intensification synthesis capabilities of GMF using a set of mass/heat exchange modules and a set of pure heat exchange modules. In this chapter, we extend GMF synthesis to:

- Enhance intra-module representation by coupling with Orthogonal Collocation,
- Formally incorporate heat integration considerations with process intensification.

5.1 GMF synthesis model with Orthogonal Collocation

To further enhance the intra-module representation and to more accurately estimate module design parameters of a highly compact and abstract mass/heat exchange module, Proios and Pistikopoulos [1] proposed to couple the Orthogonal Collocation (OC) method [2] with GMF in analogy to the OC method for multistage separation systems [3]. As depicted in Fig. 5.1, a minimum essential set of nc interior collocation points are used to discretize each mass/heat exchange module in addition to 2 exterior collocation points. These collocation points correspond to a total of M_e "intra-segments" related by the roots of Hahn polynomials following orthogonality conditions. If we denote the location of each interior collocation point as $s_{e,j}$, then $s_{e,j} = s_j(M_e)$. Liquid and vapor flowrates, enthalpies, and other stream property variables also need to be approximated at these collocation points by introducing intra-module physical constraints, such as mass and energy balances, etc. Additional physical constraints are imposed to link the intra-module GMF/OC representation with the inlet and outlet streams from the GMF module as presented in Section 4.2. In this way, only continuous variables are introduced to describe the physical phenomena at the collocation points. The combinatorial size of the resulting GMF/OC synthesis problem remains compact in terms of binary variables. The detailed application of GMF/OC for process intensification synthesis will be illustrated in Chapter 11 for extractive separation with material selection. The modeling constraints in a GMF/OC model are detailed in what follows.

80 Synthesis and Operability Strategies for Computer-Aided Modular PI

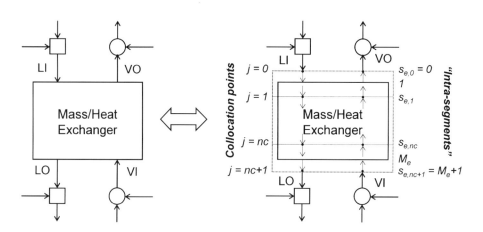

FIGURE 5.1 GMF and GMF/OC representations.

- Collocation point location calculation

$$s_{e,0} = 1 \tag{5.1}$$

$$s_{e,j} = s_j(M_e) \tag{5.2}$$

$$s_{e,nc_e+1} = (M_e + 1) \tag{5.3}$$

where j is the set for collocation points, $j = 1, ..., nc_e$. For example, if 1 interior collocation point is selected (i.e., $nc_e = 1$), then:

$$s_{e,0} = 1 \tag{5.4}$$

$$s_{e,1} = s_1(M_e) = \frac{M_e + 1}{2} \tag{5.5}$$

$$s_{e,2} = (M_e + 1) \tag{5.6}$$

- Lagrange polynomial weighting functions

$$W_{j'}^V(s_e) - \prod_{z=1, z \neq j'}^{nc_e+1} \frac{s - s_{e,j}}{s_{e,j'} - s_{e,z}} = 0 \tag{5.7}$$

$$W_{j'}^L(s_e) - \prod_{z=1, z \neq j'}^{nc_e} \frac{s - s_{e,j}}{s_{e,j'} - s_{e,z}} = 0 \tag{5.8}$$

where $W_j^V(s_e)$ and $W_j^L(s_e)$ are the Lagrange polynomial weighting functions to the number of "intra-segments" per module respectively on the vapor and liquid sides.

Chapter 5 • Enhanced GMF for process synthesis, intensification, & heat integration   81

- Component mass and energy balances

$$\sum_{j'=1}^{nc_e+1} W_{j'}^V(s_{e,j}+1) fc_i^V(s_{e,j'}) - fc_i^V(s_{e,j})$$
$$+ \sum_{j'=0}^{nc_e} W_{j'}^L(s_{e,j}-1) fc_i^L(s_{e,j'}) - fc_i^L(s_{e,j'}) = 0 \tag{5.9}$$

$$\sum_{j'=1}^{nc_e+1} W_{j'}^V(s_{e,j}+1) f^V(s_{e,j'}) h^V(s_{e,j'}) - f^V(s_{e,j}) h^V(s_{e,j})$$
$$+ \sum_{j'=0}^{nc_e} W_{j'}^L(s_{e,j}-1) f^L(s_{e,j'}) h^L(s_{e,j'}) - f^L(s_{e,j'}) h^L(s_{e,j'}) = 0 \tag{5.10}$$

where $fc_i^V(s_{e,j})$ and $fc_i^L(s_{e,j})$ are respectively the vapor and liquid flowrates for component i at collocation point $s_{e,j}$, $f^V(s_{e,j})$ and $f^L(s_{e,j})$ are respectively the total vapor and liquid flowrates at collocation point $s_{e,j}$.

- Molar fraction summations

$$\sum_i \frac{fc_c^V(s_{e,j})}{f^V(s_{e,j})} - 1 = 0 \tag{5.11}$$

$$\sum_i \frac{fc_c^L(s_{e,j})}{f^L(s_{e,j})} - 1 = 0 \tag{5.12}$$

- Driving force constraints

$$G1_i(s_{e,j}) \times G2_i(s_{e,j}) \leq 0 \tag{5.13}$$

$$G1_i(s_{e,j}) = fc_c^L(s_{e,j}) - fc_c^L(s_{e,j}-1) \tag{5.14}$$

$$G2_i(s_{e,j}) = \ln \frac{\gamma_i(s_{e,j}) P_i^{sat}(s_{e,j}) \frac{fc_c^L(s_{e,j})}{f^L(s_{e,j})}}{\phi_i(s_{e,j}) P(s_{e,j}) \frac{fc_c^V(s_{e,j})}{f^V(s_{e,j})}}$$
$$+ \sum_i \sum_k \left[\frac{\nu_{ik} \Delta G_i^f}{RT^L(s_{e,j})} + \nu_{ik} \ln \left(\phi_i(s_{e,j}) P(s_{e,j}) \frac{fc_c^V(s_{e,j})}{f^V(s_{e,j})} \right) \right] \frac{\partial \epsilon_k}{\partial n_i^L(s_{e,j})} \tag{5.15}$$

- Equations to link the interior collocation points with the inlet/outlet streams of the mass/heat module

$$f_e^{VO} - \sum_i \left(\sum_{j=1}^{nc_e+1} W_j^V(1) fc_i^V(s_{e,j}) \right) = 0 \tag{5.16}$$

$$f_e^{VO} x_{e,i}^{VO} - \sum_{j=1}^{nc_e+1} W_j^V(1) fc_i^V(s_{e,j}) = 0 \tag{5.17}$$

$$f_e^{VO}h_e^{VO} - \sum_{j=1}^{nc_e+1} W_j^V(1)f^V(s_{e,j})h^V(s_{e,j}) = 0 \qquad (5.18)$$

$$f_e^{LI} - f^L(s_{e,0}) = 0 \qquad (5.19)$$

$$f_e^{LI}x_{e,i}^{LI} - fc_i^L(s_{e,0}) = 0 \qquad (5.20)$$

$$f_e^{LI}h_e^{LI} - f^L(s_{e,0})h^L(s_{e,0}) = 0 \qquad (5.21)$$

5.2 GMF synthesis model with heat integration

A pure heat exchange module (Fig. 5.2) – regardless for utility heating, utility cooling, or heat integration between process streams – can be partitioned as a hot side and a cold side. Heat will be transferred from the hot side stream to the cold side stream if the temperature gradients suffice.

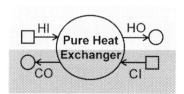

FIGURE 5.2 GMF pure heat exchange module.

From a modeling point of view, a process stream, either an initial feed stream or a stream coming out of a mass/heat exchange module, can be directed to either the hot side or the cold side of the heat exchanger. In other words, the stream property as a hot stream or a cold stream will not be declared *in priori*. To ensure the heat transfer feasibility between the stream pair, the following mathematical formulations in Eq. (5.22) and Eq. (5.23) are used to determine if temperature gradient is satisfied on both sides of the heat exchange module:

$$T^{CO} - T^{HI} + EMTA \leq M(1-y) \qquad (5.22)$$

$$T^{CI} - T^{HO} + EMTA \leq M(1-y) \qquad (5.23)$$

where T stands for the temperatures of process or utility streams – i.e. for utility heater using steam: $T^{HI} = T^{HO} = T^{steam}$; for utility cooler using cooling water: $T^{CI} = T^{CO} = T^{water}$; for heat integration between process streams, the temperatures are determined by the energy balance around the heat integration block. Furthermore, $EMTA$ stands for Exchange Minimum Temperature Approach – for example, if EMTA is assigned as 10 K, that means heat transfer can only take place between the two process streams with a temperature

Chapter 5 • Enhanced GMF for process synthesis, intensification, & heat integration   83

difference larger than 10 K. M is a sufficiently large and positive constant to activate or de-activate the modeling constraint based on the selection of y. y is a binary variable, which can take the value of 1 or 0 to denote the existence or not of this heat exchange module.

To give a more detailed example of how the binary variable for the pure heat exchange module and the big M constant are working:

- **If y equals to 1**, then this pure heat exchange module exists. In the feasibility constraints, the right hand sides turn out to be 0. Eqs. (5.22)–(5.23) are now equivalent to:

$$T^{CO} - T^{HI} + EMTA \leq 0 \tag{5.24}$$

$$T^{CI} - T^{HO} + EMTA \leq 0 \tag{5.25}$$

More intuitively, the temperature difference between the cold side and the hot side is greater than the exchange minimum temperature

- **If y takes the other value as 0**, then this pure heat exchange module does not exist. In this case, the feasibility constraints turn out to be:

$$T^{CO} - T^{HI} + EMTA \leq M \tag{5.26}$$

$$T^{CI} - T^{HO} + EMTA \leq M \tag{5.27}$$

Note that M is a very large positive constant. Thus, these two inequality constraints always hold true given any temperature combinations. In other words, no active constraints or physical phenomena are taking place.

To formally incorporate heat integration considerations into GMF process intensification synthesis, the GMF modular representation is extended to explicitly include a *Heat Integration (HI)* block [4]. As illustrated in Fig. 5.3, the *HI* block is a particular type of the pure heat exchange module in which heat exchange takes place between two process streams (i.e., without utility streams). Therefore, the HI blocks are modeled in a consistent manner with the GMF utility heat exchange modules. Additionally, a binary variable, i.e. $y_{e,ea}^{HI}$, is introduced to denote the existence or not of the HI block between mass/heat exchange modules e and ea. An indicative list of the modeling constraints around the HI block are summarized below while the associated variables are marked in Fig. 5.3 to illustrate their physical meanings:

Hot side

- Mass balances

$$f_{e,ea}^{XH} - f_{e,ea}^{VX} = 0 \tag{5.28}$$

$$f_{e,ea}^{XL} + f_{e,ea}^{XLL} + f_{e,pjiq}^{XP} - f_{e,ea}^{XH} = 0 \tag{5.29}$$

$$f_{e,ea}^{XH} x_{e,ea,i}^{XH} - f_{e,ea}^{VX} x_{ea,i}^{VO} = 0 \tag{5.30}$$

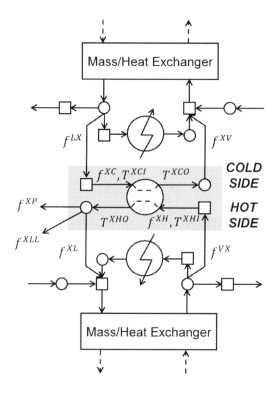

FIGURE 5.3 GMF modular building blocks with heat integration. (Reproduced from [4,5].)

- Summation of molar fractions

$$\sum_i x^{XH}_{e,ea,i} = 1 \qquad (5.31)$$

- Energy balances

$$f^{XH}_{e,ea} h^{XHI}_{e,ea} - f^{VX}_{e,ea} h^{VO}_{ea} = 0 \qquad (5.32)$$

$$Q^{HI} - f^{XH}_{e,ea}(h^{XHI}_{e,ea} - h^{XHO}_{e,ea}) = 0 \qquad (5.33)$$

- Phase defining constraints

$$\sum_i (\gamma^{XHO}_{ei} P^{sat,XHO}_{ei} x^{XH}_{ei})/(\phi^{XHO}_{ei} P_{tot}) \leq 1 \qquad (5.34)$$

- Thermodynamic properties

$$h^{XHI}_{e,ea} = f(x^{XH}_{e,ea,i}, T^{XHI}_{e,ea}) \qquad (5.35)$$

$$h^{XHO}_{e,ea} = f(x^{XH}_{e,ea,i}, T^{XHO}_{e,ea}) \qquad (5.36)$$

$$\gamma^{XHO}_{ei} = f(x^{XH}_{e,ea,i}, T^{XHO}_{e,ea}, P_{tot}) \qquad (5.37)$$

Chapter 5 • Enhanced GMF for process synthesis, intensification, & heat integration   85

$$\phi_{ei}^{XHO} = f(x_{e,ea,i}^{XH}, T_{e,ea}^{XHO}, P_{tot}) \tag{5.38}$$

$$P_{ei}^{sat,XHO} = f(x_{e,ea,i}^{XH}, T_{e,ea}^{XHO}, P_{tot}) \tag{5.39}$$

- Mixed-integer structural constraints

$$f_{e,ea}^{XH} - y_{e,ea}^{HI} F^{max} \leq 0 \tag{5.40}$$

$$f_{e,ea}^{VX} - y_{e,ea}^{HI} F^{max} \leq 0 \tag{5.41}$$

$$f_{e,ea}^{XL} - y_{e,ea}^{HI} F^{max} \leq 0 \tag{5.42}$$

$$f_{e,ea}^{XLL} - y_{e,ea}^{HI} F^{max} \leq 0 \tag{5.43}$$

$$f_{e,pliq}^{XP} - y_{e,ea}^{HI} F^{max} \leq 0 \tag{5.44}$$

$$Q_{e,e}^{HI} - y_{e,ea}^{HI} Q^{max} \leq 0 \tag{5.45}$$

Cold side

- Mass balances

$$f_{e,ea}^{XC} - f_{e,ea}^{LX} = 0 \tag{5.46}$$

$$f_{e,ea}^{XV} + f_{e,ea}^{XVV} + f_{e,pvap}^{XP} - f_{e,ea}^{XC} = 0 \tag{5.47}$$

$$f_{e,ea}^{XC} x_{e,ea,i}^{XC} - f_{e,ea}^{LX} x_{e,i}^{LO} = 0 \tag{5.48}$$

- Summation of molar fractions

$$\sum_i x_{e,ea,i}^{XC} = 1 \tag{5.49}$$

- Energy balances

$$f_{e,ea}^{XC} h_{e,ea}^{XCI} - f_{e,ea}^{LX} h_e^{LO} = 0 \tag{5.50}$$

$$Q^{HI} - f_{e,ea}^{XC} (h_{e,ea}^{XCO} - h_{e,ea}^{XCI}) = 0 \tag{5.51}$$

- Phase defining constraints

$$\sum_i (x_{ei}^{XCO} \phi_{ei}^{XCO} P_{tot}) / (\gamma_{ei}^{XCO} P_{ei}^{sat,XCO}) \leq 1 \tag{5.52}$$

- Thermodynamic properties

$$h_{e,ea}^{XCI} = f(x_{e,ea,i}^{XC}, T_{e,ea}^{XCI}) \tag{5.53}$$

$$h_{e,ea}^{XCO} = f(x_{e,ea,i}^{XC}, T_{e,ea}^{XCO}) \tag{5.54}$$

$$\gamma_{ei}^{XCO} = f(x_{e,ea,i}^{XC}, T_{e,ea}^{XCO}, P_{tot}) \tag{5.55}$$

$$\phi_{ei}^{XCO} = f(x_{e,ea,i}^{XC}, T_{e,ea}^{XCO}, P_{tot}) \tag{5.56}$$

$$P_{ei}^{sat,XCO} = f(x_{e,ea,i}^{XC}, T_{e,ea}^{XCO}, P_{tot}) \tag{5.57}$$

86 Synthesis and Operability Strategies for Computer-Aided Modular PI

- Mixed-integer structural constraints

$$f_{e,ea}^{XC} - y_{e,ea}^{HI} F^{max} \leq 0 \qquad (5.58)$$

$$f_{e,ea}^{LX} - y_{e,ea}^{HI} F^{max} \leq 0 \qquad (5.59)$$

$$f_{e,ea}^{XC} - y_{e,ea}^{HI} F^{max} \leq 0 \qquad (5.60)$$

$$f_{e,ea}^{XCC} - y_{e,ea}^{HI} F^{max} \leq 0 \qquad (5.61)$$

$$f_{e,p_{vap}}^{XP} - y_{e,ea}^{HI} F^{max} \leq 0 \qquad (5.62)$$

Heat integration feasibility constraints

$$T^{XCO} - T^{XHI} + EMTA \leq M(1 - y_{e,ea}^{HI}) \qquad (5.63)$$

$$T^{XCI} - T^{XHO} + EMTA \leq M(1 - y_{e,ea}^{HI}) \qquad (5.64)$$

The value for the parameter M can be estimated via:

$$M = (T_h^{BP_{max}} - T_l^{BP_{min}}) + EMTA \qquad (5.65)$$

where $T_h^{BP_{max}}$ is the boiling point for the heaviest component in the stream mixture at the maximum allowed column pressure, $T_l^{BP_{min}}$ is the boiling point for the lightest component in the stream mixture at the minimum allowed column pressure.

5.3 Motivating example: GMF synthesis, intensification, and heat integration of a ternary separation system

In this section, a ternary separation process synthesis problem is presented as developed by Georgiadis and Pistikopoulos [5]. We showcase how GMF can be applied to systematically generate different distillation-based design alternatives, including simple distillation column sequences, heat-integrated column sequences, and thermally-coupled distillation systems.

5.3.1 Process description

A ternary mixture of benzene, toluene, and o-xylene is to be separated into three high purity product streams. The process data are based on the problem provided by Chavez et al. [6]. The feed stream consists of 0.2 mol/mol benzene, 0.4 mol/mol toluene, and 0.4 mol/mol o-xylene at a flowrate of 1000 kmol/h. The feed is at bubble point at atmospheric pressure with a temperature of 383.2 K. The mixture phase behavior can be accurately described using ideal liquid-vapor equilibrium and the other physical property data are adapted from Reid [7].

The production specifications are set as what follow: (i) a benzene product $P1$ with $x_{P1,Benzene} \geq 0.95$ mol/mol, (ii) a toluene product S with $x_{S,Toluene} \geq 0.95$ mol/mol, and (iii)

Chapter 5 • Enhanced GMF for process synthesis, intensification, & heat integration   87

a o-xylene product $P2$ with $x_{P2,o-Xylene} \geq 0.95$ mol/mol. The optimization objective is set to minimize operating cost, consisting of heating and cooling utility costs, as formulated in Eq. (5.66). The utility availability and cost data are respectively: (i) steam – temperature at 500 K, 137.27 \$/kW·year, (ii) cooling water – temperature at 300 K, 26.19 \$/kW·year.

$$\text{Operating Cost (\$/year)} = \sum_e C_{cw} \times Qc_e + \sum_e C_{steam} \times Qh_e \tag{5.66}$$

where Qc_e and Qh_e are respectively the heating and cooling load in the pure heat exchange modules. The utility cost data are used for C_{cw} and C_{steam}.

5.3.2 Synthesis of simple distillation column sequences

For illustration purpose, we consider GMF synthesis with a simplified superstructure network and expand it step-by-step for simple distillation column sequences, heat-integrated distillation column sequences, and complex distillation sequences. However, it should be noted that they can be systematically captured under a unified representation with full structural combinations, as will be demonstrated in Chapter 12 with a case study on methyl methacrylate purification.

For the synthesis of simple distillation column sequences, the GMF simplified superstructure representation is depicted in Fig. 5.4a. From the superstructure, three plausible structural alternatives for simple distillation column sequences can be generated by activating or deactivating the modules and streams, namely the indirect sequence (Fig. 5.4b), direct sequence (Fig. 5.4c), and three-column configuration (Fig. 5.4d). The separation is considered to take place at a fixed pressure of 1 atm.

By executing the GMF synthesis problem, namely the physical and structural models as discussed in Chapter 4, the direct sequence (Fig. 5.4c) is identified as the optimal process design with a minimum operating cost of 2.37×10^6 \$/year. The corresponding heat duties and product conditions in the optimal design are summarized in Table 5.1.

Table 5.1 Optimization for simple distillation sequences.

	Heat Duties (kW)		
Qc_1	6335.3		
Qh_2	6633.4		
Qc_5	7783.3		
Qh_6	7936.5		
	Product Streams		
	F (kmol/hr)	$x_{P1,Benzene}$	$x_{S,Toluene}$	$x_{P2,o-Xylene}$
$P1$	210.5	0.95	0.05	0
S	393.3	0	0.95	0.05
$P2$	396.2	0	0.05	0.95

88 Synthesis and Operability Strategies for Computer-Aided Modular PI

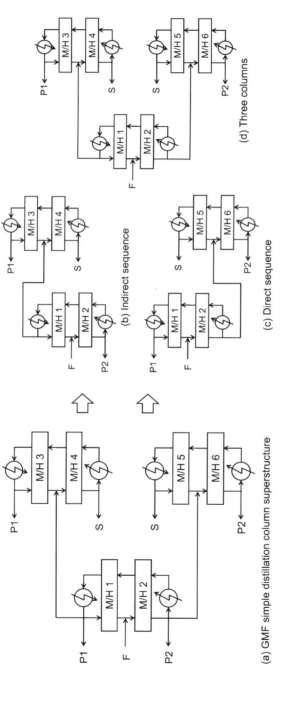

FIGURE 5.4 GMF simple distillation column superstructure.

5.3.3 Synthesis of heat-integrated distillation column sequences

For this ternary separation problem, the GMF superstructure representation with heat integration blocks are depicted in Fig. 5.5. The heat integration opportunities are considered between the heating and cooling modules associated with different distillation columns. Moreover, the column pressures are no longer fixed at atmospheric pressure but treated as optimization variables with a lower bound as 1 atm and an upper bound as 10 atm. This is to fully exploit the potential for energy savings in heat-integrated distillation systems. Accordingly, the pressure and temperature of the initial feed stream are adjusted. The feed stream is assigned to be at bubble point with a pressure equal to that of its destination mass/heat exchange module.

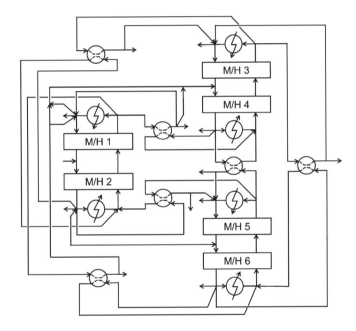

FIGURE 5.5 GMF heat-integrated distillation column superstructure. (Reproduced from [5].)

By solving the GMF synthesis problem simultaneously with heat integration considerations, an optimal process solution is obtained as shown in Fig. 5.6. It is a variant of the direct column sequence with full heat integration between the distillate from the 1st column and the bottom product from the 2nd column. In other words, the condenser of the 1st column and the reboiler of the 2nd column are replaced by the heat integration block. This process features an operating cost of 1.39×10^6 \$/year, which is 41.3% less than the direct column sequence without heat integration. The optimal operating conditions and product details are summarized in Table 5.2.

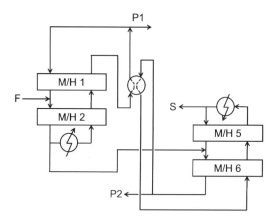

FIGURE 5.6 Optimal heat-integrated distillation column sequence.

Table 5.2 Optimization for heat-integrated distillation sequence.

	Operating Conditions		
Qh_2 (kW)	8488.4		
Qc_5 (kW)	10635.8		
P_1, P_2 (atm)	5.78		
P_5, P_6 (atm)	1		

	Product Streams			
	F (kmol/hr)	$x_{P1,Benzene}$	$x_{S,Toluene}$	$x_{P2,o-Xylene}$
P1	191	0.95	0.05	0
S	394.75	0.04	0.95	0.01
P2	414.25	0	0.05	0.95

5.3.4 Synthesis of ternary complex distillation sequence

To incorporate the thermal coupling considerations, the GMF superstructure representation can be extended to Fig. 5.7. The optimal process design in this case is the Petlyuk column (i.e., fully thermally coupled), as depicted in Fig. 5.8. Around 30% operating cost savings are reported comparing to the optimal direct column sequence without thermal coupling. The optimal heat duties and product details are given in Table 5.3.

To validate GMF representation of the resulting Petlyuk column configuration, we again perform the optimization on a consistent basis using a rigorous tray-by-tray model developed by Viswanathan and Grossmann [8]. The design parameters for the tray-by-tray model are adapted from Chavez et al. [6] as presented in Table 5.4. For the comparison on energy consumption, the optimal reboiler heat duty is 10302.7 kW from GMF synthesis while 12462.5 kW from tray-by-tray modeling. The optimal condenser heat duty is 9849.1 kW from GMF synthesis while 11919.4 kW from tray-by-tray modeling. Moreover, Fig. 5.9 shows the comparison between the o-xylene composition profile and the temperature profile in the main column. As can be noted, GMF can capture the major trends of the mass/heat transfer taking place in the Petlyuk column and identify the optimal thermal coupling process scheme.

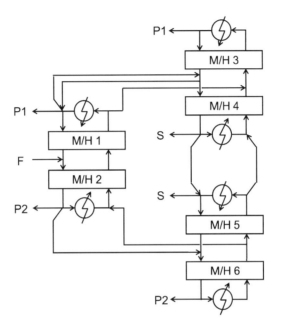

FIGURE 5.7 GMF complex distillation Column superstructure. (Reproduced from [5].)

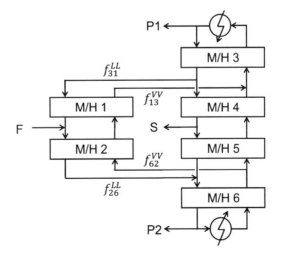

FIGURE 5.8 Optimal ternary complex distillation column sequence.

Table 5.3 Optimization for complex distillation sequences.

	Heat Duties (kW)		
Qh_6	10302.7		
Qc_3	9849.1		
	Process Streams		
	F (kmol/hr)	$x_{P1,Benzene}$	$x_{S,Toluene}$	$x_{P2,o-Xylene}$
P1	210.5	0.95	0.049	0.001
S	389.2	0	0.95	0.05
P2	400.3	0	0.05	0.95
f_{31}^{LL}	273.6	0.159	0.814	0.027
f_{13}^{VV}	675.7	0.236	0.755	0.009
f_{62}^{VV}	669.4	0	0.779	0.221
f_{26}^{LL}	1267.3	0	0.578	0.422

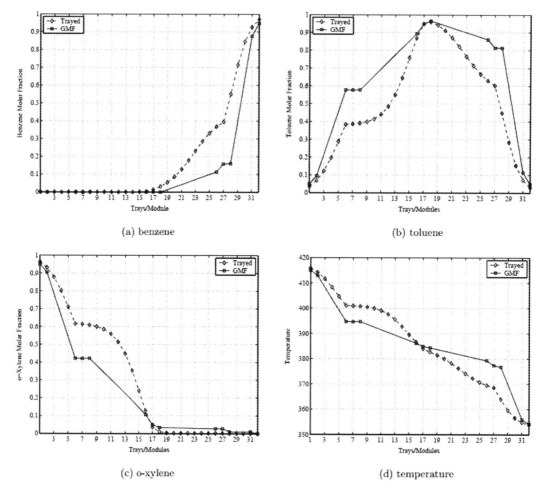

(a) benzene

(b) toluene

(c) o-xylene

(d) temperature

FIGURE 5.9 GMF validation with tray-by-tray model for Petlyuk column (reproduced from [5]).

Table 5.4 Design parameters for tray-by-tray modeling of Petlyuk column.

Prefractionator	Number of trays	20
	Feed tray location	9
Main Column	Number of trays	32
	Upper feed tray location	27
	Lower feed tray location	6
	Side feed tray location	17

References

[1] P. Proios, E.N. Pistikopoulos, Hybrid generalized modular/collocation framework for distillation column synthesis, AIChE Journal 52 (2006) 1038–1056.

[2] P. Seferlis, A. Hrymak, Optimization of distillation units using collocation models, AIChE Journal 40 (1994) 813–825.

[3] W.E. Stewart, K.L. Levien, M. Morari, Simulation of fractionation by orthogonal collocation, Chemical Engineering Science 40 (1985) 409–421.

[4] P. Proios, N.F. Goula, E.N. Pistikopoulos, Generalized modular framework for the synthesis of heat integrated distillation column sequences, Chemical Engineering Science 60 (2005) 4678–4701.

[5] M.C. Georgiadis, E.N. Pistikopoulos, Energy and Process Integration, Begell House, 2006.

[6] R. Chavez, J. Seader, T.L. Wayburn, Multiple steady-state solutions for interlinked separation systems, Industrial & Engineering Chemistry Fundamentals 25 (1986) 566–576.

[7] R.C. Reid, J.M. Prausnitz, B.E. Poling, The properties of gases and liquids, 1987.

[8] J. Viswanathan, I.E. Grossmann, Optimal feed locations and number of trays for distillation columns with multiple feeds, Industrial & Engineering Chemistry Research 32 (1993) 2942–2949.

6

Steady-state flexibility analysis

6.1 Basic concepts

A generalized process model can be defined as shown in Eq. (6.1):

$$h(d, x, z, \theta) = 0$$
$$g(d, x, z, \theta) \leq 0 \qquad \forall \theta \in T \tag{6.1}$$

where d is the vector of design variables (e.g., size, structure), x is the vector of state variables (e.g., temperature, flowrate), z is the vector of control variables (i.e., the degrees of freedom that can be manipulated during operation), and θ is the vector of uncertain parameters, T gives the range of uncertain parameters. The equality constraints $h(d, x, z, \theta) = 0$ describe heat, mass balances, sizing equations, etc. And the inequality constraints $g(d, x, z, \theta) \leq 0$ define process specification, operational constraints, etc. The region of feasible operation is determined by the set of inequality constraints as shown in Fig. 6.1.

In what follows we consider a simplified case where the dimension of equality constraints equals to the dimension of state variables x, thus x can be explicitly expressed as functions of d, z, θ, i.e. $x = x(d, z, \theta)$. Then we also define $g(d, x(d, z, \theta), z, \theta) = f(d, z, \theta) \leq 0$. The generalization of this case, namely when $dim\{h\} \neq dim\{x\}$, will be discussed later.

6.2 Problem definition

The problems in process flexibility can be generally classified as the follows. In this chapter, we introduce the flexibility test problem, flexibility index problem, and multiperiod design of flexible process systems. We aim to provide a fundamental understanding for the flexibility analysis of modular process intensification systems to be introduced in Chapter 9. The readers are referred to the recent textbook by Grossmann [1] for more theoretical details on this topic.

a. **Flexibility analysis:**
   Given a fixed design d and a range of uncertain parameters θ,

   - Flexibility test problem – to determine if the design is feasible at all θ in the range [2]
   - Flexibility index problem – to determine how flexible the design is with respect to θ [3]
   - Stochastic flexibility analysis – if the uncertainty parameters θ are described by probability distribution [4]

Synthesis and Operability Strategies for Computer-Aided Modular Process Intensification
https://doi.org/10.1016/B978-0-32-385587-7.00016-6
Copyright © 2022 Elsevier Inc. All rights reserved.

96 Synthesis and Operability Strategies for Computer-Aided Modular PI

- Dynamic flexibility analysis – if the uncertainty parameters $\theta(t)$ are functions of time t [5]

b. Design and synthesis of flexible process

- Multiperiod design – to determine a design d that is optimal and feasible for *finite points* of the uncertain parameters θ [6]
- Design under uncertainty – to determine a design d that is optimal and feasible for *all* θ points in the specified range [7]

6.2.1 Flexibility test

Given a design d and a specified range of uncertain parameter $T = \{\theta \mid \theta^L \leq \theta \leq \theta^U\}$, the flexibility test problem [2] answers the question if the design is feasible to operate at all θ in the given uncertainty range.

Starting with the simplified process model:

$$f(d, z, \theta) \leq 0 \qquad \forall \theta \in T \tag{6.2}$$

Considering the control variables z as the degrees of freedom to be adjusted during operation, we can define a new function $\psi(d, \theta)$ for given d operated at θ:

$$\psi(d, \theta) = \min_{z} \max_{j \in J_f} f_j(d, z, \theta) \tag{6.3}$$

where J_f is the index set for the inequality constraints $f_j(d, z, \theta) \leq 0$. In this way, $\psi(d, \theta) \leq 0$ indicates feasible operation for the given design d at uncertainty θ – since there exists a vector of control variables z with which all the inequality constraints are satisfied, namely $f_j(d, z, \theta) \leq 0, \forall j \in J_f$. On the other hand, $\psi(d, \theta) > 0$ indicates infeasible operation for the given design d at uncertainty θ – since there exists constraint violation(s) no matter how the control variables z are adjusted, namely $f_j(d, z, \theta) > 0, \exists j \in J_f$.

$\psi(d, \theta) \leq 0$ can be re-formulated as the following optimization problem:

$$\psi(d, \theta) = \min_{u, z} u$$
$$\text{s.t.} \quad f_j(d, z, \theta) \leq u \qquad j \in J_f \tag{6.4}$$

To determine if the operation is feasible for all θ in T, it suffices to consider the largest value of $\psi(d, \theta)$ over T, which gives:

$$X(d) = \max_{\theta \in T} \psi(d, \theta) \tag{6.5}$$

Substituting for $\psi(d, \theta)$,

$$X(d) = \max_{\theta \in T} \min_{z} \max_{j \in J_f} f_j(d, z, \theta) \tag{6.6}$$

Therefore, the flexibility text problem originally described by Eq. (6.2) with flexibility requirement $\forall \theta \in T$ is equivalent to the following tri-level max-min-max problem:

$$X(d) = \max_{\theta \in T} \min_{z} \max_{j \in J_f} f_j(d, z, \theta) \tag{6.7}$$

Or the following bi-level optimization problem:

$$X(d) = \max_{\theta \in T} \psi(d, \theta)$$
$$\text{s.t.} \quad \psi(d, \theta) = \min_{z} \max_{j \in J_f} f_j(d, z, \theta) \tag{6.8}$$

If $X(d) \leq 0$, the design d is feasible over the entire set T of uncertain parameters, as shown in Fig. 6.1a. Otherwise if $X(d) > 0$, the design d is infeasible for T, as shown in Fig. 6.1b.

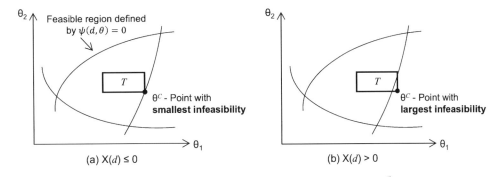

FIGURE 6.1 Geometric interpretation of the flexibility test problem.

6.2.2 Flexibility index

The flexibility index problem [3] aims to answer the question what is the maximum uncertainty set that can be handled for the given process design. In this case, the uncertainty set T is defined as:

$$T(\delta) = \{\theta \mid \theta^N - \delta \Delta \theta^- \leq \theta^N \leq \theta^N + \delta \Delta \theta^+\}, \quad \delta \geq 0 \tag{6.9}$$

where δ is defined as the flexibility index, θ^N is the vector of nominal parameter values.

From a geometric point of view as illustrated in Fig. 6.2, the flexibility index can be obtained via the following steps:

- Assume that uncertain parameters (i.e., θ_1, θ_2) vary independently,
- Select base points as nominal parameter values (i.e., θ_1^N, θ_2^N),
- Determine the maximum range of parameter variation within the region of feasible operation, by constructing a hyper-rectangle centered at the nominal points,
- Calculate the flexibility index based on parameter variations.

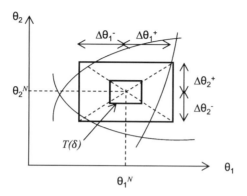

FIGURE 6.2 Geometric interpretation of the flexibility index problem.

Mathematically, the flexibility index F and the maximum uncertainty set $T(F)$ can be determined by the following optimization problem:

$$F = max \ \delta$$
$$\text{s.t.} \quad T(\delta) \subseteq R \tag{6.10}$$

To interpret the region of feasible operation (R) with flexibility test, Eq. (6.10) can be formulated as:

$$F = max \ \delta$$
$$\text{s.t.} \quad X(d) = \max_{\theta \in T} \min_{z} \max_{j \in J_f} f_j(d, z, \theta) \leq 0 \tag{6.11}$$
$$T(\delta) = \{\theta \mid \theta^N - \delta \Delta \theta^- \leq \theta^N \leq \theta^N + \delta \Delta \theta^+\}, \quad \delta \geq 0$$

6.3 Solution algorithms

In this section we discuss the solution methods for the optimization formulations of the flexibility test problem (Eq. 6.8) and the flexibility index problem (Eq. 6.11).

6.3.1 Vertex enumeration

For both the flexibility test problem and the flexibility vertex problem, a key task is to detect the critical points θ^C – i.e. the point with smallest feasibility (if flexible) or the point with largest infeasibility (if not flexible). It has been theoretically proved that, the critical points θ^C lie as a vertex of the uncertain parameter set T (or $T(\delta)$) if the process inequality constraints have any one of the following properties:

i. $f_j(d, z, \theta)$ is linear in z and θ
ii. $f_j(d, z, \theta)$ is convex in z and θ
iii. $f_j(d, z, \theta)$ is 1-dimensional quasi-convex in θ and jointly convex in z

The examples when θ^C lies at a vertex of T can be found in Fig. 6.1, while Fig. 6.3 shows a case when θ^C is not a vertex given a nonconvex region of feasible operation. For more information on the theoretical proof of the properties, readers are recommended to the relevant papers by Grossmann and co-workers [8,9].

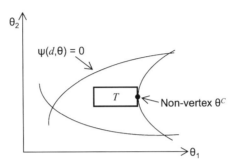

FIGURE 6.3 Non-vertex critical points.

In this regard, flexibility analysis can be performed by vertex enumeration for the (i)–(iii) classes of processes. The corresponding algorithms are summarized below. However, it is worth noting that the number of vertex calculations is an exponential function of the number of parameters (i.e., 2^P). Therefore, the enumeration only suits the process systems with small number of uncertain parameters (e.g., with only θ_1 and θ_2, $2^2 = 4$ vertex calculations are needed).

Flexibility Test

Step 1. Let θ^k denote the kth vertex for the specified uncertain parameter set T, $k \in V$. For each θ^k, solve:

$$\psi(d, \theta^k) = \min_{u, z} u$$
$$\text{s.t.} \quad f_j(d, z, \theta) \leq u \quad j \in J_f$$
(6.12)

Step 2. Set $X(d) = \max_{k \in V} \psi(d, \theta^k)$:
If $X(d) \leq 0$, the design is feasible; if $X(d) > 0$, infeasible.

Flexibility Index

Step 1. For each vertex direction $\Delta \theta^k$, solve:

$$\delta^k = \max_{\delta, z} \delta$$
$$\text{s.t.} \quad f_j(d, z, \theta) \leq u \quad j \in J_f$$
$$\theta = \theta^N + \delta \Delta \theta^k$$
(6.13)

Step 2. Set $F = \min_{k \in V} \delta^k$.

6.3.2 Active set strategy

To avoid the 2^P vertex searches and to capture non-vertex critical points, Grossmann and Floudas [10] proposed the Active Set Strategy which recast the flexibility test (or index) problem, in the form of max-min-max formulation, to a single maximization problem.

In what follows, we take the flexibility test problem as an example to detail the mathematical derivation. Since $\psi(d, \theta)$ is an optimization problem, it can be represented through its Kuhn-Tucker conditions, i.e.

$$\psi(d, \theta) = \min_{u, z} u$$
$$\text{s.t.} \quad f_j(d, z, \theta) \leq u \qquad j \in J_f \tag{6.14}$$

Kuhn-Tucker Conditions:

$$u: \quad 1 - \sum_{j \in J_f} \lambda_j = 0$$
$$z: \quad \sum_{j \in J_f} \lambda_j \frac{\partial f_j}{\partial z} = 0$$
$$\lambda_j \geq 0 \tag{6.15}$$
$$f_j(d, z, \theta) - u \leq 0$$
$$\lambda_j(f_j(d, z, \theta) - u) = 0 \qquad j \in J_f$$

where λ_j is the non-negative Lagrange multiplier corresponding to f_j.

The complementarity conditions introduce nonlinearity to this problem with:

$$\lambda_j(f_j(d, z, \theta) - u) = 0$$

However, it can be observed that:

- if $f_j(d, z, \theta) - u = 0$, then $\lambda \geq 0$,
- if $f_j(d, z, \theta) - u < 0$, then $\lambda = 0$.

This implies a binary choice to denote if a constraint f_j is active or not.

Therefore, the nonlinearity can be removed by explicitly modeling the selection of active constraints. To this purpose, we define "slack variables" as $s_j = u - f_j(d, z, \theta) \geq 0$. An additional set of binary variables y_j are also defined as:

$$y_j = \begin{cases} 1 & \text{if } j^{\text{th}} \text{ constraint active} \\ 0 & \text{otherwise} \end{cases} \tag{6.16}$$

The complementarity conditions $\lambda_j(f_j(d, z, \theta) - u) = 0$ can thus be replaced by:

$$s_j - U(1 - y_j) \leq 0$$
$$\lambda_j - y_j \leq 0$$

$$\sum_{j \in J_f} y_j = n + 1 \qquad (6.17)$$

$$s_j \geq 0$$

$$\lambda_j \geq 0$$

$$j \in J_f$$

where U is a large parameter as upper bound. The summation of y_j for all $j \in J_f$ equals to $n + 1$, because at the critical point there are $n + 1$ active constraints to make the optimization problem degrade to a system of equations with zero degree of freedom.

From Eq. (6.17), we can have:

- when $y_j = 1$, i.e. the jth constraint is active, then $s_j = 0, 0 \leq \lambda_j \leq 1$
- when $y_j = 0$, i.e. the jth constraint is inactive, then $\lambda_j = 0, 0 \leq s_j \leq U$

By substituting the Kuhn-Tucker conditions (Eq. 6.15) with complementarity conditions (Eq. 6.17), a single mixed-integer programming (MIP) problem is obtained for flexibility text as shown in Eq. (6.18). In a similar way, the flexibility index problem can also be reformulated as a single MIP problem as given in Eq. (6.19).

Flexibility Test

$$\begin{aligned}
\max_{\theta, u, z, \lambda_j, s_j, y_j} \quad & u \\
\text{s.t.} \quad & 1 - \sum_{j \in J_f} \lambda_j = 0 \\
& \sum_{j \in J_f} \lambda_j \frac{\partial f_j}{\partial z} = 0 \\
& s_j + f_j(d, z, \theta) - u = 0 \\
& s_j - U(1 - y_j) \leq 0 \qquad (6.18) \\
& \lambda_j - y_j \leq 0 \\
& \sum_{j \in J} y_j = n + 1 \\
& y_j = 0, 1 \\
& s_j \geq 0, \lambda_j \geq 0 \\
& j \in J_f, \qquad \theta \in T \\
& u \in R^1
\end{aligned}$$

102 Synthesis and Operability Strategies for Computer-Aided Modular PI

Flexibility Index

$$F = \min_{\theta,\delta,z,\lambda_j,s_j,y_j} \delta$$

$$\text{s.t.} \quad 1 - \sum_{j \in J_f} \lambda_j = 0$$

$$\sum_{j \in J_f} \lambda_j \frac{\partial f_j}{\partial z} = 0$$

$$s_j + f_j(d, z, \theta) = 0$$

$$s_j - U(1 - y_j) \leq 0$$

$$\lambda_j - y_j \leq 0 \tag{6.19}$$

$$\sum_{j \in J} y_j = n + 1$$

$$\theta^N - \delta\Delta\theta^- \leq \theta \leq \theta^N + \delta\Delta\theta^+$$

$$\delta \geq 0$$

$$y_j = 0, 1$$

$$s_j \geq 0, \lambda_j \geq 0$$

$$j \in J_f$$

6.3.3 Remarks

1. If the uncertain parameters are correlated, i.e. not independent of each other:
In this case, the correlation equations can be explicitly included in the process model formulation. Otherwise, the correlated parameters can be treated as uncertainty instead of the original set of parameters. For example, in Eq. (6.20), the correlated parameter θ is the new parameter, while T_1 and T_2 are regarded as state variables.

$$T_1 = 10(1 + \theta)$$
$$T_2 = 20(1 + \theta) \tag{6.20}$$

2. If the equations or state variables cannot be eliminated:
In this case, the flexibility index (or test) problem can be formulated as Eq. (6.21) in the complete form:

$$F = \min_{\theta,\delta,x,z,\lambda_j,\mu_j,s_j,y_j} \delta$$

$$\text{s.t.} \quad h_i(d, x, z, \theta) = 0$$

$$f_j(d, x, z, \theta) + s_j = 0$$

$$1 - \sum_{j \in J_f} \lambda_j = 0$$

$$\sum_{i \in J_h} \mu_i \frac{\partial h_i}{\partial z} + \sum_{j \in J_f} \lambda_j \frac{\partial f_j}{\partial z} = 0$$

$$\sum_{i \in J_h} \mu_i \frac{\partial h_i}{\partial x} + \sum_{j \in J_f} \lambda_j \frac{\partial f_j}{\partial x} = 0$$

$$s_j - U(1 - y_j) \leq 0 \tag{6.21}$$

$$\lambda_j - y_j \leq 0$$

$$\sum_{j \in J_f} y_j = n_z + 1$$

$$\theta^N - \delta \Delta \theta^- \leq \theta \leq \theta^N + \delta \Delta \theta^+$$

$$\delta \geq 0$$

$$y_j = 0, \ 1$$

$$s_j \geq 0, \ \lambda_j \geq 0$$

$$i \in J_h, \ j \in J_f$$

3. Analytical expressions for the optimal boundary of the region of operation:
The boundary of the region of operation can be calculated via:

$$\psi^k(d, \theta) = \sum_{i \in J_h} \mu_i h_i(d, x, z, \theta) + \sum_{j \in J_A^k} \lambda_j^k f_j(d, z, \theta) \tag{6.22}$$

where J_A^k is the Active Set of f_j constraints at the kth vertex. The number of active constraints for each active set is $n_z + 1$.

6.4 Design and synthesis of flexible processes

In this section, we consider process designs with flexibility considerations for a specified range of uncertain parameters. An iterative multiperiod optimization problem can be formulated to this purpose [11,12]. As depicted in Fig. 6.4, considering N periods of operation each of length W_i and operating condition θ_i, a multiperiod optimization approach aims to determine an optimal design d (including the process structure and equipment sizing parameters) which is feasible to operate during all the N periods. The objective function is formulated as Eq. (6.23):

$$C = C(d) + \sum_{i=1}^{N} W_i \phi_i(d, z_i, \theta_i) \tag{6.23}$$

where $C(d)$ calculates the investment cost, $W_i \phi_i(d, z_i, \theta_i)$ gives the operating cost for each operation period i.

The iterative algorithm works as follows: Starting from an initial process structure (e.g., the cost-optimal process design for nominal operation condition), flexibility test is performed to determine the critical operating conditions under a specified range of uncertain

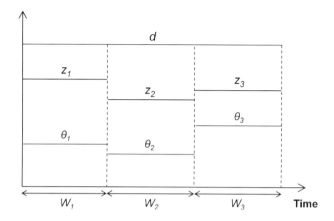

FIGURE 6.4 Multiperiod formulation.

parameters (i.e., the operating conditions with the largest violation from nominal operating condition). Then the optimization problem is extended to include all these critical operating conditions via a multiperiod representation approach by introducing an additional set of operation "periods". Specifically, design variables (e.g., heat exchange area, equipment diameter) are reformulated as continuous variables for *all periods of operation*, while the other variables (e.g., stream flows, compositions, temperatures) are regarded as continuous variables for *each period of operation*. Thus, the resulting optimization model will identify the most promising process option with desired flexibility performance. The detailed steps are as follow:

Step 1 Specification of flexibility target:

  i. Determine an initial network based on nominal operation condition (e.g., the cost-optimal structure)
  ii. Specify an expected range of uncertain parameters

Step 2 For the current network, solve a flexibility test problem:

  i. If the current configuration satisfies the flexibility target, stop. Otherwise, go to (ii)
  ii. Identify critical operating conditions (i.e., periods) as per the maximum constraint violation

Step 3 Multiperiod synthesis:

  i. Formulate a multiperiod optimization model based on the nominal and/or critical operating periods identified in Step 2
  ii. Solve to obtain "new" network structure. Then go to Step 2.

6.5 Tutorial example: flexibility analysis of heat exchanger network

6.5.1 Process description

Consider the following heat exchanger network as illustrated in Fig. 6.5. The process considers two hot process streams (i.e. H1, H2) and two cold process streams (i.e. C1, C2). The inlet/outlet temperatures and heat capacities of the process streams are summarized in Fig. 6.5. The utility available for this process includes cooling water with a heat duty of Q_c. There exist four uncertain parameters, respectively as the inlet temperatures of process streams H1, H2, C1, and C2. In what follows, we perform flexibility analysis of this heat exchanger network with: (i) vertex enumeration, and (ii) active set strategy.

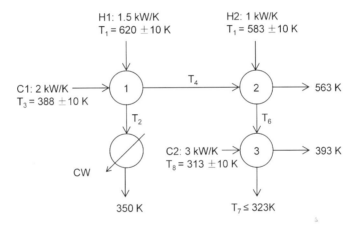

FIGURE 6.5 Heat exchanger network.

6.5.2 Problem solution

Step 1: Energy balance equations

The energy balances of the above heat exchanger network can be described by Eqs. (6.24)–(6.27). The classification of uncertain parameters and process variables are shown in Table 6.1.

Heat exchanger 1

$$1.5(T_1 - T_2) = 2(T_4 - T_3) \tag{6.24}$$

Heat exchanger 2

$$1(T_5 - T_6) = 2(563 - T_4) \tag{6.25}$$

Heat exchanger 3

$$1(T_6 - T_7) = 3(393 - T_8) \tag{6.26}$$

106 Synthesis and Operability Strategies for Computer-Aided Modular PI

Utility exchanger

$$Q_c = 1.5(T_2 - 350) \tag{6.27}$$

Table 6.1 Uncertain parameters and process variables for the heat exchanger network.

Uncertain Parameters	T_1	T_3	T_5	T_8	
Process Variables	T_2	T_4	T_6	T_7	Q_c

Step 2: Heat transfer feasibility constraints

In this example, we consider the Heat Recovery Approximation Temperature as 0 K. Then the following constraints need to be satisfied for feasible heat exchange between hot and cold process streams:

Heat exchanger 1

$$T_2 - T_3 \geq 0 \tag{6.28}$$

Heat exchanger 2

$$T_6 - T_4 \geq 0 \tag{6.29}$$

Heat exchanger 3

$$T_7 - T_8 \geq 0 \tag{6.30}$$

$$T_6 - 393 \geq 0 \tag{6.31}$$

$$T_7 \leq 323 \tag{6.32}$$

Step 3: State variable elimination

The state variables T_2, T_4, T_6, and T_7 can be eliminated from Eqs. (6.24)–(6.27).

Elimination of T_2

From Eq. (6.27), an explicit function of T_2 can be obtained in terms of the control variable Q_c:

$$T_2 = Q_c/1.5 + 350 \tag{6.33}$$

Elimination of T_4

From Eqs. (6.24) and (6.33), T_4 can be derived as a function of the control variable Q_c and the uncertain parameters:

$$T_4 = -0.5Q_c + 0.75T_1 + T_3 - 262.5 \tag{6.34}$$

Elimination of T_6

In a similar way, from Eqs. (6.25) and (6.34), T_6 can be derived as:

$$T_6 = -Q_c + 1.5T_1 + 2T_3 + T_5 - 1651 \tag{6.35}$$

Elimination of T_7

Finally, from Eqs. (6.26) and (6.35), T_7 can be derived as:

$$T_7 = -Q_c + 1.5T_1 + 2T_3 + T_5 + 3T_8 - 2830 \tag{6.36}$$

Inequalities

The inequality constraints Eqs. (6.28)–(6.32) can be rewritten as functions only of the control variables Q_c and uncertain parameters T_2, T_4, T_6, and T_7 in the form of $f_j(z, \theta) \leq 0$, $\forall i \in \{1, ..., 5\}$:

$$f_1 = -Q_c/1.5 + T_3 - 350 \leq 0 \tag{6.37}$$

$$f_2 = -0.5Q_c - 0.75T_1 - T_3 - T_5 + 1388.5 \leq 0 \tag{6.38}$$

$$f_3 = Q_c - 1.5T_1 - 2T_3 - T_5 - 2T_8 + 2830 \leq 0 \tag{6.39}$$

$$f_4 = Q_c - 1.5T_1 - 2T_3 - T_5 + 2044 \leq 0 \tag{6.40}$$

$$f_5 = Q_c + 1.5T_1 + 2T_3 + T_5 + 3T_8 - 3153 \leq 0 \tag{6.41}$$

Step 4: Flexibility index calculation

Method 1: Vertex enumeration scheme

With 4 uncertain parameters, the number of vertices is $k = 2^4 = 16$. Eq. (6.42) presents the flexibility index problem for each vertex k:

$$
\begin{aligned}
\delta^k = \max_{\delta, Q_c} \ & \delta \\
\text{s.t.} \quad f_1 = & -Q_c/1.5 + T_3^k - 350 \leq 0 \\
f_2 = & -0.5Q_c - 0.75T_1^k - T_3^k - T_5^k + 1388.5 \leq 0 \\
f_3 = & \ Q_c - 1.5T_1^k - 2T_3^k - T_5^k - 2T_8^k + 2830 \leq 0 \\
f_4 = & \ Q_c - 1.5T_1^k - 2T_3^k - T_5^k + 2044 \leq 0 \\
f_5 = & \ Q_c + 1.5T_1^k + 2T_3^k + T_5^k + 3T_8^k - 3153 \leq 0 \\
& -Q_c \leq 0 \\
T_1^k = & \ 620 \pm 10\delta \\
T_3^k = & \ 388 \pm 10\delta \\
T_5^k = & \ 583 \pm 10\delta \\
T_8^k = & \ 313 \pm 10\delta
\end{aligned}
\tag{6.42}
$$

The solutions at the 16 vertices are summarized in Table 6.2. The flexibility index can be calculated based on the vertex enumeration results as given by Eq. (6.43), which is $\delta = 0.5$ for this case.

$$\delta = \min_k \delta^k \tag{6.43}$$

108   Synthesis and Operability Strategies for Computer-Aided Modular PI

Table 6.2 Flexibility index solutions at vertices.

k	T_1	T_3	T_5	T_8	δ^k
1	$620 + 10$	$388 + 10$	$583 + 10$	$313 + 10$	1
2	$620 + 10$	$388 + 10$	$583 + 10$	$313 - 10$	∞
3	$620 + 10$	$388 + 10$	$583 - 10$	$313 + 10$	0.5
4	$620 + 10$	$388 + 10$	$583 - 10$	$313 - 10$	2.8
5	$620 + 10$	$388 - 10$	$583 + 10$	$313 + 10$	1
6	$620 + 10$	$388 - 10$	$583 + 10$	$313 - 10$	5.6
7	$620 + 10$	$388 - 10$	$583 - 10$	$313 + 10$	0.5
8	$620 + 10$	$388 - 10$	$583 - 10$	$313 - 10$	1.4
9	$620 - 10$	$388 + 10$	$583 + 10$	$313 + 10$	1
10	$620 - 10$	$388 + 10$	$583 + 10$	$313 - 10$	1.4
11	$620 - 10$	$388 + 10$	$583 - 10$	$313 + 10$	0.5
12	$620 - 10$	$388 + 10$	$583 - 10$	$313 - 10$	0.7
13	$620 - 10$	$388 - 10$	$583 + 10$	$313 + 10$	1
14	$620 - 10$	$388 - 10$	$583 + 10$	$313 - 10$	0.93
15	$620 - 10$	$388 - 10$	$583 - 10$	$313 + 10$	0.5
16	$620 - 10$	$388 - 10$	$583 - 10$	$313 - 10$	0.56

Method 2: Active set strategy

Applying Eq. (6.19) to this heat exchanger network problem:

$$F = \min_{z,y,\delta} \delta$$

$$\text{s.t.} \quad f_1 = -Q_c/1.5 + T_3 - 350 + s_1 = 0$$

$$f_2 = -0.5Q_c - 0.75T_1^k - T_3^k - T_5^k + 1388.5 + s_2 = 0$$

$$f_3 = Q_c - 1.5T_1^k - 2T_3^k - T_5^k - 2T_8^k + 2830 + s_3 = 0$$

$$f_4 = Q_c - 1.5T_1^k - 2T_3^k - T_5^k + 2044 + s_4 = 0$$

$$f_5 = Q_c + 1.5T_1^k + 2T_3^k + T_5^k + 3T_8^k - 3153 + s_5 = 0$$

$$\lambda_1 + \lambda_2 + \lambda_3 + \lambda_4 + \lambda_5 = 1$$

$$-\lambda_1/1.5 + 0.5\lambda_2 + \lambda_3 + \lambda_4 - \lambda_5 = 0$$

$$s_1 - M(1 - y_1) \leq 0$$

$$s_2 - M(1 - y_2) \leq 0$$

$$s_3 - M(1 - y_3) \leq 0$$

$$s_4 - M(1 - y_4) \leq 0 \qquad (6.44)$$

$$s_5 - M(1 - y_5) \leq 0$$

$$\lambda_1 - y_1 \leq 0$$

$$\lambda_2 - y_2 \leq 0$$

$$\lambda_3 - y_3 \leq 0$$

$$\lambda_4 - y_4 \le 0$$

$$\lambda_5 - y_5 \le 0$$

$$y_1 + y_2 + y_3 + y_4 + y_5 = 1 + 1$$

$$620 - 10\delta \le T_1 \le 620 + 10\delta$$

$$388 - 10\delta \le T_3 \le 388 + 10\delta$$

$$583 - 10\delta \le T_5 \le 583 + 10\delta$$

$$313 - 10\delta \le T_8 \le 313 + 10\delta$$

$$Q_c \ge 0, \; \delta \ge 0, \; \lambda_j \ge 0, \; s_j \ge 0, \; y_j \in \{0, 1\}, \forall j \in J_f$$

The solution to Eq. (6.44) provides:

- the active constraints: f_2 and f_5
- the critical point: $T_1 = 615$ K, $T_3 = 383$ K, $T_5 = 578$ K, $T_8 = 318$ K
- the flexibility index: $F = \delta_{min} = 0.5$
- the value of control variable at the critical point: $Q_c = 67.5$ kW

We briefly showcase how to analytically derive the expressions for the optimal boundary of the region of operation. In this problem with only 1 control variable, each active set includes 2 constraints. The active sets are given in Table 6.3 obtained from the vertex enumeration strategy.

Table 6.3 Active sets for the heat exchanger network problem.

Active Set	Constraints
AS_1	f_1, f_2
AS_2	f_1, f_3
AS_3	f_1, f_4
AS_4	f_5, f_2
AS_5	f_5, f_3
AS_6	f_5, f_4

Taking AS_4 as am example, which is the active set corresponding to vertex $k = 7$:

$$f_2 = 0.5Q_c - 0.75T_1 - T_3 - T_5 + 1388.5 = 0$$
$$f_5 = -Q_c + 1.5T_1 + 2T_3 + T_5 + 3T_8 - 3153 = 0$$

(6.45)

Based on Eq. (6.22), the following system of equations can be obtained:

$$\sum_{j \in J_f} \lambda_j = 1$$

$$\sum_{j \in J_f} \lambda_j \frac{\partial f_j}{\partial z} = 0$$

$$\longrightarrow \quad \lambda_2 + \lambda_5 = 1 \tag{6.46}$$

$$0.5\lambda_2 - \lambda_5 = 0$$

$$\longrightarrow \quad \lambda_2 = \frac{2}{3}, \quad \lambda_5 = \frac{1}{3}$$

The $\psi^7(T_1, T_3, T_5, T_8)$ function is now derived as:

$$\psi^7(T_1, T_3, T_5, T_8) = \lambda_2 f_2 + \lambda_5 f_5$$

$$\longrightarrow \quad \psi^7(T_1, T_3, T_5, T_8) = T_8 - \frac{376}{3} \tag{6.47}$$

References

[1] I.E. Grossmann, Advanced Optimization for Process Systems Engineering, Cambridge University Press, 2021.

[2] K.P. Halemane, I.E. Grossmann, Optimal process design under uncertainty, AIChE Journal 29 (1983) 425–433.

[3] R.E. Swaney, I.E. Grossmann, An index for operational flexibility in chemical process design. Part I: formulation and theory, AIChE Journal 31 (1985) 621–630.

[4] E. Pistikopoulos, T. Mazzuchi, A novel flexibility analysis approach for processes with stochastic parameters, Computers & Chemical Engineering 14 (1990) 991–1000.

[5] V.D. Dimitriadis, E.N. Pistikopoulos, Flexibility analysis of dynamic systems, Industrial & Engineering Chemistry Research 34 (1995) 4451–4462.

[6] D.K. Varvarezos, I.E. Grossmann, L.T. Biegler, An outer-approximation method for multiperiod design optimization, Industrial & Engineering Chemistry Research 31 (1992) 1466–1477.

[7] E.N. Pistikopoulos, M.G. Ierapetritou, Novel approach for optimal process design under uncertainty, Computers & Chemical Engineering 19 (1995) 1089–1110.

[8] I.E. Grossmann, M. Morari, Operability, resiliency, and flexibility: process design objectives for a changing world, 1983.

[9] I.E. Grossmann, B.A. Calfa, P. Garcia-Herreros, Evolution of concepts and models for quantifying resiliency and flexibility of chemical processes, Computers & Chemical Engineering 70 (2014) 22–34.

[10] I.E. Grossmann, C.A. Floudas, Active constraint strategy for flexibility analysis in chemical processes, Computers & Chemical Engineering 11 (1987) 675–693.

[11] C.A. Floudas, I.E. Grossmann, Synthesis of flexible heat exchanger networks for multiperiod operation, Computers & Chemical Engineering 10 (1986) 153–168.

[12] K. Papalexandri, E. Pistikopoulos, A multiperiod MINLP model for the synthesis of flexible heat and mass exchange networks, Computers & Chemical Engineering 18 (1994) 1125–1139.

7

Inherent safety analysis

In this chapter, we introduce several representative approaches for process safety analysis. Note that a comprehensive process safety assessment requires very detailed information on equipment and flowsheet design, control and protection layers, plant layout, etc. Herein we focus on analyzing inherent safety performance which is impacted by process design using rigorous steady-state or dynamic models. In Chapter 9, we will further integrate these approaches with the process intensification synthesis and design approaches to systematically generate inherently safer process systems.

7.1 Dow Chemical Exposure Index

The Dow Chemical Exposure Index (CEI) [1] has been widely adapted in the chemical process industry to analyze the relative acute health hazard potential to people in the neighborhood from possible incident of chemical release. The method aims to provide a relative ranking between different process alternatives, but not to define if a particular design is absolutely safe or not. As highlighted in the guide, the resulting index value depends on the consequence of a possible incident but independent on the failure frequency. In what follows, we will briefly outline the key steps to calculate the Dow CEI. Of particular interest is the impact of process design variables (e.g., material properties, mass inventory, temperature) on the safety evaluation. Readers are highly recommended to study the original DOW CEI guide [1] to learn about the full analysis procedure.

Given the following process information (which are available at the conceptual design stage): (i) Process flowsheet design information consisting of vessels, major piping, and chemical inventories; and (ii) Physical and chemical properties of the materials which are present in the process. The steps for CEI calculation are explained below:

Step 1: Define possible chemical release incidents

This step aims to determine which scenario has the ***largest potential airborne release*** of acutely toxic chemicals. An instantaneous or continuous release can be considered based on the incident. Several scenarios are listed in the Dow CEI guide as reference points, which includes: (i) release due to process pipes rupture, (ii) full bore rupture of hoses, (iii) release from pressure relief devices, (iv) release from vessels, (v) tank overflows and spills, and (vi) other scenarios based on specialized hazard analysis studies.

Step 2: Determine ERPG-2/EEPG-2

ERPG stands for the "Emergency Response Planning Guidelines" values published by the American Industrial Hygiene Association. EEPG is the Dow equivalent if an ERPG value

---

Synthesis and Operability Strategies for Computer-Aided Modular Process Intensification
https://doi.org/10.1016/B978-0-32-385587-7.00017-8
Copyright © 2022 Elsevier Inc. All rights reserved.

112   Synthesis and Operability Strategies for Computer-Aided Modular PI

for a certain material is not provided. Particularly, the ERPG-2 or EEPG-2 used in CEI is the "maximum airborne concentration below which it is believed that nearly all individuals could be exposed for up to one hour without experiencing or developing irreversible or other serious health effects or symptoms that could impart their abilities to take protective action." [1] For example, the ERPG-2 for Ammonia is 150 ppm based on the ERPG Table updated in 2016.

Step 3: Determine the airborne release quantity

The airborne quantity considered in the CEI calculation refers to the total quantity of material entering the atmosphere over time via vapor, liquid flashing, or pool evaporation. Fig. 7.1 presents the flowchart to quantify the airborne formation in different cases.

Step 4: CEI and hazard distance calculation

The Chemical Exposure Index is defined via Eq. (7.1) (in SI units):

$$\text{CEI} = 655.1 \sqrt{\frac{AQ}{\text{ERPG-2}}} \tag{7.1}$$

where AQ is the airborne quantity in kg/s, ERPG-2 is in mg/m^3.

Moreover, the Hazard Distance (HD) is defined as the distance to the ERPG-2 concentration, which can be calculated via Eq. (7.2) (in SI units):

$$\text{HD} = 6551 \sqrt{\frac{AQ}{\text{ERPG-2}}} \tag{7.2}$$

7.2 Dow Fire and Explosion Index

The Dow Fire and Explosion Index (F&EI) [2] is another safety analysis approach which is widely accepted in the chemical process industry. It provides an index to evaluate the risk of individual process unit losses under potential fires and explosions. Similar to the Dow Chemical Exposure Index, Dow F&EI also outputs the relative ranking of process alternatives. It also aims to serve as a guide for adding fire protection methods. However, the detailed equipment design with protection layers, plant layout optimization, and instrumental considerations are beyond the scope of this book. We will focus our discussion on the evaluation of process inherent hazards, while readers are referred to the original F&EI guide for more comprehensive information. The steps for Dow F&EI calculation (up to obtaining the Area of Exposure value) are explained below.

Step 1: Select pertinent process units

This step aims to determine which process units should be included in this F&EI calculation. These process unites, known as Pertinent Process Units, should affect loss prevention from the following perspectives: (i) chemical energy potential, (ii) quantity of

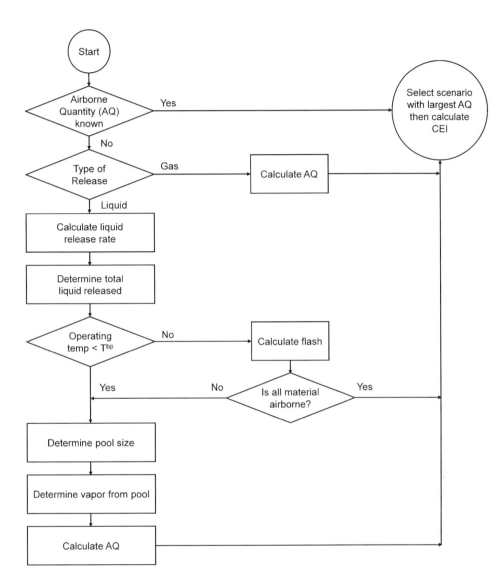

FIGURE 7.1 The workflow for airborne quantity calculation (reproduced from [1]).

hazardous material present, (iii) capital density, (iv) process temperature and pressure, (v) historical data on fire and explosion incident, and (vi) critical units to plant operation.

Step 2: Determine material factor

The material factor (MF) measures the intrinsic rate of potential energy release from a fire or explosion incident. The F&EI guide lists the MF values for a given set of compounds and materials. A table is also provided in the guide to determine MF values based on NFPA N_F and N_R values, as well as material flammability and combustibility properties (e.g., flash point, boiling point). Note that, for mixtures without corresponding testing data, an approximation can be made to use the MF of the compound which has the highest MF value. The component should take up more than 5% of the mixture concentration. It should also be highlighted that the MF defines the material hazard under ambient pressure and temperature. Certain adjustments are required to recognize the fact that – fire and explosion hazards can increase significantly with a rise in temperature. More detailed adjustments can be found in the F&EI guide.

Step 3: Determine Process Unit Hazards Factor

Process Unit Hazards Factor (F_3) is obtained by multiplying the General Process Hazards Factor (F_1) and the Special Process Hazards Factor (F_2).

Step 3.1: General Process Hazards

A list of six items are considered under General Process Hazards, which can affect the **magnitude of a loss incident**: (a) exothermic chemical reaction, (b) endothermic processes, (c) material handling and transfer, (d) enclosed or indoor process units, (e) access to emergency equipment, and (f) drainage and spill control. These items require careful evaluation of for each Pertinent Process Unit under the **most hazardous normal operating conditions**, since they have been reported to have a significant impact on fire and explosion incidents from a historical point of view. Different penalty factors are assigned to different scenarios. For example, for the consideration of exothermic chemical reactions:

- Mild Exotherms – e.g., hydrogenation, hydrolysis, isomerization, sulfonation, neutralization – a penalty of 0.30
- Moderate Exotherms – e.g., alkylation, esterification, addition reactions, oxidation, polymerization, condensation) – a penalty of 0.50
- Critical-to-Control Exotherms – e.g., halogenation – a penalty of 1.00
- Particularly Sensitive Exotherms – e.g., nitration – a penalty of 1.25

Step 3.2: Special Process Hazards

Special process hazards are considered to increase the **probability of a loss incident**. In other words, these items have been reported as the **major causes of fire and explosion incidents**. A list of twelve items can be found in this category. We highlight three of them which are related to process pressures and mass inventories:

- Sub-Atmospheric Pressure – for the unit in which air leakage can result in flammable mixtures formation – a penalty of 0.50
- Relief Pressure – for the unit in which operating pressure is above atmospheric – the relationship between the penalty factor Y and the pressure (X in psig) is given by Eq. (7.3):

$$Y = 0.16109 + \frac{1.61503X}{1000} - 1.42879(\frac{X}{1000})^2 + 0.5172(\frac{X}{1000})^3 \tag{7.3}$$

- Quantity of Flammable and Unstable Liquid or Gas Materials – which adds the penalty based on the quantity of material that might cause a fire or reactive chemical event if

Chapter 7 • Inherent safety analysis   115

spilled – the relationship between the penalty factor Y and the process energy content (X in BTU $\times 10^9$) is given by Eq. (7.4):

$$\log Y = 0.17179 + 0.42988(\log X) - 0.37244(\log X)^2 \\ + 0.17712(\log X)^3 - 0.029984(\log X)^4 \tag{7.4}$$

The penalty factors under each item of the General Process Hazards and Special Process Hazards are respectively summed up to define the value of F_1 and F_2. Furthermore, the Process Unit Hazards Factor F_3 equals to the product (not sum) of F_1 and F_2 due to the compounding impact between F_1 and F_2.

Step 4: Determine F&EI

The Fire and Explosion Index is defined as the product of the Process Unit Hazards Factor (F_3) and the Materials Factor (MF). The degree of hazard reflected by the corresponding F&EI range is summarized in Table 7.1.

Table 7.1   Degree of hazard for F&EI.

F&EI Range	Degree of Hazard
1–60	Light
61–96	Moderate
97–127	Intermediate
128–158	Heavy
>159	Severe

Step 5: Calculate the radius or the area or exposure

The Fire and Explosion Index can be further correlated with the Radius of Exposure or the Area of Exposure via Eqs. (7.5) and (7.6). The radius or the area is that can be exposed to a fire or explosion due to the Process Unit being evaluated (as the center of the area).

$$R = 0.84 \times FEI \quad \text{(in ft)} \tag{7.5}$$

$$Area = \pi R^2 \tag{7.6}$$

7.3 Safety Weighted Hazard Index

Different with the Dow CEI and F&EI which originated from industry, the Safety Weighted Hazard Index [3] was proposed by academic research groups expertized in process safety. SWeHI, defined by Eq. (7.7), considers two major components: (i) B – which quantitatively measures the potential damage caused by an incident in the investigated unit or plant, (ii) A – which quantified the credits of add-on safety and control measures. The value of SWeHI gives the radius of area under moderate hazard (i.e. 50% probability of fatality or damage). The inherent safety performance of a unit or plant can be either improved by increasing A (i.e., enhancing protection procedures) or decreasing B (i.e., reduce inherent

116 Synthesis and Operability Strategies for Computer-Aided Modular PI

process hazards). The workflow for SWeHI calculation is summarized in Fig. 7.2 and explained below, with particular emphasis on the quantification of B term which is related to the process design and operating characteristics.

$$SWeHI = B/A \tag{7.7}$$

Quantification of A

Factor A quantifies various measures which are adapted in a unit or plant to control the damage potential (e.g., emergency resource planning, disaster management plan) and to reduce the event frequency of occurrence (e.g., control system, detection devices, human error).

Quantification of B

Based on the different types of hazards present in the process, B provides two types of methodologies respectively to quantify: (i) $B1$ – fire and explosion hazards, and (ii) $B2$ – toxic and corrosive hazards.

Quantification of B1

The process units are classified to five categories and the calculation of $B1$ is defined separately for each category: (i) storage units, (ii) units involving physical operations (e.g., heat/mass transfer, phase change), (iii) units involving chemical reactions, (iv) transportation units, and (v) other units such as furnaces, boilers, etc. Let us take the "units involving chemical reactions" as an example to illustrate the $B1$ methodology.

For a unit in which chemical reaction takes place, $B1$ is defined via Eqs. (7.8) and (7.9):

$$\text{Hazard Potential} = (F1 \times pn1 + F \times pn2 + F4 \times pn9 \times pn10)$$
$$\times pn3 \times pn4 \times pn5 \times pn6 \times pn7 \times pn8 \tag{7.8}$$

$$B1 = 4.76 \times \text{Hazard Potential}^{1/3} \tag{7.9}$$

where $F1$, F (determined by $F2$ and $F3$), and $F4$ are energy factors which account for physical and chemical energy, $pn1$-$pn10$ are penalty factors considering different process aspects. More specifically:

a. Energy factors
   - $F1$ accounts for chemical energy via combustion energy

$$F1 = 0.1M \times (Hc)/K \tag{7.10}$$

   where M gives the released mass of chemical (kg) or mass release rate (kg/s), Hc denotes heat of combustion (kJ/mol), K is a constant as 3.148
   - $F2$ and $F3$ account for physical energy

$$F2 = 1.304 \times 10^{-3} \times PP \times V \tag{7.11}$$

$$F3 = 1.0 \times 10^{-3} \times 1/(T + 273) \times (PP - VP)^2 \times V \tag{7.12}$$

Chapter 7 • Inherent safety analysis 117

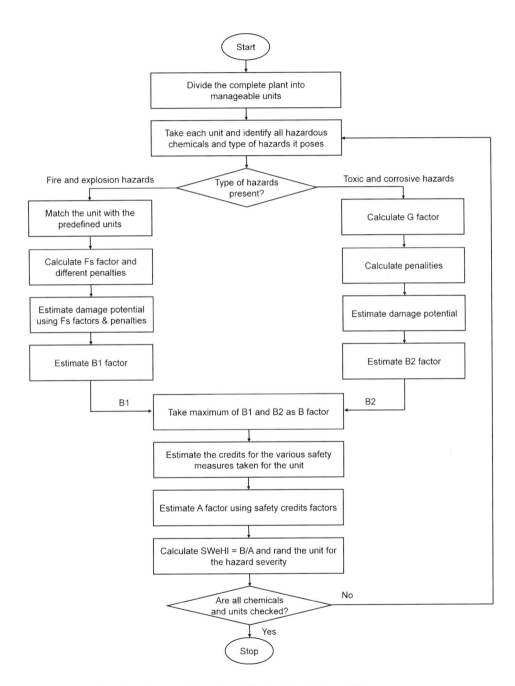

FIGURE 7.2 The workflow for Safety Weighted Hazard Index (adapted from [3]).

118   Synthesis and Operability Strategies for Computer-Aided Modular PI

where PP is processing pressure (kPa), VP denotes vapor pressure (kPa), V represents chemical volume (m³), and T gives temperature (°C)

- $F4$ accounts for energy release from chemical reactions

$$F4 = M \times Hrxn/K \tag{7.13}$$

where $Hrxn$ is the heat of reaction (kJ/kg)

b. Penalty

- $pn1$ accounts for temperature effects

$$pn1 = \begin{cases} 1.45 & \text{if fire point > temperature > flash point} \\ 1.75 & \text{if 0.75 autoignition temperature > temperature > fire point} \\ 1.95 & \text{if temperature > 0.75 autoignition temperature} \\ 1.1 & \text{elsewise} \end{cases}$$

$$\tag{7.14}$$

- $pn2$ accounts for pressure effects
  - if $VP > AP$ and $PP > AP$

$$pn2 = 1 + ((PP - VP)/PP) \times 0.6 \tag{7.15}$$
$$F = F2 + F3 \tag{7.16}$$

  Otherwise:

$$pn2 = 1 + ((PP - VP)/PP) \times 0.4 \tag{7.17}$$
$$F = F2 \tag{7.18}$$

  - if $AP > VP$ and $PP > AP$

$$pn2 = 1 + ((PP - VP)/PP) \times 0.2 \tag{7.19}$$
$$F = F3 \tag{7.20}$$

  Otherwise:

$$pn2 = 1.1 \tag{7.21}$$
$$F = F3 \tag{7.22}$$

- $pn3$ accounts for the mass quantity of chemicals present in the unit/plant (a figure is presented in Khan et al. [3] to find the penalty under different NF/NR values)
- $pn4$ accounts for the effect of reactivity and flammability

$$pn4 = \text{Maximum}[1, 0.30 \times (NF + NR)] \tag{7.23}$$

- $pn5$ accounts for the effect of the location of the nearest hazardous unit (a figure is presented in Khan et al. [3] to find the penalty)

Chapter 7 • Inherent safety analysis   119

- $pn6$ accounts for the unit density

$$pn6 = (1+\% \text{ space occupied by the unit in an area of 30 m} \\ \text{radius from the unit}/100) \tag{7.24}$$

- $pn7$ accounts for the frequency of external factors such as earthquake and hurricane
- $pn8$ accounts for the vulnerability of the surroundings to accident based on historical data
- $pn9$ accounts for the severity of different reaction types (e.g., $pn9 = 1.95$ for nitration reaction, $pn9 = 1.60$ for oxidation reaction)
- $pn10$ accounts for the impact of undesirable side reaction (e.g., $pn10 = 1.65$ for autocatalytic reaction)

Quantification of B2

$B2$ is derived based on relevant transport phenomena and empirical models with the assumption of dispersion occurring under slightly stable atmospheric conditions. $B2$ is computed via the following Eq. (7.25):

$$B2 = a(G \times pnr1 \times pnr2 \times pnr3 \times pnr4 \times pnr5 \times pnr6 \times pnr7)^{b} \tag{7.25}$$

where $a = 25.35$ and $b = 0.425$ are constants, G is defined as the core factor which accounts for difference release scenarios, $pnr1$-$pnr7$ are penalties. More specifically,

a. Core factor G:

$$G = S \times m \tag{7.26}$$

where m is the anticipated release rate (kg/s), the value of S is determined based on the NFPA rank and the release conditions (i.e., in the form of liquid, liquefied gas, gas, or solid). A table of values is provided in Khan et al. [3] for S values under different cases.

b. Penalty

- $pnr1$ accounts for temperature effects
  - if chemical is flammable

$$pn1 = \begin{cases} 1.45 & \text{if fire point} > \text{temperature} > \text{flash point} \\ 1.75 & \text{if 0.75 autoignition temperature} > \text{temperature} > \text{fire point} \\ 1.95 & \text{if temperature} > 0.75 \text{ autoignition temperature} \end{cases} \tag{7.27}$$

  - or if chemical is toxic or corrosive

$$pn1 = \begin{cases} 1.55 & \text{if temperature} > 4\times \text{ambient temperature} \\ 1.35 & \text{if temperature} > 2\times \text{ambient temperature} \end{cases} \tag{7.28}$$

  - or

$$pnr1 = 1.1 \tag{7.29}$$

120 Synthesis and Operability Strategies for Computer-Aided Modular PI

- $pnr2$ accounts for pressure effects
  - if $VP > AP$

$$pn2 = \begin{cases} h1(PP) & \text{if } PP > 3.0 \times AP \\ 1.3 & \text{elsewise} \end{cases} \tag{7.30}$$

  - or if $PP < VP$

$$pnr2 = -h2(pp) \qquad (\text{provided}, PP < 0.3 \times AP) \tag{7.31}$$

  - otherwise,

$$pnr2 = 1.2 \tag{7.32}$$

  where the functions $h1(PP)$ and $h2(PP)$ are plotted in [3]
- $pnr3$ accounts for the effect of vapor density on dispersing

$$pnr3 = 1.2 \times \text{Vapor density / air density} \tag{7.33}$$

- $pnr4$ accounts for the toxicity of the release chemical

$$pnr4 = \text{Maximum}(1, \ 0.6 \times NH) \tag{7.34}$$

  where NH is the NFPA ranking for health hazards
- $pnr5$ accounts for population density
- $pnr6$ and $pnr7$ are estimated the same as $pn6$ and $pn7$

7.4 Quantitative risk assessment

In addition to the above safety indices, Quantitative Risk Assessment (QRA) provides another method to evaluate process safety performance. The value of risk in a process is defined as the product of consequence severity and equipment failure frequency. Equipment failure frequency can be obtained via the average historical data for equipment failure frequencies. Table 7.2 presents the failure frequencies for plate heat exchangers.

Table 7.2 Failure frequency data for plate heat exchangers. (Adapted from Handbook of Failure Frequencies [4].)

Type of failure	Failure frequency (per heat exchanger per year)		
	$P < 5$ bar	5 bar $\leq P < 8$ bar	8 bar $\leq P$
Small leak $0 < d \leq 25$ mm $d_{eq} = 10$ mm	4.6×10^{-3}	7.0×10^{-3}	1.8×10^{-2}
Medium leak $25 < d \leq 50$ mm $d_{eq} = 35$ mm	2.0×10^{-3}	3.0×10^{-3}	7.2×10^{-3}
Rupture	5.5×10^{-6}	8.3×10^{-6}	2.0×10^{-5}

Chapter 7 • Inherent safety analysis   121

To quantify the consequence severity, we showcase in this section the indication number approach documented in *Guidelines for Quantitative Risk Assessment* [5]. The indication number, A, is defined as Eq. (7.35).

$$A = \frac{Q \times O_1 \times O_2 \times O_3}{G} \tag{7.35}$$

where Q gives the quantity (kg) of substance present in the unit or plant, O_i are the factors for process conditions, and G is the limit value (kg). More specifically,

a. Q considers the total amount of the materials

b. O_1 takes into account the type of installation

$$O_1 = \begin{cases} 1 & \text{process installation} \\ 0.1 & \text{storage installation} \end{cases} \tag{7.36}$$

c. O_2 takes into account the positioning of installation and the presence of provisions. For example,

$$O_2 = \begin{cases} 1 & \text{outdoor installation} \\ 0.1 & \text{indoor installation} \end{cases} \tag{7.37}$$

d. O_3 takes into account the process conditions and the amount of released substance in gas phase as presented in Table 7.3. Note that: i. P_{sat} is saturation pressure (absolute) at process temperature, ii. $X = 4.5 \times P_{sat} - 3.5$, iii. Δ is assigned as 0–3 based on substance boiling point to account for liquid pool evaporation, and iv. O_3 is limited to the range of 0.1–10.

Table 7.3 Quantification of O_3. (Adapted from Stoffen [5].)

Phase conditions of released substance	O_3
Gas phase	10
Liquid phase	
$P_{sat} \geq 3$ bar	10
1 bar $< P_{sat} < 3$ bar	$X + \Delta$
$P_{sat} \leq 1$ bar	$P_i + \Delta$
Solid phase	0.1

e. G is the limit value which quantifies the hazardous properties of the substance

- Limit value for toxicity is determined using: (i) LC_{50} (rat, inh, 1h in mg/m^3 / i.e. Lethal Concentration 50%, rat, inhalation, 1 hour), (2) the substance phase condition at 25 degree C (i.e., vapor, liquid, or solid), and (iii) the boiling point if in liquid. A table is provided in [5] – for example, if a substance is gaseous with $LC_{50} \leq 100$, the corresponding limit value is 3; however, if the gaseous has a LC_{50} between 100 and 500, the limit value is 30.

- Limit value is 10000 kg for all flammable substances, i.e. for which the process temperature equal to or higher than the flashpoint.
- Limit value for explosive substances is calculated as the quantity of substance (in kg) which has an energy release equivalent to 1000 kg TNT (with an explosive energy of 4600 kJ/kg).

References

[1] J.T. Marshall, A. Mundt, Dow's chemical exposure index guide, Process Safety Progress 14 (1995) 163–170.

[2] American Institute of Chemical Engineers (AIChE) and American Institute of Chemical Engineers Staff, Dow's Fire & Explosion Index Hazard Classification Guide, Wiley, 2010.

[3] F.I. Khan, T. Husain, S.A. Abbasi, Safety weighted hazard index (SWeHI): a new, user-friendly tool for swift yet comprehensive hazard identification and safety evaluation in chemical process industrie, Process Safety and Environmental Protection 79 (2001) 65–80.

[4] Flemish Government, Handbook failure frequencies 2009 for drawing up a safety report, 2009.

[5] P.G. Stoffen, Guidelines for quantitative risk assessment, Ministerie van Volkshuisvesting Ruimtelijke Ordening en Milieu. CPR E 18 (2005).

8

Multi-parametric model predictive control

8.1 Process control basics

8.1.1 PID control basics

A control system (or controller) is the process of manipulating certain system variables $u(t)$ (referred as manipulated variables) to obtain a set of desired system outputs $y(t)$ (referred as control variables). As shown in Fig. 8.1, the controller receives reference signals y_{sp}, namely the desired setpoints for the process to meet regardless of disturbances or uncertainties $d(t)$. The setpoints can be set for product purity, production rate, process temperature, etc. Based on the control law, the controller will calculate the manipulated variables $u(t)$ and send them to the dynamic process trying to regulate the control variables $y(t)$ at the desired setpoints. The mismatches between the actual output variables $y(t)$ and their setpoints y_{sp}, i.e. the error terms $e(t)$, are also used as feedback to the controller to adjust the manipulated variables in a closed loop control system (more information on chemical process control basics can be found in the seminal textbooks [1,2,16]).

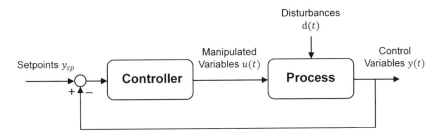

FIGURE 8.1 Process control system.

The most widely used controller in chemical industry is the Proportional-Integral-Derivative (PID) control (Fig. 8.2), which comprises three types of control considerations:

- The proportional term (P) considers how far the output variables are from their setpoints only at time t (Fig. 8.3a). Namely, the control action is calculated based on $e(t)$ for a given time t;
- The integral term (I) considers how long and how far the output variables have been away from their setpoints (Fig. 8.3b). Namely, the control action is calculated by continually summing $e(t)$;

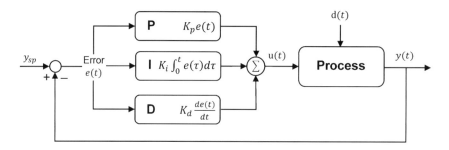

FIGURE 8.2 A PID control system.

- The derivative term (D) considers how fast the trajectory of $e(t)$ changes at time t (Fig. 8.3c). Namely, the control action is calculated as the slope of $e(t)$ for a given time t.

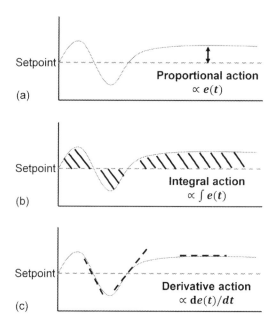

FIGURE 8.3 PID control actions.

For PID control, the correlation between the controller outputs and the error terms can be expressed as:

$$u(t) = K_p e(t) + K_i \int_0^t e(\tau)d\tau + K_d \frac{de(t)}{dt} \tag{8.1}$$

where K_p is the proportional gain, K_i is the integral gain, and K_d is the derivative gain. In some application cases, not all the above three control terms are needed, which results in P control, PI control, etc.

Compared to other model-based control strategies, PID control is easy to design, tune, and implement – mostly via trial and error approaches. Since the control law (Eq. 8.1) only relies on the error terms between process setpoints versus the measured control variables, PID control can be used merely based on plant operation data without a plant model. However, technical challenges of PID control have been observed. For example, PID control concerns mostly single-input single-output process systems, while showing insufficiency in handling multi-input multi-output systems with complex nonlinear process dynamics, process constraints, high directionality of the process gain, etc. – which are normally the case for intensified processes [3].

8.1.2 Model predictive control basics

Model predictive control (MPC) strategies have been developed to address the above challenges. Unlike PID control which calculates the manipulated variables based on the error terms, MPC predicts process output variables using a process model and explicitly considers process constraints. Moreover, optimal control actions are computed by optimizing a pre-defined objective function to measure tracking error, profit, etc. over a certain output horizon and/or input horizon of interest. Following a receding horizon policy, the manipulated variables are updated at every sampling time by repetitively solving the dynamic optimization problem online (i.e., during system operation). The main elements of model predictive control are illustrated in Fig. 8.4 and summarized as follow:

1. Starting at time t, given the measured control variables $y(t)$ and/or state variables $x(t)$
2. A constrained optimization problem (Eq. 8.2) is solved to obtain:
    i. Predicted future outputs: $y_{t+1|t}, y_{t+2|t}, ..., y_{t+N|t}$
    ii. Optimal manipulated variables: $u^*_{t|t}, u^*_{t+1|t}, u^*_{t+2|t}, ..., u^*_{t+M-1|t}$
3. Apply the first input of the sequence $u^*_{t|t}$ until time $t+1$
4. At time $t+1$ repeat the above steps.

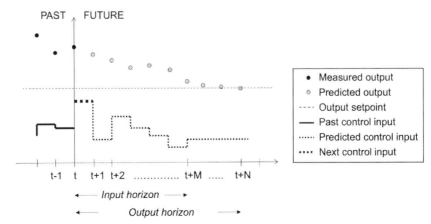

FIGURE 8.4 Model predictive control with receding horizon policies.

126   Synthesis and Operability Strategies for Computer-Aided Modular PI

The general mathematical formulation of an MPC problem is given by Eq. (8.2). Eq. (8.2a) defines the objective function. Eqs. (8.2b) and (8.2c) describe the process model which is used to describe the process dynamics and to predict the state and output variables. Eqs. (8.2d)–(8.2g) set the constraints respectively for output, state, and input variables due to operational, physical, and safety considerations, etc. Eq. (8.2h) is the state estimator. L defines the objective function, ϕ is the terminal cost function, N gives the output prediction horizon, t is the current time, $y_{t+j|t}$ and $u_{t+j|t}$ are respectively the output and input variables at j time steps after the current time t.

$$\min_{u} \quad J = \sum_{j=0}^{N-1} L(y_{t+j|t}, u_{t+j|t}, \Delta u_{t+j|t}) + \phi(y_{t+N|t}) \tag{8.2a}$$

$$\text{s.t.} \quad x_{t+j+1|t} = f(x_{t+j|t}, u_{t+j|t}), \quad j = 0, ..., N-1 \tag{8.2b}$$

$$y_{t+j|t} = g(x_{t+j|t}, u_{t+j|t}), \quad j = 0, ..., N-1 \tag{8.2c}$$

$$\underline{y} \leq y_{t+j|t} \leq \overline{y} \tag{8.2d}$$

$$\underline{x} \leq x_{t+j|t} \leq \overline{x} \tag{8.2e}$$

$$\underline{u} \leq u_{t+j|t} \leq \overline{u} \tag{8.2f}$$

$$\underline{\Delta u} \leq \Delta u_{t+j|t} \leq \overline{\Delta u} \tag{8.2g}$$

$$x_{t|t} = x(t) \tag{8.2h}$$

8.1.2.1 Tutorial example 1: quadratic programming formulation of MPC

Based on the above general formulation, let us consider a motivating example on a Constrained Linear Quadratic Regulator (CLQR) problem. The dynamic process system is defined as Eq. (8.3):

$$\dot{x}(t) = u(t)$$
$$y(t) = x(t) \tag{8.3}$$

The discrete-time state space model of the above process system can be obtained by approximating the first order derivative of x with a sampling time of $T_s = 1s$:

$$\dot{x}(t) = u(t) \Longrightarrow \frac{x(t+1) - x(t)}{T_s} = u(t) \Longrightarrow x(t+1) - x(t) = T_s u(t) \tag{8.4}$$

Considering the following MPC problem (Eq. 8.5):

$$\min_{u_{t|t}, u_{t+1|t}} \quad J = \sum_{k=0}^{1} [q(x_{t+k|t} - x_{ref})^2 + r(u_{t+k|t} - u_{ref})^2] + p(x_{t+2|t} - x_{ref})^2$$

$$\text{s.t.} \quad x_{t+k+1|t} = x_{t+k|t} + u_{t|t}, \quad k = 0, 1, 2 \tag{8.5}$$

$$-2 \leq u_{t+k|t} \leq 2, \quad k = 0, 1$$

Chapter 8 • Multi-parametric model predictive control   127

In this case, $M = N = 2$, x_{ref} is the setpoint for state variables, u_{ref} is the input when the system is at x_{ref}. Since $\dot{x} = 0 = u(t)$, we have $u_{ref} = 0$. For simplicity, we will neglect u_{ref} in the following calculations since it is always 0.

To solve Eq. (8.5), we need to reformulate the MPC problem, including the objective function and the constraints, to a known optimization format:

Step 1: Expand all state variables over the output prediction horizon (i.e. $N = 2$).

$$x_{t+1|t} = x_{t|t} + u_{t|t}$$
$$x_{t+2|t} = x_{t+1|t} + u_{t+1|t} = x_{t|t} + u_{t|t} + u_{t+1|t} \tag{8.6}$$

Step 2: The objective function in Eq. (8.5) can be rearranged as:

$$J = q(x_{t|t} - x_{ref})^2 + ru_{t|t}^2 + q(x_{t+1|t} - x_{ref})^2 + ru_{t+1|t}^2 + p(x_{t+2|t} - x_{ref})^2 \tag{8.7}$$

Substitute $x_{t+1|t}$ and $x_{t+2|t}$ obtained via Step 1:

$$J = (2q + p)(x_{t|t} - x_{ref})^2 + 2(q + p)x_{t|t}u_{t|t} - 2(q + p)x_{ref}u_{t|t} + 2px_{ref}u_{t+1|t}$$
$$+ 2pu_{t|t}u_{t+1|t} + (q + r + p)u_{t|t}^2 + (r + p)u_{t+1|t}^2 \tag{8.8}$$

Reformulate into the matrix form:

$$J = \frac{1}{2}\begin{bmatrix} u_{t|t}, u_{t+1|t} \end{bmatrix} \begin{bmatrix} 2(q + r + p), 2p \\ 2p, 2(r + p) \end{bmatrix} \begin{bmatrix} u_{t|t} \\ u_{t+1|t} \end{bmatrix}$$
$$+ \begin{bmatrix} x_{t|t}, x_{ref} \end{bmatrix} \begin{bmatrix} 2(q + p), 2p \\ -2(q + p), -2p \end{bmatrix} \begin{bmatrix} u_{t|t} \\ u_{t+1|t} \end{bmatrix} \tag{8.9}$$
$$+ (2q + p)(x_{t|t} - x_{ref})^2$$

Let $U = \begin{bmatrix} u_{t|t} \\ u_{t+1|t} \end{bmatrix}$, $H = \begin{bmatrix} 2(q + r + p), 2p \\ 2p, 2(r + p) \end{bmatrix}$, $F = \begin{bmatrix} 2(q + p), 2p \\ -2(q + p), -2p \end{bmatrix}$

Then Eq. (8.9) can be written as:

$$J = \frac{1}{2}U^T HU + \begin{bmatrix} x_{t|t}, x_{ref} \end{bmatrix} FU + (2q + p)(x_{t|t} - x_{ref})^2 \tag{8.10}$$

Step 3: Rearrange the constraints Eq. (8.5) into the matrix form:

$$\begin{bmatrix} 1, 0 \\ -1, 0 \\ 0, 1 \\ 0, -1 \end{bmatrix} \begin{bmatrix} u_{t|t} \\ u_{t+1|t} \end{bmatrix} \le \begin{bmatrix} 2 \\ 2 \\ 2 \\ 2 \end{bmatrix} \tag{8.11}$$

Let $G = \begin{bmatrix} 1, 0 \\ -1, 0 \\ 0, 1 \\ 0, -1 \end{bmatrix}$, $W = \begin{bmatrix} 2 \\ 2 \\ 2 \\ 2 \end{bmatrix}$

128 Synthesis and Operability Strategies for Computer-Aided Modular PI

As a result, the MPC problem in Eq. (8.5) can be written as Eq. (8.12), which is a **Quadratic Program** (QP) to be solved with known quadratic programming or general nonlinear programming algorithms:

$$\min_{u} \quad J = \frac{1}{2}U^T H U + \left[x_{t|t}, x_{ref}\right] F U + (2q + p)(x_{t|t} - x_{ref})^2$$

$$\text{s.t.} \quad GU \leq W \tag{8.12}$$

For a more generalized CLQR (finite horizon) problem as defined in Eq. (8.13):

$$\min_{u} \quad J = \sum_{j=0}^{N-1} x_{t+j|t}^T Q x_{t+j|t} + u_{t+j|t}^T R u_{t+j|t} + x_{t+N|t}^T P x_{t+N|t} \tag{8.13a}$$

$$\text{s.t.} \quad x_{t+j+1|t} = A x_{t+j|t} + B u_{t+j|t}, \quad j \geq 0 \tag{8.13b}$$

$$y_{t+j|t} = C x_{t+j|t} + D u_{t+j|t}), \quad j \geq 0 \tag{8.13c}$$

$$\underline{y} \leq y_{t+j|t} \leq \overline{y} \tag{8.13d}$$

$$\underline{x} \leq x_{t+j|t} \leq \overline{x} \tag{8.13e}$$

$$\underline{u} \leq u_{t+j|t} \leq \overline{u} \tag{8.13f}$$

$$\underline{\Delta u} \leq \Delta u_{t+j|t} \leq \overline{\Delta u} \tag{8.13g}$$

$$u_{t+j|t} = -K x_{t+j|t}, \quad j \geq N \tag{8.13h}$$

$$x_{t|t} = x(t) \tag{8.13i}$$

The equivalent QP problem, given $x(t)$, is shown by Eq. (8.14):

$$\min_{u} \quad J(U, x(t)) = \frac{1}{2}U^T H U + x^T(t) F U + x(t) Y x(t)$$

$$\text{s.t.} \quad GU \leq W + E x(t) \tag{8.14}$$

In sum, MPC offers the advantages to handle complex multiple-input multiple-output systems using model-based prediction by explicitly taking care of hard constraints. MPC is able to obtain optimal control actions via optimization. However, some technical challenges for MPC implementation exist such as the expensive computational demands, the need to repetitively solve online dynamic optimization problems at each time instance, etc. Besides, MPC may not be applicable for systems with fast dynamics or small sampling time, in which cases the online optimization problem cannot be solved in time.

8.2 Explicit model predictive control via multi-parametric programming

Explicit/Multi-parametric model predictive control (mp-MPC) has been proposed to reduce the computational load of MPC online optimization. mp-MPC reformulates and solves the MPC problem offline via multi-parametric programming. The solution to this

Chapter 8 • Multi-parametric model predictive control   129

problem generates the explicit control law via offline calculation, which expresses the optimal control action as explicit functions of the defined parametric set (e.g., disturbance, state variables, design variables). In this context, online efforts are only needed to map through the parameter space and to perform function evaluation. For more details on multi-parametric programming, readers are referred to the recent topical book by Pistikopoulos et al. [4].

Let us continue with the CLQR (finite horizon) problem as an illustrative example. Let $z = U + H^{-1}F^T x(t)$, Eq. (8.14) can be reformulated as:

$$\min_{z} \quad \frac{1}{2}z^T H z$$
$$\text{s.t.} \quad Gz \leq W + Sx(t) \tag{8.15}$$

where $S = E + GH^{-1}F$ and $H > 0$.

By solving the mp-QP problem (Eq. 8.15), we will obtain:

- Local optimum $z^*(x)$ – the state variables x are considered as the "parameters" which are used to calculate $z(x)$
- The corresponding $x(t)$-space where the optimality of $z^*(x)$ is valid – defined as the "Critical Region"
- Proceed iteratively to cover all $x(t)$-space

To solve Eq. (8.15), the Karush-Kuhn-Tucker (KKT) conditions of optimality are applied as shown below [5]:

$$Hz^* + G^T\lambda^* = 0, \qquad \lambda^* \in R^l$$
$$\lambda^*(G_i z^* - W_i - S_i x) = 0 \tag{8.16}$$
$$\lambda^* \geq 0, \qquad\qquad i = 1, ..., l$$

For active constraints:

$$G_i z^* - W_i - S_i x = 0, \qquad \lambda^* > 0$$

For inactive constraints:

$$G_i z^* - W_i - S_i x < 0, \qquad \lambda^* = 0$$

where λ^* is the Lagrange multiplier.

In what follows, we use a tutorial example to detail step by step how to solve Eq. (8.15) via multi-parametric programming and how to generate a map of solutions over the parametric space.

8.2.1 Tutorial example 2: multi-parametric MPC

Consider the following multi-parametric quadratic optimization (mp-QP) problem [4]:

$$
\begin{aligned}
\min \quad & \frac{1}{2}z^2 \\
\text{s.t.} \quad & z \le 1 + \frac{1}{2}\theta_1 - \frac{1}{6}\theta_2 \quad (g_1) \\
& -z \le 1 - \frac{1}{2}\theta_1 + \frac{1}{6}\theta_2 \quad (g_2) \\
& -3 \le \theta_1 \le 3 \\
& -3 \le \theta_2 \le 3
\end{aligned}
\tag{8.17}
$$

ITERATION 1:

Step 1.1: Select starting points for the parameter space.

$$
\theta^0 = [\theta_1^0, \theta_2^0]^T = [0, 0]^T
$$

Step 1.2: Substitute θ^0 to the inequality constraints g_1 and g_2, which results in a nonlinear programming problem. More specifically, a quadratic programming (QP) problem.

$$
\begin{aligned}
\min \quad & \frac{1}{2}z^2 \\
\text{s.t.} \quad & z \le 1 \\
& -z \le 1
\end{aligned}
\tag{8.18}
$$

Step 1.3: The above QP problem (Eq. 8.18) can be solved via the Karush-Kuhn-Tucker (KKT) active set strategy. More information on nonlinear optimization basics can be found in the Book Chapter 3 of Floudas [6].

The optimal solution, $z = 0$, can also be obtained via observation in this case. In this context, both the inequality constraints g_1 and g_2 are inactive.

Step 1.4: Based on the active and inactive constraints of the above optimal solution, the following sets can be defined:

- $\hat{G}, \hat{W}, \hat{S}$ – Matrices to define active constraints
- $\check{G}, \check{W}, \check{S}$ – Matrices to define inactive constraints

$$
G = \begin{bmatrix} 1 \\ -1 \end{bmatrix}, \quad W = \begin{bmatrix} 1 \\ 1 \end{bmatrix}, \quad S = \begin{bmatrix} 1/2, & -1/6 \\ -1/2, & 1/6 \end{bmatrix}, \quad H = \begin{bmatrix} 1 \end{bmatrix}
$$

$$
\hat{G} = \begin{bmatrix} 0 \end{bmatrix}, \quad \hat{W} = \begin{bmatrix} 0 \end{bmatrix}, \quad \hat{S} = \begin{bmatrix} 0 \end{bmatrix}
$$

$$
\check{G} = \begin{bmatrix} 1 \\ -1 \end{bmatrix}, \quad \check{W} = \begin{bmatrix} 1 \\ 1 \end{bmatrix}, \quad \check{S} = \begin{bmatrix} 1/2, & -1/6 \\ -1/2, & 1/6 \end{bmatrix}
$$

Chapter 8 • Multi-parametric model predictive control   131

Step 1.5: The expressions for λ and z as functions of θ:

$$\lambda^*(\theta) = -(\hat{G} \cdot H^{-1} \cdot \hat{G}^T)^{-1}(\hat{W} + \hat{S}\theta)$$
$$z^*(\theta) = H^{-1} \cdot \hat{G}^T)(\hat{G} \cdot H^{-1} \cdot \hat{G}^T)^{-1}(\hat{W} + \hat{S}\theta)$$

Substitute the *active* set matrix values:

$$\lambda^*(\theta) = 0$$
$$z^*(\theta) = 0$$

Step 1.6: The Critical Region (CR) is defined by the *inactive* set and the dual feasibility condition as shown below.

Inactive inequality constraints:

$$\begin{bmatrix} 1 \\ -1 \end{bmatrix} z^*(\theta) \leq \begin{bmatrix} 1 \\ 1 \end{bmatrix} + \begin{bmatrix} 1/2, & -1/6 \\ -1/2, & 1/6 \end{bmatrix} \begin{bmatrix} \theta_1 \\ \theta_2 \end{bmatrix}$$
$$\implies \begin{bmatrix} 1 \\ -1 \end{bmatrix} \leq \begin{bmatrix} 1 \\ 1 \end{bmatrix} + \begin{bmatrix} 1/2, & -1/6 \\ -1/2, & 1/6 \end{bmatrix} \begin{bmatrix} \theta_1 \\ \theta_2 \end{bmatrix}$$

Box constraints:

$$-3 \leq \theta_1 \leq 3$$
$$-3 \leq \theta_2 \leq 3$$

Since there are no active constraints, dual feasibility condition does not apply for this iteration. Thus, we obtain the following optimal solution of z on Critical Region CR_1 (as depicted in Fig. 8.5):

$$z = 0, \quad \text{if} \quad CR_1 = \{-1 \leq -\frac{1}{2}\theta_1 + \frac{1}{6}\theta_2 \leq 1, \ -3 \leq \theta_1, \theta_2 \leq 3\}$$

ITERATION 2:

Step 2.1: Select a new starting point θ^0, which should be outside the region of CR_1.

$$\theta^0 = [-2, \ 2]^T$$

Step 2.2: Substitute θ^0 to the inequality constraints g_1 and g_2. Again this results in a QP problem.

$$\min \quad \frac{1}{2}z^2$$
$$\text{s.t.} \quad z \leq -\frac{1}{3} \tag{8.19}$$
$$-z \leq \frac{7}{3}$$

132 Synthesis and Operability Strategies for Computer-Aided Modular PI

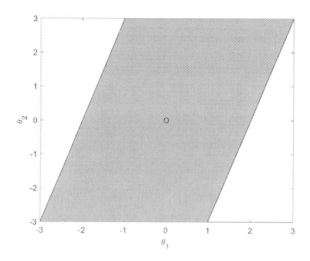

FIGURE 8.5 Critical Region for Iteration 1.

<u>Step 2.3:</u> The optimal solution, $z = \frac{1}{3}$, can be obtained via observation. In this context, g_1 is active and $\lambda_1 = \frac{1}{3}$.

<u>Step 2.4:</u> Based on the active and inactive constraints of the above optimal solution, the following sets can be defined:

$$\hat{G} = [1], \quad \hat{W} = [1], \quad \hat{S} = [1/2, -1/6]$$
$$\check{G} = [-1], \quad \check{W} = [1], \quad \check{S} = [-1/2, 1/6]$$

<u>Step 2.5:</u> The expressions for λ and z as functions of θ.

$$\lambda^*(\theta) = -(\hat{G} \cdot H^{-1} \cdot \hat{G}^T)^{-1}(\hat{W} + \hat{S}\theta)$$
$$z^*(\theta) = H^{-1} \cdot \hat{G}^T)(\hat{G} \cdot H^{-1} \cdot \hat{G}^T)^{-1}(\hat{W} + \hat{S}\theta)$$

Substitute the *active* set matrix values:

$$\lambda^*(\theta) = -1 + [-1/2, 1/6]\theta$$
$$z^*(\theta) = 1 + [1/2, -1/6]\theta$$

<u>Step 2.6:</u> The Critical Region (CR) is defined by the *inactive* set and the dual feasibility condition as shown below.

Inactive inequality constraints:

$$[-1]z^*(\theta) \leq [1] + [-1/2, 1/6]\begin{bmatrix}\theta_1\\\theta_2\end{bmatrix}$$
$$\Longrightarrow 0 \leq 2 \quad [redundant]$$

Box constraints:

$$-3 \leq \theta_1 \leq 3$$
$$-3 \leq \theta_2 \leq 3$$

Dual feasibility condition:

$$\lambda^*(\theta) = -1 + [-1/2,\ 1/6]\theta \geq 0$$

Thus, we obtain the following optimal solution of z on Critical Region CR_2 (as depicted in Fig. 8.6):

$$z = 1 + [1/2,\ -1/6]\theta, \quad \text{if} \quad CR_2 = \{1 \leq -\frac{1}{2}\theta_1 + \frac{1}{6}\theta_2,\ -3 \leq \theta_1, \theta_2 \leq 3\}$$

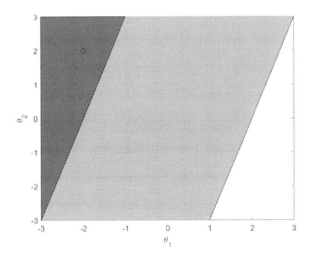

FIGURE 8.6 Critical Region for Iteration 2.

ITERATION 3:

Step 3.1: Select another new starting point θ^0, which should be outside the regions of CR_1 and CR_2.

$$\theta^0 = [2,\ -2]^T$$

134   Synthesis and Operability Strategies for Computer-Aided Modular PI

Step 3.2: Substitute θ^0 to the inequality constraints g_1 and g_2. Again this results in a QP problem.

$$\begin{aligned} \min \quad & \frac{1}{2}z^2 \\ \text{s.t.} \quad & z \le \frac{7}{3} \\ & -z \le -\frac{1}{3} \end{aligned} \tag{8.20}$$

Step 3.3: The optimal solution, $z = \frac{1}{3}$, can be obtained via observation. In this context, g_2 is active and $\lambda_2 = \frac{1}{3}$.

Step 3.4: Based on the active and inactive constraints of the above optimal solution, the following sets can be defined:

$$\begin{aligned} \hat{G} = \begin{bmatrix} -1 \end{bmatrix}, \quad \hat{W} = \begin{bmatrix} 1 \end{bmatrix}, \quad \hat{S} = \begin{bmatrix} -1/2, \ 1/6 \end{bmatrix} \\ \check{G} = \begin{bmatrix} 1 \end{bmatrix}, \quad \check{W} = \begin{bmatrix} 1 \end{bmatrix}, \quad \check{S} = \begin{bmatrix} 1/2, \ -1/6 \end{bmatrix} \end{aligned}$$

Step 3.5: The expressions for λ and z as functions of θ.

$$\begin{aligned} \lambda^*(\theta) &= -(\hat{G} \cdot H^{-1} \cdot \hat{G}^T)^{-1}(\hat{W} + \hat{S}\theta) \\ z^*(\theta) &= H^{-1} \cdot \hat{G}^T)(\hat{G} \cdot H^{-1} \cdot \hat{G}^T)^{-1}(\hat{W} + \hat{S}\theta) \end{aligned}$$

Substitute the *active* set matrix values:

$$\begin{aligned} \lambda^*(\theta) &= -1 + \begin{bmatrix} 1/2, \ -1/6 \end{bmatrix}\theta \\ z^*(\theta) &= -1 + \begin{bmatrix} 1/2, \ -1/6 \end{bmatrix}\theta \end{aligned}$$

Step 3.6: The Critical Region (CR) is defined by the *inactive* set and the dual feasibility condition as shown below.

Inactive inequality constraints:

$$\begin{bmatrix} 1 \end{bmatrix}z^*(\theta) \le \begin{bmatrix} 1 \end{bmatrix} + \begin{bmatrix} 1/2, \ -1/6 \end{bmatrix}\begin{bmatrix} \theta_1 \\ \theta_2 \end{bmatrix}$$

$$\implies -2 \le 0 \quad [redundant]$$

Box constraints:

$$\begin{aligned} -3 \le \theta_1 \le 3 \\ -3 \le \theta_2 \le 3 \end{aligned}$$

Dual feasibility condition:

$$\lambda^*(\theta) = -1 + \begin{bmatrix} 1/2, \ -1/6 \end{bmatrix}\theta \ge 0$$

Thus, we obtain the following optimal solution of z on Critical Region CR_3 (as depicted in Fig. 8.7):

$$z = -1 + [1/2, -1/6]\theta, \quad \text{if} \quad CR_3 = \{-1 \leq \frac{1}{2}\theta_1 - \frac{1}{6}\theta_2, \ -3 \leq \theta_1, \theta_2 \leq 3\}$$

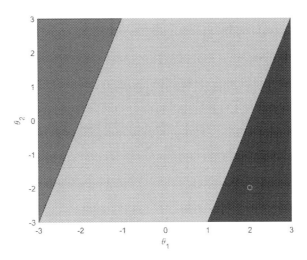

FIGURE 8.7 Critical Region for Iteration 3.

FINAL SOLUTION:

$$z(\theta) = \begin{cases} 0, & \text{if } -1 \leq -\frac{1}{2}\theta_1 + \frac{1}{6}\theta_2 \leq 1, \ -3 \leq \theta_1, \theta_2 \leq 3 \\ 1 + [1/2, -1/6]\theta, & \text{if } 1 \leq -\frac{1}{2}\theta_1 + \frac{1}{6}\theta_2, \ -3 \leq \theta_1, \theta_2 \leq 3 \\ -1 + [1/2, -1/6]\theta, & \text{if } -1 \leq \frac{1}{2}\theta_1 - \frac{1}{6}\theta_2, \ -3 \leq \theta_1, \theta_2 \leq 3 \end{cases}$$

8.3 The PAROC framework

Based on the explicit/multi-parametric model predictive control strategy, we present the PAROC framework which stands for "PARametric Optimization and Control". It provides a unified framework and software platform for the design, operational optimization, and advanced model-based control of process systems [7,8]. The PAROC platform can be accessed via http://paroc.tamu.edu/ and more information can be found in the book by Burnak et al. [9]. As shown in Fig. 8.8, the PAROC framework consists of the following steps to generate optimal process designs with optimal control actions in dynamic operations:

Step 1 – High fidelity dynamic modeling and analysis

High fidelity dynamic models (Eq. 8.21), based on first-principles and correlations (e.g., mass and energy balances), are critical to the accuracy of process dynamic behavior

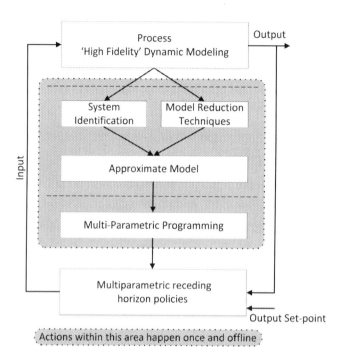

FIGURE 8.8 The PAROC framework (adapted from Pistikopoulos et al. [7]).

and the validity of operational analysis. It normally consists of a system of (Partial) Differential-Algebraic Equations, with continuous and binary design and operating variables which can be manipulated or optimized via the following control and optimization steps. This modeling task takes place in gPROMS ModelBuilder.

$$\frac{d}{dt}x(t) = f(x(t), u(t), Y(t), d(t), De)$$
$$y(t) = g(x(t), u(t), Y(t), d(t), De)$$
(8.21)

where $x(t)$ is the vector of state variables, $u(t)$ is the vector of manipulated variables, $Y(t)$ is the vector of binary variables, $d(t)$ is the vector of disturbances, De is the vector of design variables, and $y(t)$ is the vector of control variables.

Step 2 – Model approximation

Due to the high nonlinearity and complexity of the high fidelity models, an approximation step is needed to reduce the computational requirement and thus to enable the application of advanced optimization approach in a later controller design step. A number of approaches can be applied to simplify the model to a *linear discrete-time state-space model* (Eq. 8.22) while preserving its accuracy, such as: (i) Monte Carlo-based model-reduction [10], (ii) model reduction using machine learning techniques [11], and (iii)

Chapter 8 • Multi-parametric model predictive control   137

statistical methods adapted in the System Identification Toolbox in MATLAB®.

$$\overline{x}_{k+1} = A\overline{x}_k + Bu_k + Cd_k \qquad (8.22a)$$

$$y_k = D\overline{x}_k \qquad (8.22b)$$

where the matrices A, B, C, D are used to define the state-space model, the index k denotes the current time instant, $\overline{x}$ is the vector of identified state variables, u is the vector of manipulated variable, y represents the vector of control variables, and d is the vector of disturbances. De will also be included as design variables if for design-aware controller design. The pseudo-states $\overline{x}$ no longer provide physical meanings as the original state variables x due to the model reduction. However, the other variables retain the physical meanings as in the original high fidelity model.

Step 3 – Design-aware explicit/multi-parametric model predictive control

In deriving explicit model predictive control schemes, the objective is to express the optimal control actions as explicit functions of the parameters of the system. In the seminal paper by Bemporad et al. [12], parameters ($\theta \in \Theta$), which are incorporated in the control law, include the states ($\overline{x}$), outputs (y) and disturbances of the system (d). The parameter space has been more recently extended to include time-invariant design variables (De) by Diangelakis et al. [8]. Consider a generalized CLQR (finite horizon) problem for setpoint tracking as described by Eq. (8.23): QR_k, R_k are the weights of the controller, P is derived from the solution of the discrete time Riccati equation, N and M are the output and control horizons respectively and ϵ takes into account the mismatch between the process and the developed approximate model.

$$\min_{u} \quad J = x_N^T P x_N + \sum_{k=1}^{N-1} \left(\left(y_k - y_k^R \right)^T QR_k \left(y_k - y_k^R \right) \right)$$

$$+ \sum_{k=0}^{M-1} \left(u_k - u_k^R \right)^T R_k \left(u_k - u_k^R \right)$$

$$\text{s.t.} \quad x_{k+1} = Ax_k + Bu_k + Cd_k \qquad (8.23)$$

$$y_k = Dx_k + \epsilon$$

$$\underline{x} \le x_k \le \overline{x}$$

$$\underline{u} \le u_k \le \overline{u}$$

$$\underline{y} \le y_k \le \overline{y}$$

It was proved that the above problem can be exactly reformulated into a multi-parametric quadratic programming problem (mp-QP). The resulting mp-QP has the following form as shown in Eq. (8.24) (recall that in tutorial example 1, we presented the

138   Synthesis and Operability Strategies for Computer-Aided Modular PI

reformulation for a 1-D system).

$$\begin{aligned}
\min_{u} \quad & f(u,\theta) = \frac{1}{2}u^T Q u + u^T H^T \theta + \theta^T Q_\theta \theta + c_u^T u + c_\theta^T \theta + c_c \\
\text{s.t.} \quad & g_i(u,\theta) := A_i u \le b_i + F_i \theta \\
& h_j(u,\theta) := A_j u = b_j + F_j \theta \\
& u_c \in \mathbb{R}^n, \theta \in \Theta := \left\{ \theta \in \mathbb{R}^m \mid C R^A \theta \le C R^b \right\} \\
& Q \succ 0 \\
& i \in \mathbb{I}, j \in \mathbb{J}
\end{aligned} \tag{8.24}$$

where the indices i and j correspond to the ith and j^{th} inequality and equality constraints respectively, which belong to the sets $\mathbb{I}$ and $\mathbb{J}$. The multi-parametric solution of Eq. (8.24), via the POP toolbox [13] in MATLAB, returns a list with the optimal partitions of the parameter space, i.e. Critical Regions. Each critical region is described by a unique active set. When the parameters belong to this critical region, the optimal control law is the following:

$$\begin{aligned}
u^* &= K_i \theta^* + r_i \\
\theta^* &\in C R^i = \{ C R_i^A \theta \le C R_i^b \} \\
\theta^* &= [x_k; u_{k-1}^*; d_k; De; y_k; y_k^{SP}]
\end{aligned} \tag{8.25}$$

where u^* is the optimal control action at the parameter value θ^*, CR^i define the ith critical region, K_i and r_i define the affine expression for the ith critical region, De are the design variables. As can be noted in Eq. (8.25), the design variables De are treated as uncertain parameters and are aware by the optimal mp-MPC controller. Thus the derived design dependent mp-MPC controller can be applied for different design alternatives without a reformulation of the control problem for different designs [8].

Step 4 – Closed-loop validation
The derived design-aware explicit model predictive controller is validated against the original high fidelity model to verify whether it provides satisfactory control performance for setpoint tracking (and/or disturbance rejection) for different designs.

Step 5 – Simultaneous design and control via (mixed-integer) dynamic optimization
By creating Dynamic Link Libraries in Microsoft Visual C++, the design dependent control laws generated from the MATLAB POP toolbox are introduced into gPROMS to integrate with the original high fidelity process model. The solution of the (mixed-integer) dynamic optimization problem fully validates the optimality and control of the resulting designs. The mathematical formulation of the simultaneous design and control optimization problem, which incorporates the high-fidelity dynamic model and

Chapter 8 • Multi-parametric model predictive control   139

explicit control actions, is described as in Eq. (8.26):

$$\min_{Y, De} \quad F = \int_0^\tau P(x, y, u, Y, d, De) dt$$

$$\text{s.t.} \quad \frac{dx}{dt} = f(x, y, u, Y, d, De)$$

$$y = g(x, u, Y, d, De)$$

$$u = h(x, y, Y, d, De)$$

$$\underline{y} \le y \le \overline{y}$$

$$\underline{u} \le u \le \overline{u} \qquad\qquad (8.26)$$

$$Y \in \{0, 1\}^q$$

$$\begin{bmatrix} \underline{x}^T & \underline{d}^T \end{bmatrix}^T \le \begin{bmatrix} x^T & d^T \end{bmatrix}^T \le \begin{bmatrix} \overline{x}^T & \overline{d}^T \end{bmatrix}^T$$

$$\underline{De} \le De \le \overline{De}$$

where the multivariate functions f and g define the original high fidelity process model, h give the explicit control actions derived from mp-MPC, and F is the objective function to be minimized. The typical goal for this dynamic optimization is to achieve a specific process operating and production target by minimizing the cost of the process.

8.4 Case study: multi-parametric model predictive control of an extractive distillation column

In what follows, we present an example for the control of an ethanol-water-ethylene glycol extractive distillation column using the mp-MPC strategy.

8.4.1 Process description

The feed stream consists of 85 mol% ethanol (ETOH) and 15 mol% water (H2O) with a flow rate of 10 kmol/s at 351.3 K, 1 atm (i.e., saturated liquid feed). Ethylene glycol is utilized in this system as solvent with a flowrate of 7.2 kmol/s. The separation target is to obtain a liquid ethanol product with a purity of 99 mol% and a flow rate of 8 kmol/s. The NRTL equation (Eq. 8.27) is utilized to describe the nonideal liquid behavior in this system, with the binary interaction parameters given in Table 8.1 adapted from Ismail et al. [14].

$$\ln \gamma_i = \frac{\sum_{j \in c} \tau_{ji} G_{ji} x_j}{\sum_{j \in c} G_{ji} x_j} + \sum_{j \in c} \frac{G_{ij} x_j}{\sum_{l \in c} G_{lj} x_l} \tau_{ij} - \frac{\sum_{l \in c} \tau_{lj} G_{lj} x_l}{\sum_{l \in c} G_{lj} x_l}$$

$$\tau_{ij} = \frac{\Lambda_{ij}}{RT}, \qquad G_{ij} = exp(-\alpha_{ij} \tau_{ij}), \qquad \Lambda_{ii} = 0, \qquad \alpha_{ij} = \alpha_{ji} \qquad (8.27)$$

$$\Lambda_{ij} = aa_{ij} + ab_{ij} \times (T - 273.15) \text{ cal/mol}$$

Table 8.1 NRTL binary interaction parameters.

	Ethanol (1) - Water (2) - Ethylene Glycol (3) System				
ij	aa_{ij}	ab_{ij}	aa_{ji}	ab_{ji}	α_{ij}
12	−441.20	18.3280	3293.17	17.0471	0.475000
13	13527.42	−92.7391	−4351.97	53.3769	0.370400
23	1383.43	8.0409	−1445.97	−9.1506	0.185894

The extractive distillation column is illustrated in Fig. 8.9. The design and operating parameters at nominal conditions are shown in Table 8.2. Two sets of disturbance exist during the column operation: (i) a disturbance in the ETOH composition of the feed stream (0.85 ± 0.05 mol/mol), and (ii) a disturbance in the ethylene glycol solvent flowrate (6.8–7.2 kmol/s). The control objective is to maintain ethanol product purity at 0.99 mol/mol (note that the ethanol product is obtained from the column distillate). Moreover, the ETOH composition from the column bottom is required to be less than 0.05 mol/mol. Reflux ratio and reboiler heat duty are selected as the manipulated variables. Therefore, the extractive distillation column is a multi-input multi-output (MIMO) system.

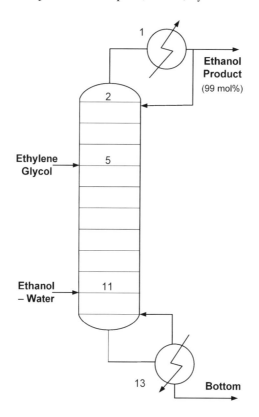

FIGURE 8.9 Ethanol-water-ethylene glycol extractive distillation column.

Chapter 8 • Multi-parametric model predictive control   141

Table 8.2   Extractive distillation nominal design.

Parameters		Values
Column Trays	Number of Trays	13
	Ethanol-Water Feed Tray	Tray 5
	Solvent Feed Tray	Tray 11
Operating Conditions	Column Pressure	1 atm
	Reflux Ratio	0.54
Top Product	Flowrate	8.13 kmol/s
	ETOH Purity	0.99 mol/mol
Bottom Product	Floware	9.07 kmol/s
	ETOH Composition	0.05 mol/mol

8.4.2 mp-MPC controller design via the PAROC framework

To meet the control objectives under disturbances, we will design a MIMO control system for the distillation column following the PAROC framework:

Step 1: high fidelity dynamic modeling

A generalized high fidelity dynamic model for distillation columns [15] is adapted here. The detailed mathematical model can be found in Appendix C, which considers:

- Dynamic mass balances and energy balances for each column tray;
- Liquid and vapor mass and energy holdups on each column tray;
- Liquid hydraulic calculation using modified Francis weir formulation;
- Sieve tray geometry design (e.g., weir height, holes area, etc.);
- Flooding and entrainment correlations to determine the minimum allowable column diameter;
- Phase equilibrium or non-equilibrium behaviors by assigning Murphree tray efficiencies.

The model statistics are given in Table 8.3. The model is built in gPROMS ModelBuilder, integrated with the MultiFlash thermodynamic package for physical property calculation. Table 8.4 presents a summary of the process variables which will be used later for controller design (e.g., manipulated variables, control variables).

Table 8.3   Model statistics.

Modeling equations	720
Initial conditions	33
Algebraic variables	687
Differential variables	33

142 Synthesis and Operability Strategies for Computer-Aided Modular PI

Table 8.4 Extractive distillation – Types of variables for controller design.

Symbol	Definition	Variables	Physical description
$x(t)$	State variables	$M_{i,k}, i = 1, ..., NC$	Molar holdups
		$k = 1, ..., N$	
$y(t)$	Control variables	x_{ETOH}^{top}	ETOH composition in top product
		x_{ETOH}^{bottom}	ETOH composition in bottom stream
$u(t)$	Manipulated variables	RR	Reflux ratio
		Q_{reb}	Reboiler heat duty
$d(t)$	Disturbances	$F^{solvent}$	Solvent flowrate
		x_{ETOH}^{feed}	ETOH composition in feed

Step 2: model approximation

Using the MATLAB System Identification Toolbox, a discrete linear state-space model is developed based on the input-output simulation data generated from the original high fidelity model. As shown in Eq. (8.28), the approximated model consists of 5 identified state variables comparing to the 33 state variables used in the original model. The values of matrices A, B, C, D, E are reported in Eq. (8.29).

$$\bar{x}_{k+1} = A\bar{x}_k + Bu_k + Cd_k$$
$$y_k = D\bar{x}_k + Eu_k \tag{8.28}$$

$$A = \begin{bmatrix} 0.9061 & -0.03299 & -0.04581 & -0.04579 & -0.04716 \\ 0.2084 & 0.873 & -0.221 & -0.2061 & -0.05922 \\ 0.1458 & 0.134 & 0.9485 & -0.1549 & -0.2723 \\ -0.009694 & 0.4836 & 0.1511 & 0.8574 & -0.5918 \\ 0.01222 & -0.05963 & -0.02946 & 0.0263 & 0.9856 \end{bmatrix}$$

$$B = \begin{bmatrix} 0.0748 & -0.02216 \\ -0.003592 & 0.002283 \\ -0.005627 & -0.001754 \\ 0.04477 & -0.02557 \\ 0.009074 & 0.002538 \end{bmatrix} \quad C = \begin{bmatrix} 0.1585 & 0.01313 \\ -0.07265 & 0.01126 \\ -0.01034 & 0.007001 \\ 0.2189 & 0.004659 \\ -0.07771 & 0.01177 \end{bmatrix} \tag{8.29}$$

$$D = \begin{bmatrix} 0.05475 & -0.06541 & 0.1235 & -0.0521 & -0.001263 \\ 0.3149 & -0.04875 & -0.02458 & 0.03864 & -0.02138 \end{bmatrix}$$

$$E = \begin{bmatrix} -6.401e-5 & -2.474e-5 \\ 0.01505 & -0.05059 \end{bmatrix}$$

Step 3: mp-MPC controller design

The controller tuning parameters for this system are presented in Table 8.5. The resulting mp-MPC problem is solved with the POP Toolbox in MATLAB [13]. The optimal control action at time k is expressed as an affine function of 13 parameters (i.e., 5 states, 2 manipulated variables at last time step, 2 disturbances, 2 output variables, and 2 output setpoints). A total of 171 Critical Regions are identified.

Table 8.5 mp-MPC controller tuning parameters.

MPC Parameters	Tuning Value
OH	2
CH	1
QR	[1E4, 0 ; 1E3, 0]
$R1$	[1E-1, 0 ; 5E-1, 0]
u_{min}	[0.3 ; 4.5]
u_{max}	[1 ; 6]
y_{min}	[0 ; 0]
y_{max}	[1 ; 1]
d_{min}	[0.8 ; 6.8]
d_{max}	[0.9 ; 7.6]

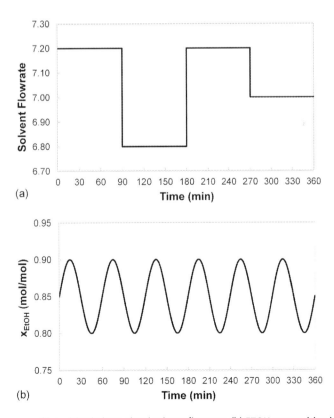

FIGURE 8.10 Disturbance profiles – (a) Ethylene glycol solvent flowrate, (b) ETOH composition in ethanol-water feed.

144 Synthesis and Operability Strategies for Computer-Aided Modular PI

Step 4: closed-loop validation

In this step, we test the above mp-MPC controller on the original high fidelity process model for disturbance rejection. The disturbance profiles for ethylene glycol solvent flowrate and ETOH composition in ethanol-water feed are depicted in Fig. 8.10. The open-loop simulation results and the closed-loop results with mp-MPC controller on are presented in Fig. 8.11. It can be concluded that the mp-MPC controllers can provide a good performance in maintaining the output setpoints in the presence of disturbances. In Chapter 13, we will present another case study to compare mp-MPC and PI control on a reactive distillation system. In Chapter 14, we will showcase simultaneous design and control via the PAROC framework.

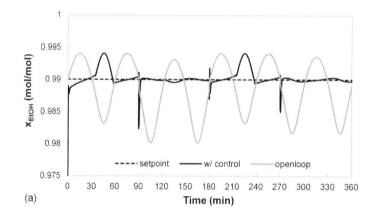

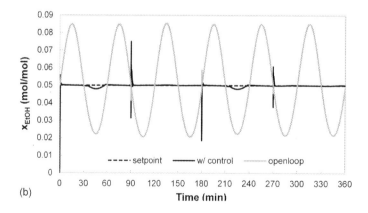

FIGURE 8.11 Output profiles – (a) ETOH composition in top ethanol product, (b) ETOH composition in bottom stream.

References

[1] G. Stephanopoulos, Chemical Process Control, Vol. 2, Prentice Hall, New Jersey, 1984.

[2] D.E. Seborg, D.A. Mellichamp, T.F. Edgar, F.J. Doyle III, Process Dynamics and Control, John Wiley & Sons, 2010.

[3] N.M. Nikačević, A.E. Huesman, P.M. Van den Hof, A.I. Stankiewicz, Opportunities and challenges for process control in process intensification, Chemical Engineering and Processing: Process Intensification 52 (2012) 1–15.

[4] E.N. Pistikopoulos, N.A. Diangelakis, R. Oberdieck, Multi-Parametric Optimization and Control, John Wiley & Sons, 2020.

[5] M.S. Bazaraa, H.D. Sherali, C. Shetty, Nonlinear Programming. Theory and Algorithms, John Wiley&Sons, Inc., New York, 1993.

[6] C.A. Floudas, Nonlinear and Mixed-Integer Optimization: Fundamentals and Applications, Oxford University Press, 1995.

[7] E.N. Pistikopoulos, N.A. Diangelakis, R. Oberdieck, M.M. Papathanasiou, I. Nascu, M. Sun, PAROC – an integrated framework and software platform for the optimisation and advanced model-based control of process systems, Chemical Engineering Science 136 (2015) 115–138.

[8] N.A. Diangelakis, B. Burnak, J. Katz, E.N. Pistikopoulos, Process design and control optimization: a simultaneous approach by multi-parametric programming, AIChE Journal 63 (2017) 4827–4846.

[9] B. Burnak, N.A. Diangelakis, E.N. Pistikopoulos, Integrated process design and operational optimization via multiparametric programming, Synthesis Lectures on Engineering, Science, and Technology 2 (2020) 1–258.

[10] R.S. Lambert, P. Rivotti, E.N. Pistikopoulos, A Monte-Carlo based model approximation technique for linear model predictive control of nonlinear systems, Computers & Chemical Engineering 54 (2013) 60–67.

[11] J. Katz, I. Pappas, S. Avraamidou, E.N. Pistikopoulos, Integrating deep learning models and multiparametric programming, Computers & Chemical Engineering 136 (2020) 106801.

[12] A. Bemporad, M. Morari, V. Dua, E.N. Pistikopoulos, The explicit linear quadratic regulator for constrained systems, Automatica 38 (2002) 3–20.

[13] R. Oberdieck, N.A. Diangelakis, M.M. Papathanasiou, I. Nascu, E.N. Pistikopoulos, POP – parametric optimization toolbox, Industrial & Engineering Chemistry Research 55 (2016) 8979–8991.

[14] S.R. Ismail, E.N. Pistikopoulos, K.P. Papalexandri, Modular representation synthesis framework for homogeneous azeotropic separation, AIChE Journal 45 (1999) 1701–1720.

[15] V. Bansal, J.D. Perkins, E.N. Pistikopoulos, A case study in simultaneous design and control using rigorous, mixed-integer dynamic optimization models, Industrial & Engineering Chemistry Research 41 (2002) 760–778.

[16] C. Kravaris, I.K. Kookos, Understanding Process Dynamics and Control, Cambridge University Press, United Kingdom, 2021.

9

Synthesis of operable process intensification systems

9.1 Problem statement

The following problem definition presents the generalized problem addressed in this chapter for the synthesis of modular process intensification systems with operability, safety, and control considerations (Fig. 9.1).

Given:

1. *Process design target*

    - A set of feed streams with given flowrates, compositions, and supply temperatures;
    - A set of desired products and specifications on their flowrates, temperatures, and/or purities;
    - A set of available heating and/or cooling utilities such as steam and cooling water with their availability, supply temperatures, and compositions;
    - A set of available mass utilities such as mass separating agents (e.g., solvents, adsorbents) and catalysts;
    - All reaction schemes and kinetics data;
    - All physical property models;
    - Cost data of feed streams, mass/heat utilities, and equipment;

2. *Flexibility target*

    - A specified range of uncertain parameters, where process flexibility is desired (e.g., feed compositions, flowrates, temperatures, heat utility flowrates, temperatures);

3. *Safety target*

    - A set of assessment criteria on inherent safety performances (e.g., toxicity, flammability, explosiveness);
    - A set of available equipment with their failure frequency data;
    - Hazardous property data (e.g., lethal concentration);

4. *Control target*

    - A set of disturbances during process operation;
    - A set of control variables with desired setpoints;
    - A set of available manipulated variables to maintain feasible operation based on degrees of freedom analysis.

Synthesis and Operability Strategies for Computer-Aided Modular Process Intensification
https://doi.org/10.1016/B978-0-32-385587-7.00019-1
Copyright © 2022 Elsevier Inc. All rights reserved.

Objective: To determine process solution(s) with

- Minimized total annualized cost consisting of capital costs, heat utility costs, and mass utility costs;
- Optimal equipment or flowsheet configuration(s) with design and operating parameters, which satisfy the desired flexibility and inherent safety criteria for both steady-state design and dynamic operation;
- Optimal control actions to achieve process specifications during dynamic operation.

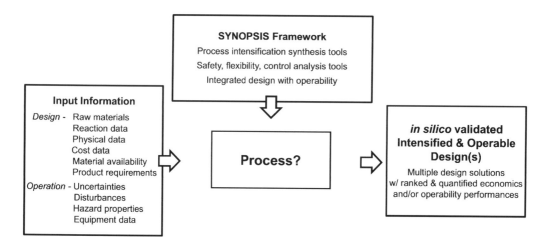

FIGURE 9.1 Problem definition.

9.2 A systematic framework for synthesis of operable process intensification systems

To address the above problem, we present the SYNOPSIS framework, standing for SYNthesis of Operable ProcesS Intensification Systems, to integrate modular process intensification synthesis, operability analysis, and dynamic control optimization. The step-wise procedure is shown in Fig. 9.2.

Step 1: process intensification synthesis representation
The first step is to represent chemical processes via the Generalized Modular Representation framework (GMF) from the phenomena level without any pre-postulation of possible unit operation-based process alternatives which may hinder the discovery of novel intensified process solutions. Just to briefly recap the GMF introduction in Chapters 3–5: Two types of modular building blocks – i.e. pure heat exchange module & multifunctional mass/heat exchange module – are employed in GMF to overcome process bottlenecks by intensifying process fundamental performance, such as improving mass/heat transfer

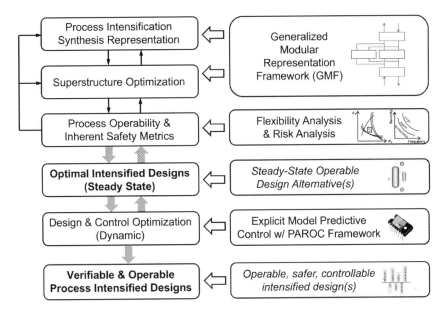

FIGURE 9.2 A systematic framework for synthesis of operable PI systems (adapted from Tian et al. [14]).

and/or shifting reaction equilibrium. Gibbs free energy-based driving force constraints are derived to characterize mass transfer feasibility, thus providing a more compact and effective synthesis representation strategy exploiting the "ultimate" thermodynamic space.

Step 2: superstructure optimization
A superstructure formulation is developed based on the GMF modular building blocks to account for all possible network configurations. The overall synthesis problem is formulated as a mixed-integer nonlinear programming (MINLP) problem which enables efficient screening of the resulting combinatorial design space. The solution of this optimization problem will identify the optimal GMF modular process alternatives with respect to a pre-defined objective function (e.g., total annualized cost, energy consumption, environmental impact). Alternative process solutions can also be generated by introducing integer cuts into synthesis model formulation, since arguably even "intermediate" solutions can provide useful information on the process.

Step 3: steady-state design with flexibility and safety
After obtaining the nominal optimal design from Step 2 as base design, flexibility and safety metrics are introduced to assess steady-state operability performance of the derived intensified structure. In this chapter, flexibility test is utilized to evaluate the functionality of the resulting design under varying operating conditions, and risk analysis is applied to indicate its inherent safety performance by accounting for process consequence severity and equipment failure frequency. If the nominal design fails the operability and/or safety

assessment, alternative design structures will be derived by incorporating flexibility and safety targets into GMF synthesis model to deliver intensified GMF modular structures with guaranteed steady-state flexibility and safety performance without compromising process specifications. However, the cost performance of the more operable and safer designs will be inferior than that of the nominal design due to the trade-offs on operability and safety. Note that other model-based operability criteria can also be incorporated, such as safety indices, structural controllability, given that their required design/operation information for assessment are readily available at this phenomena-based design stage.

Step 4: optimal intensified steady-state designs
The resulting GMF modular designs with enhanced flexibility and safety performance will be translated to equipment-based process alternatives and will be validated using steady-state simulation tools. The identification of process equipment or flowsheet is mostly based on heuristics suggested by the types of GMF module, their interconnections, and operating conditions. For example, a mass/heat exchange module for (reactive) separation can be translated to either a column section for (reactive) distillation/absorption/etc., or a (reactive) membrane unit which satisfies the performance targets given by GMF synthesis results. Rigorous steady-state design is then performed to optimize the derived process units to determine optimal design and operating parameters.

Step 5: simultaneous design and control optimization
In this step, we take the above identified intensified process solutions to dynamic analysis. First, a high-fidelity dynamic model is developed to accurately describe the process behavior which can be highly nonlinear with strong variable interactions due to task integration, etc. To ensure that the desired operational performances have been succeeded from steady-state design, the dynamic systems are then analyzed with respect to flexibility and safety performance. After the consistency check, simultaneous design and control is performed to ensure economical and smooth operation despite the influence of disturbances. It is achieved by integrating the rigorous process model and design-aware explicit controllers via mixed-integer dynamic optimization following the PAROC framework.

Step 6: verifiable and operable process intensification designs
The outcomes of this framework will be modular and/or intensified process solutions with: (i) optimal process design and operating configurations (steady-state and dynamic validated); (ii) guaranteed operability and safety performance, and (iii) optimal explicit model predictive controller design.

9.3 Steady-state synthesis with flexibility and safety considerations

The computational strategies utilized in the SYNOPSIS framework have all been detailed in the previous chapters (e.g., GMF, flexibility analysis, risk analysis, multi-parametric model

Chapter 9 • Synthesis of operable process intensification systems 151

predictive control). Note that the framework can be easily extended with other model-based process design and operability analysis strategies. However, there might be a missing link for integrated GMF synthesis with steady-state flexibility and inherent safety analysis, since the latters are typically considered for equipment-based design.

In this section, we introduce an integrated GMF-flexibility-safety synthesis approach to bridge this gap, as depicted in Fig. 9.3a. Risk analysis is treated as process constraints to set upper bounds on the process hazardous risks. Following the multiperiod optimization approach to synthesize flexible process systems (Section 6.4), the GMF-flexibility-safety synthesis approach results in an iterative optimization scheme (Fig. 9.3b) to deliver cost-optimal intensified processes with guaranteed flexibility and inherent safety performances. The resulting phenomenological flowsheets will then be validated with steady-state simulation to identify the corresponding equipment-based equipment or flowsheet configurations.

9.3.1 GMF synthesis for flexible process systems

The GMF synthesis model (Chapter 4) can be recast in the following compact mathematical form:

$$\min_{d,x,z} \quad f(d,x,z)$$
$$\text{s.t.} \quad h(\theta,d,x,z) = 0 \tag{9.1}$$
$$g(\theta,d,x,z) \leq 0 \qquad \forall \theta \in T(\theta)$$

where θ stands for the set of uncertain parameters, and $T(\theta)$ is the specified range of uncertainty where flexibility is desired; d is the set of design variables, including the diameter, height, and catalyst load of each mass/heat exchange module as well as the variables defining the network topology; x denotes state variables which describe the network operation; z is the set of control variables (i.e., degrees of freedom that can be adjusted during operation), including the heat duties of pure heat exchange modules; f is the objective function; h and g are the sets of equality and inequality constraints used in this superstructure model.

Recall the flexibility test problem [1,2] presented in Section 6.2, this original problem described by Eq. (9.1) with flexibility requirement $\forall \theta \in T(\theta)$ is equivalent to the following constrained max-min-max problem:

$$\max_{\theta \in T(\theta)} \min_{z} \max_{j \in J_f} \quad f_j(\theta,d,x,z) \leq 0$$
$$\text{s.t.} \quad h(\theta,d,x,z) = 0 \tag{9.2}$$

To synthesize flexible mass and heat networks, Papalexandri and Pistikopoulos [3,4] proposed an iterative multiperiod optimization formulation integrated with flexibility test. Herein we extend this approach for GMF synthesis of flexible process solution. As illustrated in Fig. 9.3b:

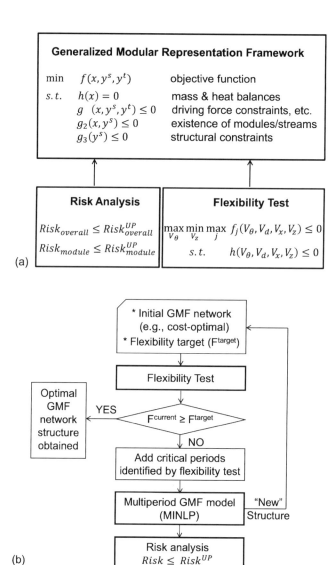

FIGURE 9.3 Steady-state GMF synthesis with flexibility and safety (adapted from Tian and Pistikopoulos [15]). (a) The integrated approach, (b) Iterative optimization scheme.

i. The algorithm starts with an initial process structure, such as the cost-optimal GMF structure at nominal operating conditions;

ii. Flexibility test is performed to determine the *critical operating conditions* under a specified range of uncertain parameters;

iii. The GMF superstructure is extended to include all these critical operating conditions via a multiperiod representation approach [5]. An additional set of operation "peri-

Chapter 9 • Synthesis of operable process intensification systems   153

ods" is introduced. Accordingly, design variables (e.g., diameter and height of mass exchange module) are reformulated as continuous variables for *all periods of operation*, while the other variables (e.g., stream flows, compositions, temperatures as well as those involved in characterizing heat and mass transfer) are regarded as continuous variables for *each period of operation*;

iv. The resulting GMF synthesis model encapsulates all plausible GMF network configurations, from which the most promising GMF design can be obtained at the minimum total annualized cost, also with the desired flexibility performance.

However, it should be noted that when applying flexibility test [1,2] to nonconvex systems (e.g., GMF), the KKT conditions used in its inner optimization problems cannot guarantee global optimality. Thus the solution obtained for the outer optimization problems might not indicate the actual flexibility of the system [6]. In this context, the resulting "flexible" design configuration need to be re-tested through the uncertainty range to ensure its actual flexibility performance.

9.3.2 Integrated GMF synthesis with risk analysis

Quantitative risk analysis (QRA) [7] evaluates the inherent safety performance of a process by accounting for equipment failure frequency (depending on the equipment type and etc.) and consequence severity (depending on the amount of substance present, their physical and hazardous properties, specific process conditions and etc.). The value of process risk is thus determined by the multiplication of these two terms, as shown in Eq. (9.3). A conservative way to evaluate the inherent safety performance is by considering the total amount of intrinsic hazard of a process at the release of the entire content in the process, as that presented in Nemet et al. [8]. Another advantage of their approach [8], based on the indication number approach introduced in Section 7.4, is that it can be readily incorporated into superstructure-based optimization model as process constraints without repetitive iteration in a manner of posterior evaluation. In this context, this approach is extended for the risk analysis of GMF-based process structures.

$$Risk = Failure\ frequency \times Severity \tag{9.3}$$

- *Equipment failure frequency* – As there is no pre-postulation of unit operations in GMF synthesis, average historical data for equipment failure frequencies, such as those documented in the *Handbook of Failure Frequencies [9]* or in the *Guideline for Quantitative Risk Assessment* [10], cannot be directly applied. Thus, similarly to the estimation of pseudo-capital module cost, the GMF pure heat exchange modules are approximated as heat exchangers, while mass/heat exchange modules are treated as process vessels (where a change in the physical properties of the substance occurs, e.g. temperature or phase [10]) if pure separation is taking place, or as reactor vessels (where a chemical change of the substances occurs [10]) if there is any reaction going on. This approximation helps to provide an initial evaluation on the process inherent safety performance

154   Synthesis and Operability Strategies for Computer-Aided Modular PI

and to penalize the intrinsic process risks. The failure frequencies for these three types of modules during the instantaneous release of the complete inventory are determined from *Guideline for Quantitative Risk Assessment* [10] as shown in Table 9.1.

Table 9.1   Failure frequencies for GMF modules.

	Heat Exchange Module	Mass/Heat Exchange Module	
		Separation	Reaction (& Separation)
$frequency/(\text{module}\cdot\text{yr})^{-1}$	5×10^{-5}	5×10^{-6}	5×10^{-6}

- **Consequence severity** – The indication number for each GMF module can be calculated via Eq. (9.4) [7,8]:

$$A_{i,e,risk} = \frac{W_{i,e} \times O_e^1 \times O_e^2 \times O_i^3}{S_{risk}} \quad \forall e \in E, risk \in RISK, i = 1, ..., NC \tag{9.4}$$

where A defines the indication number to quantify the consequence severity pertaining to a single GMF module, W represents the quantity of substances present within the module (kg), O^i are the factors accounting for process conditions, and S is the limit value (kg) measuring the hazardous properties of each substance based on their physical and toxic, explosive, or flammable properties. The determination of these parameters within the GMF framework is further elaborated below:

- Quantity of substances present (W): The mass quantity of each substance present in a mass/heat exchange module is estimated from the module's volume. The relation between mass holdup and module volume can be approximated by:

$$V = \frac{W^{liq}}{\widetilde{\rho}^{LO}} + \frac{W^{vap}}{\widetilde{\rho}^{VO}} \tag{9.5}$$

where W^{liq} and W^{vap} are liquid and vapor mass holdups in the module (kg), respectively; $\widetilde{\rho}^{LO}$ and $\widetilde{\rho}^{VO}$ are the mass densities of the module's outlet liquid and vapor stream (kg/m^3).

The module volume can be determined from its diameter and height:

$$V = \frac{\pi D^2}{4} H \tag{9.6}$$

For a liquid-vapor module where separation or reactive separation is taking place, the liquid level ($Level^{liq}$) is assumed to be 25% of module height in analogy to a distillation tray, while for a liquid-liquid module where only reaction is going on, $Level^{liq}$ is assumed to be 75% of the module height as that in reactors. Namely,

$$Level^{liq} = 25\% \times H \times y_{sep} + 75\% \times H \times (1 - y_{sep}) \tag{9.7}$$

Chapter 9 • Synthesis of operable process intensification systems   155

where y_{sep} is the binary variable which denotes the existence or not of the separation phenomena in a mass/heat exchange module. Note that if H is used as variables rather than parameters, Eq. (9.7) needs to be reformulated to avoid the nonlinearity in binary variable. Thus explicit expressions of the liquid and vapor holdups can be derived from Eq. (9.5)–(9.7):

$$W^{liq} = \frac{\pi D^2}{4} \tilde{\rho}^{LO} Level^{liq}$$

$$W^{vap} = \frac{\pi D^2}{4} \tilde{\rho}^{LO} (H - Level^{liq})$$

(9.8)

For each substance present in the module, its amounts in liquid or vapor phase are determined by:

$$W_i^{liq} = W^{liq} \tilde{x}_i^{LO}$$

$$W_i^{vap} = W^{vap} \tilde{x}_i^{VO}$$

(9.9)

where $\tilde{x}_i^{LO}$ and $\tilde{x}_i^{VO}$ are the mass fractions of component i in the module's outlet liquid and vapor stream, respectively.

In sum, the quantity of liquid/vapor substances present in a mass/heat exchange module is given by:

$$W_i^{liq} = \frac{\pi D^2}{4} \tilde{\rho}^{LO} \tilde{x}_i^{LO} Level^{liq}$$

$$W_i^{vap} = \frac{\pi D^2}{4} \tilde{\rho}^{LO} \tilde{x}_i^{VO} (H - Level^{liq}) \qquad i = 1, ..., NC$$

(9.10)

* Factors for processing conditions (O^i):
    i. O^1 describes if the installation is designed for processing or storage. As all the GMF modules are used for chemical processing, $O^1 = 1$ as per its definition in Ref. [7];
    ii. O^2 accounts for the installation's positioning. With the assumption that the resulting intensified equipment will be positioned outdoor, O^2 is selected to be 1;
    iii. O^3 provides a measurement over the amount of substance in the gas phase after its release, which is determined by its phase state and saturation pressure at process temperature – these information can be directly obtained from the GMF model.

    However, these factors only apply to the calculation of toxic and flammable risks. When explosive risk is considered, $O^1 = O^2 = O^3 = 1$.

* Limit value (S): For toxicity, S_{tox} is determined by the lethal concentration of the substance (i.e., LC_{50}(rat,inh,1h)) as well as its phase state at 298 K, as shown in Table 2.5 in Ref. [7]. The limit value for flammables, S_{flam}, is 10,000 kg. For explosive

156 Synthesis and Operability Strategies for Computer-Aided Modular PI

substances, S_{expl} is the amount of substance (kg), the explosion of which releases the same amount of energy as that of 1000 kg TNT (explosion energy 4600 kJ/kg).

Given all these constituents for risk calculation, the risk resulted by the existence of a single substance (or component) in a GMF module is determined via Eq. (9.11), which can be a good indicator to investigate the role of hazardous materials towards an inherently safer design:

$$
R_{i,e,risk}^{comp} = \sum_{fail \in FAIL} freq_{e,fail} \frac{W_{i,e}^{liq} \times O_e^1 \times O_e^2 \times O_{i,e}^3}{S_{i,risk}^{liq}} +
$$

$$
\sum_{fail \in FAIL} freq_{e,fail} \frac{W_{i,e}^{vap} \times O_e^1 \times O_e^2 \times O_{i,e}^3}{S_{i,risk}^{vap}} \qquad \forall e \in E, risk \in RISK, i = 1, ..., NC
$$

$$(9.11)$$

The risk of individual module ($R_{e,risk}^{mod}$) is determined by a summation of $R_{i,e,risk}^{comp}$ over substances:

$$
R_{e,risk}^{mod} = \sum_{i=1}^{NC} \sum_{fail \in FAIL} freq_{e,fail} \frac{W_{i,e}^{liq} \times O_e^1 \times O_e^2 \times O_{i,e}^3}{S_{i,risk}^{liq}} +
$$

$$
\sum_{i=1}^{NC} \sum_{fail \in FAIL} freq_{e,fail} \frac{W_{i,e}^{vap} \times O_e^1 \times O_e^2 \times O_{i,e}^3}{S_{i,risk}^{vap}} \qquad \forall e \in E, risk \in RISK
$$

$$(9.12)$$

And the overall risk of the process network ($R_{risk}^{overall}$) is calculated through:

$$
R_{risk}^{overall} = \sum_{e \in E} \sum_{i=1}^{NC} \sum_{fail \in FAIL} freq_{e,fail} \frac{W_{i,e}^{liq} \times O_e^1 \times O_e^2 \times O_{i,e}^3}{S_{i,risk}^{liq}} +
$$

$$
\sum_{e \in E} \sum_{i=1}^{NC} \sum_{fail \in FAIL} freq_{e,fail} \frac{W_{i,e}^{vap} \times O_e^1 \times O_e^2 \times O_{i,e}^3}{S_{i,risk}^{vap}} \qquad \forall risk \in RISK
$$

$$(9.13)$$

Risk tolerance can be set for individual module (Eq. 9.14) and/or for the overall process (Eq. 9.15).

$$
R_{e,risk}^{mod} \leq y_e R_{e,risk}^{mod,UP} \qquad \forall e \in E, risk \in RISK \tag{9.14}
$$

$$
R_{risk}^{overall} \leq R_{risk}^{overall,UP} \qquad \forall risk \in RISK \tag{9.15}
$$

However, as the risk values obtained in this approach do not provide a scaled measurement (e.g., 0-1), the risk tolerances should be set on a comparative basis. For instance, to realize an inherently safer process, each type of risk for the overall network (e.g., toxicity, flammability, explosiveness) should be reduced by at least 20% of that in the nominal (i.e., cost-optimal) structure; or to eliminate the existence of a "particularly unsafe" module in the process, the individual module risk should be less than 30% of the overall process risk.

These inherent safety targets will be presented later in more details through the case study demonstration.

9.4 Motivating example: heat exchanger network synthesis

In what follows, the SYNOPSIS framework is applied to a heat exchanger network (HEN) synthesis problem for thermal intensification, as a motivating example for proof-of-concept. An extensive case study is included in Chapter 14 to demonstrate the versatility and applicability of the framework with application to reactive separation systems.

9.4.1 Process description

This case study considers two hot streams (H1, H2), two cold streams (C1, C2), and one hot utility (HU). Given are: (i) stream flowrate data [11]; (ii) uncertain heat transfer coefficient (U_{H1-C1}), for which the flexibility of the network is desired; (iii) disturbance and control objective, for which controller design is essential; and (iv) stream toxicity, represented by LC_{50} (i.e., lethal concentration, 50%) of each substance, and equipment failure data for four types of heat exchanger (HE), namely double pipe HE (DP), plate and frame HE (PF), fixed plate shell and tube HE (SF), and U-tube shell and tube HE (UT), which necessitates inherent safety analysis [8]. The objective is to synthesize a heat exchanger network with minimized total annual cost and desired operability, safety, and control performances.

9.4.2 Steady-state synthesis with flexibility and safety considerations

The superstructure optimization problem for steady-state HEN synthesis is formulated using the GMF heat exchange modules allowing for both utility heat exchange and heat integration. A cost-optimal HEN design without flexibility or safety considerations (i.e., nominal design) is first obtained as a base case (Fig. 9.4).

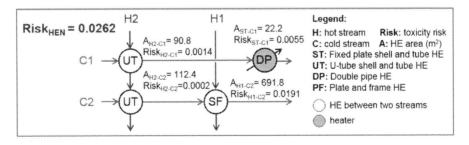

FIGURE 9.4 HEN synthesis – Nominal design.

Regarding this nominal design, feasibility analysis identifies two periods of operation, respectively characterized by the lower and upper bounds of the uncertain parameter U_{H1-C1}. To obtain an inherently safer design, the overall HEN toxicity risk is constrained to be 25% less than that of the nominal design. This results in the change of H1-C2 exchanger type from SF to UT (Fig. 9.5a), as UT has much higher area density to significantly reduce the amount of hazards contained in the equipment.

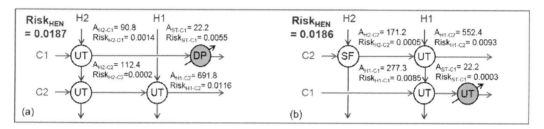

FIGURE 9.5 HEN synthesis – Operable and inherently safer designs. (a) Design 1, (b) Design 2.

However, as the individual risk of H1-C2 HE takes up more that 75% of the overall process risk in this case, the risk tolerance is further decreased by constraining individual HE toxicity risk to be less than 50% of overall risk. As a result, a different network configuration is synthesized in order to render H1-C2 HE a lower risk level by relieving its heat exchange burden (Fig. 9.5b). These two safely operable HENs are exported as Designs 1 and 2 for the next step dynamic analysis.

9.4.3 Dynamic modeling and mp-MPC controller design

The dynamic HEN is described by a Partial Differential Algebraic Equation (PDAE) model based on gPROMS Process Model Library for Heat Exchange. The network configuration is fixed for each design as per Fig. 9.5, and heat exchanger areas are used as design variables. To ensure the consistency going from steady-state synthesis to dynamic simulation, the dynamic model is validated to match its steady-state synthesis analogue.

In the derived HENs, bypass flowrate and heat utility duties are used as the manipulated variables and the outlet temperature of stream H1 and C1 are the control variables. The inlet temperature of stream H2 is considered as a disturbance to the operation. The mp-MPC controller design, following the PAROC framework [12], is performed for each of the designs. The mp-MPC problems are formulated using corresponding linear state-space model approximated by the System Identification Toolbox of MATLAB®. Via POP toolbox in MATLAB, the problem of Design 1 is solved for an output horizon of 2 and a control horizon of 2 resulting in 118 critical regions in solution map, while that of Design 2 is solved for an output horizon of 2 and a control horizon of 1 resulting in 89 critical regions. Given random disturbances deviating within ±10 K per second, the controllers are tested against the high fidelity model for closed-loop validation, indicating the agreement between the outputs and setpoints. See Fig. 9.6.

9.4.4 Simultaneous design and control

The dynamic optimization problem is then formulated and solved for the minimal TAC under each HEN configuration. The obtained results are shown in Table 9.2. Up to this point, two HENs are designed with different levels of operability, control, and safety. While the final construction decision depends on the trade-off between the desired operability behavior and economic performance, this framework has demonstrated the potential for generation and comparison of various operable design alternatives.

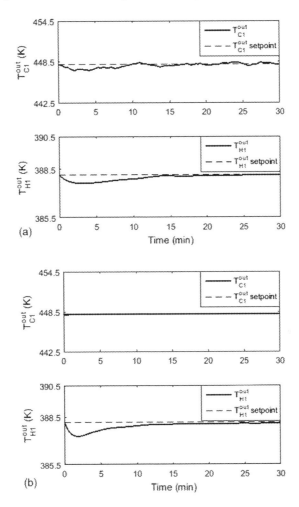

FIGURE 9.6 HEN synthesis – Closed-loop validation of mp-MPC controller. (a) Design 1, (b) Design 2 (Reproduced from [13]).

Table 9.2 HEN synthesis – Dynamic optimization results (adapted from [13]).

	Heat Exchanger Area (m^2)				Investment Cost (k$)	Operating Cost (k$/yr)
	H1 – C1	H1 – C2	H2 – C1	H2 – C2		
Design 1	/	604.4	125.6	99.9	528.4	7357.0
Design 2	221.4	522.2	/	165.7	557.4	8593.9

References

[1] K.P. Halemane, I.E. Grossmann, Optimal process design under uncertainty, AIChE Journal 29 (1983) 425–433.

[2] I.E. Grossmann, C.A. Floudas, Active constraint strategy for flexibility analysis in chemical processes, Computers & Chemical Engineering 11 (1987) 675–693.

[3] K.P. Papalexandri, E.N. Pistikopoulos, An MINLP retrofit approach for improving the flexibility of heat exchanger networks, Annals of Operations Research 42 (1993) 119–168.

[4] K.P. Papalexandri, E.N. Pistikopoulos, Synthesis and retrofit design of operable heat exchanger networks. 1. Flexibility and structural controllability aspects, Industrial & Engineering Chemistry Research 33 (1994) 1718–1737.

[5] K. Papalexandri, E. Pistikopoulos, A multiperiod MINLP model for the synthesis of flexible heat and mass exchange networks, Computers & Chemical Engineering 18 (1994) 1125–1139.

[6] I.E. Grossmann, B.A. Calfa, P. Garcia-Herreros, Evolution of concepts and models for quantifying resiliency and flexibility of chemical processes, Computers & Chemical Engineering 70 (2014) 22–34.

[7] R.A. Freeman, CCPS guidelines for chemical process quantitative risk analysis, Plant/Operations Progress 9 (1989) 231–235, https://doi.org/10.1002/prsb.720090409.

[8] A. Nemet, J.J. Klemeš, I. Moon, Z. Kravanja, Safety analysis embedded in heat exchanger network synthesis, Computers & Chemical Engineering 107 (2017) 357–380.

[9] Flemish Government, Handbook failure frequencies 2009 for drawing up a safety report, 2009.

[10] P.G. Stoffen, Guidelines for quantitative risk assessment, Ministerie van Volkshuisvesting Ruimtelijke Ordening en Milieu. CPR E 18 (2005).

[11] E. Kotjabasakis, B. Linnhoff, Flexible heat exchanger network design: comments on the problem definition and on suitable solution techniques, in: IChemE. Symposium Serie No 105, 1987.

[12] E.N. Pistikopoulos, N.A. Diangelakis, R. Oberdieck, M.M. Papathanasiou, I. Nascu, M. Sun, PAROC – an integrated framework and software platform for the optimisation and advanced model-based control of process systems, Chemical Engineering Science 136 (2015) 115–138.

[13] Y. Tian, M.S. Mannan, Z. Kravanja, E.N. Pistikopoulos, Towards the synthesis of modular process intensification systems with safety and operability considerations – Application to heat exchanger network, in: Computer Aided Chemical Engineering, Vol. 43, Elsevier, 2018, pp. 705–710.

[14] Y. Tian, I. Pappas, B. Burnak, J. Katz, E.N. Pistikopoulos, A Systematic Framework for the synthesis of operable process intensification systems – Reactive separation systems, Computers & Chemical Engineering 134 (2020) 106675.

[15] Y. Tian, E.N. Pistikopoulos, Synthesis of operable process intensification systems – steady-state design with safety and operability considerations, Industrial & Engineering Chemistry Research 58 (15) (2018) 6049–6068.

PART 3

Case studies

10

Envelope of design solutions for intensified reaction/separation systems

Given the enriched design space including both conventional and innovative process options, a key open question remains on how to derive an envelope of possible design solutions prior to establishing any process alternatives, hereby to identify the ultimate bounds of process improvements that can be achieved as well as to rapidly screen the design space with respect to productivity, economics, and/or safety objectives.

A classic development to identify process boundaries is the attainable region (AR) theory for reactor network synthesis [1,2], where the entire physically realizable outlet conditions can be characterized for a given set of feed conditions and reactions with prescribed kinetics independent of actual reactor design. The AR theory has also been extended to systems involving reaction, mixing, and separation. Feinberg and Ellison [3] introduced an optimization-based approach, namely the continuous flow stirred tank reactor (CFSTR) equivalence principle, to explore the molar productivity limits of a given chemistry in any steady-state reactor-mixer-separator (RMS) systems. Hereafter, we refer to this approach as "Feinberg Decomposition (FD)". Frumkin and Doherty [4,5] further extended the FD approach to characterize selectivity bounds in RMS systems. Despite the indispensable boundary information given by the CFSTR principle, this approach cannot generate candidate process alternatives at, or within, the derived bounds. On the other hand, process synthesis approaches have also been applied to identify the performance limits of a reactive separator network as presented in da Cruz and Manousiouthakis [6]. However, there is no guarantee that the obtained performance limits can serve as the absolute "upper bound" of process improvements as those derived from AR approaches. Thus it remains an open question – **how close can actual intensified designs, e.g. derived by the PI synthesis methods, approach these ultimate attainable region bounds.**

In this chapter, we introduce how an envelope of design solutions based on the FD theory can be effectively integrated with a phenomenological process intensification synthesis strategy using thermodynamic driving force constraints, taking the Generalized Modular Representation Framework (GMF) as an example [10].

Synthesis and Operability Strategies for Computer-Aided Modular Process Intensification
https://doi.org/10.1016/B978-0-32-385587-7.00021-X
Copyright © 2022 Elsevier Inc. All rights reserved.

10.1 The Feinberg Decomposition

We first provide a brief overview on the Feinberg Decomposition approach before applying it to identify design boundaries in the subsequent sections.

The Continuous Flow Stirred Tank Reactor (CFSTR) Equivalence Principle, proposed by Feinberg and Ellison [3], allows one to decompose any arbitrary steady-state reactor-mixer-separator (RMS) system with total reaction volume $V > 0$ into a new system which has the same effluent molar flow rates as the original system. The new system comprises only CFSTRs as reactors with no more than $\mathcal{R} + 1$ in number (where $\mathcal{R}$ is the number of linearly independent reaction), each of which coupled to a perfect separator system as shown in Fig. 10.1. Thus, the CFSTR principle, hereafter referred as the Feinberg Decomposition (FD), provides a unified CFSTR-based representation which can capture any and every steady-state combined reaction/separation system with equivalent production rates under given kinetics and capacity resource, regardless known process designs or "out-of-the-box" innovative ones.

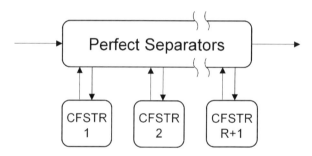

FIGURE 10.1 Feinberg Decomposition with Continuous Flow Stirred Tank Reactors. (Reproduced from Tian and Pistikopoulos [10].)

The new CFSTR-based system retains key design and operation characteristics from the original RMS system [3]: (i) the total reaction volume of CFSTRs is no greater than the reactive volume V in the original RMS system, and (ii) the mixture states within these CFSTRs, such as temperatures, pressures, and compositions, are no more extreme than those within the original system. Moreover, an optimization-based formulation [7] is utilized in the FD approach to assess bounds with respect to a specific process performance (e.g., productivity, selectivity) without having to first mapping out the AR or its boundary as that in conventional geometrically-based AR approaches. Frumkin and Doherty [5] further extended the FD optimization formulation as a nonlinear programming problem (NLP) for RMS systems with relatively larger number of components. The full NLP formulation, which can be found in Appendix D, is adapted in this chapter to explore the olefin metathesis design boundaries as detailed in the next section.

Chapter 10 • Envelope of design solutions for intensified reaction/separation systems 165

10.2 Case study: olefin metathesis

10.2.1 Process description

The metathesis of 2-pentene (C_5H_{10}) to form 2-butene (C_4H_8) and 3-hexene (C_6H_{12}) (Eq. 10.1) is an equilibrium-limited reaction. The reaction takes place in liquid phase at atmospheric pressure and can be described with ideal vapor-liquid equilibrium [8]. The reaction kinetics are given in Eq. (10.2) adapted from Okasinskin and Doherty [9].

$$2C_5H_{10} \rightleftharpoons C_4H_8 + C_6H_{12} \tag{10.1}$$

$$r = k_f (x_{C_5H_{10}}^2 - \frac{x_{C_4H_8} x_{C_6H_{12}}}{K_{eq}}) \quad (\text{h}^{-1})$$

$$k_f = 1.0661 \times 10^5 e^{(-3321.2/T(K))} \quad (\text{h}^{-1}) \tag{10.2}$$

$$K_{eq} = 0.25$$

This process has been widely investigated in literature especially featuring the use of reactive distillation as a promising design solution to enhance the conversion of pentene by removing the low boiling component butene from liquid reactive mixture. However, these design studies were mostly driven by cost minimization. In a more recent work, da Cruz and Manousiouthakis [6] applied the Infinite Dimensional Steady State (IDEAS) framework to identify the performance limits by minimizing reactive holdup.

Herein, we revisit this example with FD and GMF to investigate its design envelope. The synthesis task is to produce 50 kmol/h of 0.98 mol/mol butene and 50 kmol/h of 0.98 mol/mol hexene at 1 atm, given as raw material a saturated liquid stream of 100 kmol/h pure pentene.

10.2.2 Design boundaries via Feinberg Decomposition

We characterize the design space under a certain production task using the FD theory. Since the pentene metathesis reaction takes place in liquid phase at fixed atmospheric pressure, the available design parameters are reaction temperature (T) and reactive volume (V). Reaction kinetics for this chemistry are valid between 277.15 K and 340.15 K, which provide the largest feasible temperature range. To identify the design boundaries in the space of Temperature-Volume (T-V), the following parametric studies are performed with the NLP model provided in Appendix D with specific focus on Eqs. (D.1k) and (D.1l):

- Step 1: The total reactive volume is increased from 0 m^3 at a step size of 0.1 m^3 to maximize butene production, while CFSTR temperatures are constrained between 277.15 K and 340.15 K. The results show that **at least 1.68 m^3 reactive volume is required to produce butene at a rate of 49 kmol/h**. Moreover, it is worth noting that the two CFSTRs are both operated at 340.15 K, which indicates that the reaction is favored by higher temperature operation. This gives the upper design/operation limit as shown in Fig. 10.2.

- Step 2: The maximum allowable reaction temperature (T^{max}) is varied from 277.15 K to 340.15 K at a step size of 2 K and the butene production rate is constrained to be greater than 49 kmol/h. Then the optimization problem is set to minimize reactive volume to identify the minimum V required at different operation temperatures, which results in the lower design/operation limit as shown in Fig. 10.2.

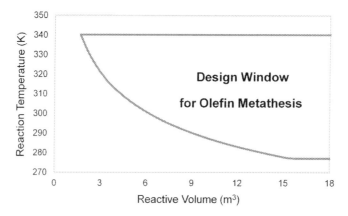

FIGURE 10.2 Temperature-Volume design boundaries for olefin metathesis. (Adapted from Tian and Pistikopoulos [10].)

The resulting T-V design space is depicted in Fig. 10.2, which provides boundary information on:

- The ultimate minimum reactive volume required for the desired production task (i.e., the leftmost vertex at 1.68 m³)
- For a given reactive volume, the range of operation temperatures
- For a given reaction temperature, the candidate reactive volumes.

10.2.3 Design boundaries via Generalized Modular Representation Framework

In what follows, we apply the Generalized Modular Representation Framework (GMF) to identify the performance limits for the olefin metathesis process and to generate corresponding design alternatives.

10.2.3.1 Design 1

Design 1 is shown in Fig. 10.3. Note that the shaded mass/heat exchange module (referred as "M/H") denotes a reactive module, and the blank M/H module is for pure separation. It corresponds to a GMF solution with an objective function to minimize reactive volume, which provides a valid underestimation of equipment size determined by reaction kinetics while independent of separation schemes. It features 3.77 m³ reactive volume, which is 20.6% less than the minimal reactive volume given in Ref. [6]. This configuration comprises two GMF mass/heat exchange modules: a reactive separation module (i.e., M/H 1) and a pure separation module (i.e., M/H 2). The pure pentene feed stream is partially vaporized before entering M/H 1 to provide the energy for reaction and separation taking place later in M/H 1. In M/H 1, the metathesis reaction takes place to form bentene and hexene from pentene, coupled with separation which vaporizes the produced butene to give the desired product P1 (50 kmol/h, C_4H_8 98 mol%). However, the heavier components pentene and hexene are retained in liquid phase (200 kmol/h, C_5H_{10} 0.708 mol/mol, C_6H_{12} 0.290 mol/mol, C_5H_{10} 0.002 mol/mol) and then sent to M/H 2. M/H 2 separates the above liquid mixture to a stream of desired Product 2 (50 kmol/h, C_6H_{12} 0.98 mol/mol), and also recycle the excessive pentene reactant back to M/H 1 for reaction.

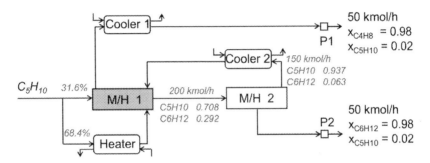

FIGURE 10.3 GMF Design 1 – Optimal solution for reactive volume minimization. (Adapted from Tian and Pistikopoulos [10].)

The detailed stream temperature, flowrate, and molar composition data for GMF Design 1 is summarized in Table 10.1. The pressures in each design are fixed to 1 atm. For brevity, we only include outlet stream information in the tables below, while the inlet stream conditions can be readily obtained via the combinatorial structure of the GMF design solution.

Table 10.1 GMF Design 1 – Flowsheet data. (Adapted from Tian and Pistikopoulos [10].)

Module ID	T (K)	F (kmol/h)	$x_{C_5H_{10}}$	$x_{C_4H_8}$	$x_{C_6H_{12}}$
Liquid Outlet Stream					
M/H 1	316.0	200.0	0.708	0	0.292
M/H 2	310.0	50.0	0.020	0	0.980
Cooler 1	277.2	50.0	0.020	0.980	0
Cooler 2	277.2	150.0	0.937	0	0.063
Vapor Outlet Stream					
M/H 1	278.2	50.0	0.020	0.980	0
M/H 2	339.5	150.0	0.937	0	0.063
Heater	340.1	68.4	1	0	0

10.2.3.2 Design 2

The results of Design 2 are presented in Fig. 10.4 and Table 10.2, as an intermediate design solution for reactive volume minimization. It has a reactive volume of 4.41 m^3, 17.6% better than the optimal result in Ref. [6]. It features a reactive distillation type of design as have been reported in our previous GMF works. The "column section" includes two pure separation modules in the rectification section, one reactive separation modules in the reaction zone, and another pure separation module in the stripping section. The pentene feed stream is fed to the reactive module.

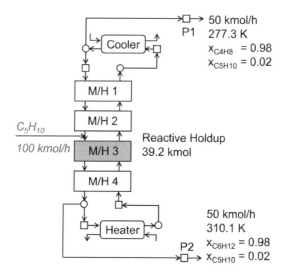

FIGURE 10.4 GMF Design 2 – Intermediate solution for reactive volume minimization. (Adapted from Tian and Pistikopoulos [10].)

Table 10.2 GMF Design 2 – Flowsheet data. (Adapted from Tian and Pistikopoulos [10].)

Module ID	T (K)	F (kmol/h)	$x_{C_5H_{10}}$	$x_{C_4H_8}$	$x_{C_6H_{12}}$
Liquid Outlet Stream					
M/H 1	292.1	51.0	0.996	0.004	0
M/H 2	321.1	100.0	0.508	0.002	0.490
M/H 3	309.8	200.0	0.754	0.001	0.245
M/H 4	310.1	200.0	0.020	0	0.980
Cooler	277.2	50.2	0.02	0.98	0
Vapor Outlet Stream					
M/H 1	310.6	50.2	0.020	0.980	0
M/H 2	298.7	101.0	0.513	0.487	0
M/H 3	326.6	150.0	0.345	0.328	0.327
M/H 4	340.1	150.0	0.999	0.001	0
Heater	340.1	150.0	0.020	0	0.980

10.2.3.3 Design 3

Design 3 is depicted in Fig. 10.5 and the numerical results are summarized in Table 10.3, which is the GMF optimal solution to minimize operating cost accounting for separation energy consumption. The design configuration comprises three mass/heat exchange modules, as well as two pure heat exchange modules. M/H 1 and M/H 2 perform both reaction and separation tasks. Moreover, the pentene feed stream is split before fed to M/H 2 and M/H 3.

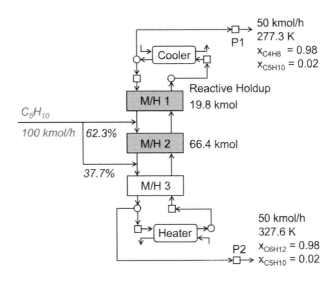

FIGURE 10.5 GMF Design 3 – Operating cost minimization (adapted from Tian and Pistikopoulos [10]).

Table 10.3 GMF Design 3 – Flowsheet data. (Adapted from Tian and Pistikopoulos [10].)

Liquid Outlet Stream						
Module ID	T (K)	F (kmol/h)	$x_{C_5H_{10}}$	$x_{C_4H_8}$	$x_{C_6H_{12}}$
M/H 1	326.2	95.0	0.136	0	0.864
M/H 2	304.7	124.6	0.604	0	0.396
M/H 3	286.6	70.0	0.016	0	0.984
Cooler	277.2	70.0	0.016	0.984	0

Vapor Outlet Stream						
Module ID	T (K)	F (kmol/h)	$x_{C_5H_{10}}$	$x_{C_4H_8}$	$x_{C_6H_{12}}$
M/H 1	278.0	70.0	0.016	0.984	0
M/H 2	326.3	145.0	0.115	0.329	0.556
M/H 3	321.8	112.3	0.998	0	0.002
Heater	339.3	20.0	0.016	0	0.984

10.2.3.4 Remarks

The identification of the three GMF design solutions (i.e., Design 1, 2, 3) in the FD design boundaries is illustrated in Fig. 10.6. Design 1 and 2 have provided two candidate process solutions on the FD lower boundary. Moreover, the trade-off between separation energy consumption and reactive equipment volume can be clearly seen from the design envelope. However, further investigations are needed to identify if there exists a feasible process configuration corresponding to the FD leftmost vertex at the ultimate minimal reactive volume.

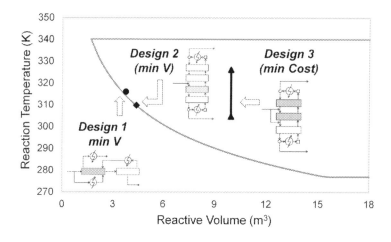

FIGURE 10.6 GMF design solutions in comparison with FD boundaries (adapted from Tian and Pistikopoulos [10]).

Another GMF design obtained during optimization is shown in Fig. 10.7 with a reactive volume of 2.20 m^3. Two mass/heat exchange modules are selected as well as two pure heat exchange modules. Reaction only takes place in M/H 1. However, the liquid inlet and outlet streams of M/H 1 turn out to be exactly the same in flowrates, compositions, and temperature. This deactivates the driving force constraints (Driving force = $G1_i \times G2_i \geq 0$) in M/H 1 by having $G1_i = f^{LO}x_i^{LO} - f^{LI}x_i^{LI}$ always equals to 0. Thus, M/H 1 has a most "intensified" and "de-bottlenecked" reactive holdup where the reaction mixture only contains pentene (i.e., pure reactant), with no limitation of reaction equilibrium since the products butene and pentene are all taken away by the vapor streams.

However, this design requires in situ, instantaneous, and perfect separation of products from the reactants to totally shift reaction equilibrium – which is doubtful if possible to find or create such separators in reality to meet this ideal separation scheme. More important is that it is *not a thermodynamically feasible design* due to the deactivation of driving force constraints. If comparing with the FD boundaries (Fig. 10.8), this design also *lies outside the FD boundaries* which cross-validates the consistency and accuracy of FD and GMF from the other way around.

Chapter 10 • Envelope of design solutions for intensified reaction/separation systems 171

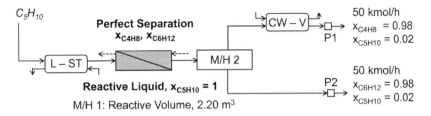

FIGURE 10.7 A thermodynamically infeasible design for reactive volume minimization. (Adapted from Tian and Pistikopoulos [10].)

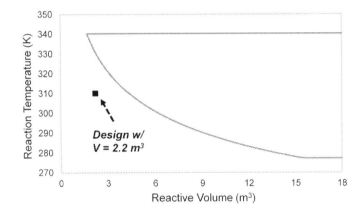

FIGURE 10.8 A thermodynamically infeasible design in comparison with FD boundaries. (Adapted from Tian and Pistikopoulos [10].)

In sum, we have discussed how to develop the design envelope of a combined reaction/separation process for olefin metathesis. The attainable region-based Feinberg Decomposition theory provides the potential to characterize the ultimate design boundaries under given kinetics regardless process design configurations. The incorporation of FD approach into process intensification synthesis methods, such as the Generalized Modular Representation Framework, can: (i) provide thermodynamic/kinetic verification to the abstract phenomena-based representation, and (ii) assist rapid screening of the combinatorial PI design space. Based on the discussions in this chapter, it will be worthy to further explore the following two open questions: (i) For the ultimate minimum reactive volume identified in Fig. 10.2, will there be a practical design to realize the miniaturization potential? How can this serve as a driving force pushing for innovative design? and (ii) How to formulate the FD design boundaries into the GMF driving force constraints to systematically intensify the process performance towards the ultimate bounds? This may provide a promising way to identify the "global" optimal process solution using GMF from mechanistic perspective compared to seeking global optimization techniques from numerical point of view.

References

[1] F. Horn, Attainable and non-attainable regions in chemical reaction technique, in: Third European Symposium on Chemical Reaction Engineering, Pergamon Press, London, 1964, pp. 1–10.

[2] D. Ming, D. Glasser, D. Hildebrandt, B. Glasser, M. Metgzer, Attainable Region Theory: An Introduction to Choosing an Optimal Reactor, John Wiley & Sons, 2016.

[3] M. Feinberg, P. Ellison, General kinetic bounds on productivity and selectivity in reactor-separator systems of arbitrary design: principles, Industrial & Engineering Chemistry Research 40 (2001) 3181–3194.

[4] J.A. Frumkin, L. Fleitmann, M.F. Doherty, Ultimate reaction selectivity limits for intensified reactor–separators, Industrial & Engineering Chemistry Research (2018).

[5] J.A. Frumkin, M.F. Doherty, Ultimate bounds on reaction selectivity for batch reactors, Chemical Engineering Science 199 (2019) 652–660.

[6] F.E. da Cruz, V.I. Manousiouthakis, Process intensification of reactive separator networks through the ideas conceptual framework, Computers & Chemical Engineering 105 (2017) 39–55.

[7] Y. Tang, M. Feinberg, Carnot-like limits to steady-state productivity, Industrial & Engineering Chemistry Research 46 (2007) 5624–5630.

[8] R.C. Reid, J.M. Prausnitz, B.E. Poling, The properties of gases and liquids, 1987.

[9] M.J. Okasinski, M.F. Doherty, Design method for kinetically controlled, staged reactive distillation columns, Industrial & Engineering Chemistry Research 37 (1998) 2821–2834.

[10] Y. Tian, E.N. Pistikopoulos, Toward an envelope of design solutions for combined/intensified reaction/separation systems, Industrial & Engineering Chemistry Research 59 (2020) 11350–11354.

11

Process intensification synthesis of extractive separation systems with material selection

A key enabling factor to achieve process intensification goals is the development of advanced materials (e.g., ionic liquids, molecularly porous materials). The integration of advanced materials in intensified equipment and processes requires a multi-scale consideration to bridge the gap between material performance evaluation, equipment optimization, and process integration in a simultaneous or hierarchical manner. From PSE point of view, material selection has been one of the core topics for process synthesis and design. Particularly with the recent advances in data analytics and computer-aided molecular design methods [1], there is an increasing interest to integrate model-based solvent selection and design strategies with process synthesis to identify simultaneously the optimal process design with the optimal mass separating agent. An indicative list of publications in this area is provided in Table 11.1, which apply mostly to extractive separation systems.

In this chapter, we showcase the application of Generalized Modular Representation Framework (GMF), integrated with Orthogonal Collocation (OC), for process design, synthesis, and intensification of extractive separation systems with solvent selection. We particularly investigate how the selection of materials can directly impact on the fundamental mass and/or heat transfer performances in a phenomenological synthesis framework.

11.1 Problem statement

The generalized design problem addressed in this chapter for extractive separation processes with solvent selection can be defined as follows:

Given:

- A multi-component feed stream with minimum-/maximum- boiling azeotropes or low relative volatility mixtures and given compositions, flow rates, supply temperatures;
- A set of desired products with specifications on flow rates, purities;
- A set of available solvents which can facilitate the homogeneous separation with given availability, temperatures, and compositions;
- A set of available heating and cooling utilities with given availability and temperatures;
- All physical property models and parameters (e.g., activity coefficient model, Antoine coefficients, heat capacity coefficients);
- Cost data of solvents, heating and cooling utilities.

Synthesis and Operability Strategies for Computer-Aided Modular Process Intensification
https://doi.org/10.1016/B978-0-32-385587-7.00022-1
Copyright © 2022 Elsevier Inc. All rights reserved.

174 Synthesis and Operability Strategies for Computer-Aided Modular PI

Table 11.1 Literature review on integrated material selection/design and process synthesis – an indicative list (reproduced from Tian and Pistikopoulos [18]).

Methods	Authors	Process Applications	Problem Formulation
Computer-aided molecular design	Hostrup et al. [2]	Solvent design for acetone-chloroform separation and solvent substitutes for acetic acid-water separation	Hybrid methods for solvent design, screening, and process synthesis
	Papadopoulos and Linke [3]	Solvent design for liquid-liquid extraction and gas-absorption processes	Multi-objective optimization
	Folić et al. [4]	Solvent design for a solvolysis reaction	Stochastic optimization
	Lek-utaiwan et al. [5]	Simultaneous solvent selection and extractive distillation column design for C8-Aromatics mixture separation	Hierarchical process design and optimization framework
	Valencia-Marquez et al. [6]	Simultaneous design of ionic liquid solvent and extractive distillation column for ethanol-water separation	Disjunctive mixed-integer nonlinear programming
	Medina-Herrera et al. [7]	Solvent selection for extractive distillation processes with safety considerations	Hierarchical framework, Genetic algorithm
	Zhou et al. [8]	Optimal solvent design for extractive distillation processes	Multi-objective optimization-based hierarchical framework
Group contribution	Marcoulaki and Kokossis [9]	Design of novel solvents for ethanol-water extraction, etc.	Simulated annealing
	Chen et al. [10]	Integrated ionic liquid and process design for azeotropic separation	Mixed-integer nonlinear programming
Molecular clustering	Papadopoulos and Linke [11]	Synthesis of solvents for cyclohexane-benzene extractive distillation, etc.	Clustering, Multi-objective optimization
Use of thermodynamic models (e.g., NRTL, UNIFAC)	Ismail et al. [12]	Simultaneous solvent selection and process synthesis for homogeneous azeotropic separation processes	Mixed-integer nonlinear programming
	Waltermann et al. [13]	Hierarchical solvent selection and process design with energy integration for azeotropic distillation	Mixed-integer nonlinear programming

The *objectives* are: (i) to minimize total annualized cost (TAC), (ii) to synthesize optimal process solution(s), conventional or intensified, for the specified separation problem, and (iii) to identify the optimal solvent choice integrated with design.

11.2 Case study: ethanol-water separation

We discuss two case studies for the design of ethanol-water separation systems. Case Study 1 showcases the GMF and GMF/OC approaches to obtain ethanol product with 99 mol% purity using methanol and ethylene glycol as the solvent candidates. Case Study 2 investigates the use of an ionic liquid solvent, i.e. [EMIM][OAc], to obtain a high-purity ethanol product with 99.8 mol% purity and a water product with 99 mol% purity. Via the case stud-

Chapter 11 • PI synthesis of extractive separation systems with material selection   175

ies, we aim to: (i) demonstrate the applicability and effectiveness of the GMF and GMF/OC synthesis approaches to systematically generate extractive separation process solutions with improved cost performance and integrated solvent selection considerations, and (ii) evaluate the design feasibility and economic optimality on using a specific ionic liquid solvent in ethanol-water extractive separation process.

11.2.1 Case Study 1: ethylene glycol vs. methanol

The minimum boiling azeotrope composition of ethanol (EtOH) and water (H2O) is at EtOH 0.8943 mol/mol, 351.15 K, 1 atm. Herein, we consider a case study in which ethylene glycol (EG) and methanol (MeOH) are available as solvent candidates. The feed stream consists of 0.85 mol/mol EtOH and 0.15 mol/mol H2O with a flow rate of 10 kmol/s at 351.3 K, 1 atm (i.e., saturated liquid feed). The separation target is to obtain a liquid ethanol product with a purity of 0.99 mol/mol and a flow rate of 8 kmol/s. The system is operated at constant atmospheric pressure. The NRTL equation (Eq. 11.1) is utilized and integrated to the GMF model formulation. The NRTL binary interaction parameters for the ternary systems (i.e., Ethanol-Water-Ethylene Glycol and Ethanol-Water-Methanol) are adapted from Ismail et al. [12] (Table 11.2). The pseudo capital cost calculation is calculated and the utility cost data are respectively $26.19/(kW·yr) for cooling water and $137.27/(kW·yr) for steam. The synthesis objective is to determine a cost-optimal design and to identify the corresponding solvent which can satisfy the product specifications.

$$\ln \gamma_i = \frac{\sum_{j\in c} \tau_{ji} G_{ji} x_j}{\sum_{j\in c} G_{ji} x_j} + \sum_{j\in c} \frac{G_{ij} x_j}{\sum_{l\in c} G_{lj} x_l} \tau_{ij} - \frac{\sum_{l\in c} \tau_{lj} G_{lj} x_l}{\sum_{l\in c} G_{lj} x_l}$$

$$\tau_{ij} = \frac{\Lambda_{ij}}{RT}, \qquad G_{ij} = exp(-\alpha_{ij}\tau_{ij}), \qquad \Lambda_{ii} = 0, \qquad \alpha_{ij} = \alpha_{ji}$$

$$\Lambda_{ij} = aa_{ij} + ab_{ij} \times (T - 273.15) \ \text{cal/mol}$$

(11.1)

Table 11.2 Case Study 1 – NRTL binary interaction parameters. (Reproduced from Tian and Pistikopoulos [18].)

	Ethanol (1) - Water (2) - Ethylene Glycol (3) System				
ij	aa_{ij}	ab_{ij}	aa_{ji}	ab_{ji}	α_{ij}
12	−441.20	18.3280	3293.17	17.0471	0.475000
13	13527.42	−92.7391	−4351.97	53.3769	0.370400
23	1383.43	8.0409	−1445.97	−9.1506	0.185894
	Ethanol (1) - Water (2) - Methanol (4) System				
ij	aa_{ij}	ab_{ij}	aa_{ji}	ab_{ji}	α_{ij}
12	206.7	0	5270.3	0	0.4
14	0	0	0	0	0.4
24	3641.5	0	−788.20	0	0.4

176 Synthesis and Operability Strategies for Computer-Aided Modular PI

11.2.1.1 GMF synthesis

The GMF synthesis superstructure considers for a maximum of sixteen pure heat exchange modules and eight mass/heat exchange modules. It results in a mixed-integer nonlinear programming (MINLP) model with 4864 continuous variables, 478 binary variables, and 4377 equality/inequality constraints. The objective is set to minimize operating cost (including heating and cooling utility cost).

Fig. 11.1(a) shows the optimal separation system design identified by GMF, with an operating cost of 6.34×10^7 per year. Numbered in a descending order, the design consists of one pure heat exchange module for cooling, three mass/heat exchange modules for ethanol-water-ethylene glycol separation, as well as another pure heat exchange module for heating. Ethylene glycol, as the heavier solvent compared to methanol, is selected and the desired ethanol product is obtained from the distillate stream. This modular solution will be later verified and compared with rigorous equipment-based simulation.

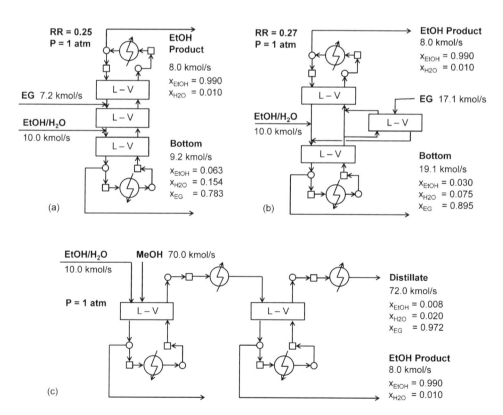

FIGURE 11.1 Case Study 1 – GMF synthesis. (a) Optimal solution, (b) Alternative 1, (c) Alternative 2. (Reproduced from Tian and Pistikopoulos [18].)

Chapter 11 • PI synthesis of extractive separation systems with material selection   177

An intermediate solution is generated by introducing integer cuts. As shown in Fig. 11.1(b), Alternative 1 selects EG as solvent and features an operating cost of 6.98×10^7 per year. Compared to the optimal solution in Fig. 11.1(a), one of the modules is placed as a "side-module" coupled with the other two modules featuring a partially thermally coupled process design. However, this design requires a larger reflux flow rate and higher operating cost than the optimal design.

Another process alternative is generated by specifying MeOH as solvent, featuring an operating cost of 11.4×10^7 per year. In this context, ethanol is the intermediate component in the MeOH-EtOH-H2O mixture with respect to volatility. As depicted in Fig. 11.1(c), two sequential separation steps are included in this process. Namely, the first mass/heat exchange module separates water from the liquid outlet stream, while the vapor outlet stream is sent to the next module to purify ethanol from methanol.

11.2.1.2 GMF/OC synthesis
Herein, we apply the GMF/OC method to design EtOH-H2O separation system(s) by minimizing TAC, which accounts for the pseudo module capital cost, heating and cooling utility cost. The selection of the number of collocation points (i.e., *nc*) in each mass/heat exchange module is critical to balance the modular representation accuracy and the computational time. In light of this, we perform a parametric analysis to discretize each module with different numbers of interior collocation points and to optimize the process with respect to TAC. The results are summarized in Table 11.3. As can be noted, the OC discretization scheme with 2 interior collocation points provides the optimal TAC value. Moreover, the resulting optimal GMF/OC modular structure remains consistent with that identified using GMF (Fig. 11.1a). This is due to the fact that, operating cost is the dominant driving force for cost optimization since it takes up a major part of TAC in distillation systems.

Table 11.3   Case Study 1 – GMF/OC synthesis results. (Reproduced from Tian and Pistikopoulos [18].)

nc	1	2	3
M/H 1	10.0	2.9	3.3
M/H 2	2.0	6.0	3.0
M/H 3	10.0	2.4	3.0
Reflux ratio	0.30	0.25	0.27
Reboiler duty ($\times 10^5$ kW)	3.86	3.70	3.87
Condenser duty ($\times 10^5$ kW)	4.12	4.03	4.05
TAC ($/year)	7.10×10^7	6.76×10^7	7.01×10^7

Note: *M/H* stands for mass/heat exchange module.
Modules are numbered in a descending order.

11.2.1.3 Equipment-based process simulation
The GMF and GMF/OC design solutions obtained in the previous sections will be translated to equipment-based process designs and then validated using Aspen Plus simulation. Taking the optimal GMF/OC solution as an example (i.e., the case of $nc = 2$ in Table 11.3),

178 Synthesis and Operability Strategies for Computer-Aided Modular PI

it can be identified as an extractive distillation column for ethanol-water separation using ethylene glycol as solvent. To determine the column design parameters, the value of GMF/OC "intra-segments" in each mass/heat exchange module can be used as the approximated number of distillation trays in each column section. Thus, Module M/H 1 is translated to 3 column trays, M/H 2 to 6 trays, M/H 3 to 2 trays. The resulting design solution features a 13-tray extractive distillation column (note that condenser is numbered as the first tray and reboiler the last tray). According to the GMF structural combination shown in Fig. 11.1(a), the EG solvent is introduced to the 5th tray (Module M/H 2) and the EtOH-H2O mixture stream is fed onto the 11th tray (Module M/H 3). These design parameters, together with the operating parameters (e.g., reflux ratio) reported in the GMF/OC synthesis results, are used as an initial guess to set up the equipment-based simulation. The Aspen simulation flowsheet using RADFRAC module is depicted in Fig. 11.2(a) and the temperature profile comparison between Aspen simulation and GMF synthesis is presented in Fig. 11.2(b). It can be noted that both GMF and GMF/OC synthesis succeed in capturing the major trend of the temperature profile, while the GMF/OC results provide more detailed design and operation information within each column section. The design and operating parameters can be further adjusted to improve product specifications and/or to minimize reboiler/condenser duties.

To validate if the GMF/OC design with 2 interior collocation points retains cost optimality at the stage of equipment-based design, the other GMF/OC solutions and GMF intermediate solutions are also translated and validated using Aspen simulation. For the GMF/OC solutions identified by $nc = 1$ and $nc = 3$, the equipment translation follows the same step with that of $nc = 2$. The resulting design and operating parameters are summarized in Table 11.4. It is consistent with the GMF/OC synthesis results that the these two process alternatives have higher energy consumption rates than the optimal design identified by $nc = 2$.

Table 11.4 Case Study 1 – Identification of GMF solutions to equipment-based designs (reproduced from Tian and Pistikopoulos [18]).

	GMF/OC designs			GMF Alternative 1
	$nc=1$	$nc=2$ (optimal)	$nc=3$	
Number of trays	24	13	11	13 (main) + 6 (side)
EtOH-H2O feed location	16	11	8	9 (main)
Solvent feed location	12	5	4	1 (side)
Reflux ratio	0.90	0.50	0.90	0.37
Condenser duty ($\times 10^5$ kW)	5.96	4.70	5.96	4.30
Reboiler duty ($\times 10^5$ kW)	6.38	5.12	6.37	6.34
EtOH product purity (mol%)	99.1	99.0	99.0	99.0
EtOH product flowrate (kmol/s)	8.0	8.0	8.0	8.0

Notes: The number of trays for Aspen simulation include condenser and reboiler.
Trays are numbered in a descending order.
main – main column, *side* – side column.

Chapter 11 • PI synthesis of extractive separation systems with material selection   179

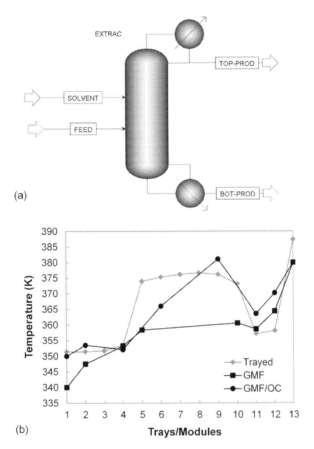

FIGURE 11.2 Case Study 1 – Equipment-based validation for GMF optimal solution.
(a) Aspen simulation, (b) Temperature profile (reproduced from Tian and Pistikopoulos [18]).

The two GMF intermediate process solutions (Fig. 11.1b, 11.1c) are translated as shown in Fig. 11.3. The Alternative 1 using ethylene glycol as solvent is identified as a partially thermally coupled column sequence, in which the side-module corresponds to the prefractionator and the other two GMF modules to the main column. The simulation results are detailed in Table 11.4. Using the reboiler duty as an indicative comparison metric, Alternative 1 requires 23.8% more heating utility consumption than the optimal design due to the less "intensified" scheme with two integrated columns instead of a single column.

For Alternative 2 which uses methanol as solvent, GMF suggests two separation steps in sequence to obtain ethanol product, which are translated to two sequential distillation columns as depicted in Fig. 11.3b. Due to the low relative volatility of methanol/ethanol, large reflux ratios are required in the equipment-based design to first break the ethanol/water azeotrope to separate ethanol/methanol from the first column distillate stream without loss of ethanol from the bottom stream, and then to separate the ethanol product from the second column bottom stream while avoiding methanol contamination. This

180   Synthesis and Operability Strategies for Computer-Aided Modular PI

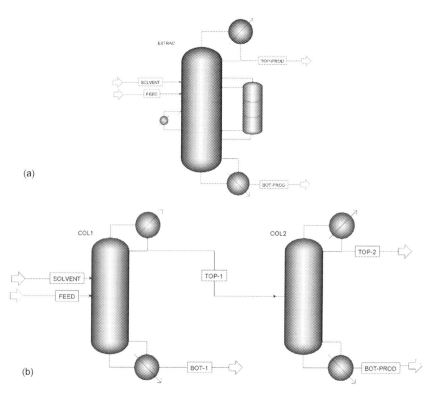

FIGURE 11.3 Case Study 1 – Equipment-based validation for GMF alternative solutions. (a) Alternative 1, (b) Alternative 2 (reproduced from Tian and Pistikopoulos [18]).

results in an internal column flow rate larger than 100 kmol/s to meet the product specifications, which is not feasible from equipment design point of view. Thus this process alternative is not considered for further design and comparison.

Just to highlight again that the design space predicted by GMF using the Gibbs free energy-based driving force constraints is a superset of that for equipment-based extractive distillation systems. This is due to the facts that: (i) GMF allows more degrees of freedom for stream variables – e.g., each module inlet/outlet stream can have different temperatures while satisfying the overall energy balance, (ii) mass transfer feasibility is characterized between liquid inlet stream and vapor outlet stream, as well as between liquid outlet stream and vapor inlet stream. Both the two factors can result in a more "powerful" separation performance in a GMF mass/heat exchange module compared to a single distillation tray. It is also proved that if the driving force constraints are set between the liquid and vapor outlet streams while enforcing uniform temperature for module outlet streams, a mass/heat exchange module is equivalent to a distillation tray assuming liquid-vapor phase (in)equilibrium. Although these modeling considerations introduce thermodynamic approximations compared to rigorous tray-by-tray modeling, more advantages are offered at this phenomena-based design stage to rapidly screen the design space and

Chapter 11 • PI synthesis of extractive separation systems with material selection   181

to evaluate the potential performance improvements by enabling: (i) compact physics-based representation for general reaction/separation systems with conventional type of mass/heat transfer (including but not limited to extractive separation), (ii) identification of performance limits (e.g., cost, energy consumption) regardless the use of intensified/conventional units, and (iii) compact model size and reduced computational load.

11.2.2 Case Study 2: ethylene glycol vs. [EMIM][OAc]

In Case Study 2, we examine the use of ionic liquids solvent for ethanol-water separation, compared to ethylene glycol as the conventional solvent. Ionic liquids are an emerging class of solvents in extractive separation owing to their advantages such as negligible vapor pressure, high boiling point, thermal and chemical stability, etc. However, the industrial application of ionic liquids is still limited due to the high material cost, high viscosity, and lack of physical property data. Thus, it highlights the need to systematically and quantitatively study the trade-off between process efficiency and cost by integrating solvent selection with process design intensification.

1-ethyl-3-methylimidazolium acetate ([EMIM][OAc]) has been identified as an efficient ionic liquid solvent for several azeotropic separation systems, including ethanol-water [14,15], acetone-methanol [16], isopropyl alcohol dehydration [17]. In addition to significantly increasing the selectivity between the azeotropic pair, [EMIM][OAc] also has a relatively lower viscosity than other ionic liquid solvents. However, it should be noted that the degradation temperature of [EMIM][OAc] is at 433.15 K, which normally requires vacuum pressure operation.

The ethanol-water feed stream is considered to comprise 0.80 mol/mol EtOH and 0.20 mol/mol H2O at 200 kmol/h, 351.3 K, 1 atm. The separation target is to obtain a liquid ethanol product at 160 kmol/h with 0.998 mol/mol purity and a liquid water product with 0.99 mol/mol purity. Two solvent candidates are available: (i) ethylene glycol (EG), and (ii) [EMIM][OAc] (denoted as "IL" hereafter). The NRTL binary interaction parameters for the Ethanol-Water-Ethylene Glycol system are used consistently as that in Case Study 1, while the set of parameters for the Ethanol-Water-IL system is given in Table 11.5 [15]. The ionic liquid solvent is assumed to be non-volatile, while this assumption can be easily relaxed if corresponding vapor pressure parameters are available. For potential solvent make-up streams, the cost of IL solvent is assumed to be much higher than that of the EG solvent by adding a penalized cost term in the objective function. The synthesis objective is again to determine a cost-optimal design and to identify the corresponding solvent which can satisfy the aforementioned product specifications.

The initial guess consists of four pure heat exchange modules and six mass/heat exchange modules. All the interconnecting liquid and vapor streams exist between these modules. The GBD solution converges in 25 iterations and the optimal GMF configuration is shown in Fig. 11.4 which features a total annualized cost of 1.10×10^6 per year. Regarding the GMF module selection, to accommodate the low volatility of IL which cannot be vaporized in the GMF pure heat exchange module, a modified type of mass/heat exchange module "HU-L-V" is added in which additional heating duties can be introduced

Table 11.5 Case Study 2 – NRTL parameters for ethanol (1), water (2), and [EMIM][OAc] (3). (Reproduced from Tian and Pistikopoulos [18].)

ij	a_{ij}	b_{ij}	a_{ji}	b_{ji}	α_{ij}
12	−91.8057	1.6016	1162.4117	1.0213	0.4
13	−1023.3691	0	−662.9004	0	0.3
23	−823.8181	0	−1672.7723	0	0.3

to a mass/heat exchange module in analogy to a flash column. In this context, the optimal GMF process solution in Fig. 11.4 can be interpreted as two unit operations: (i) an extractive distillation column – which consists of three "liquid-vapor (L-V)" mass/heat exchange modules for separation, one "HU-L-V" mass/heat exchange module, and one pure heat exchange module for process stream cooling, and (ii) a solvent recovery flash column – which consists of one "HU-L-V" mass/heat exchange module to separate water and IL respectively via vapor and liquid outlet streams, as well as one pure heat exchange module to cool the vapor stream to obtain liquid H2O product. To minimize the solvent cost, [EMIM][OAc] is all recovered from the solvent recovery flash column (i.e., no solvent make-up stream is needed).

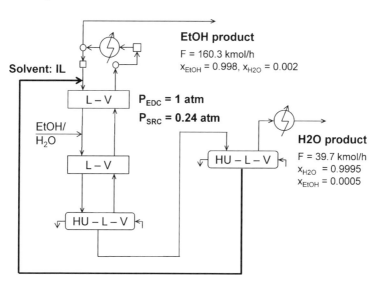

FIGURE 11.4 Case Study 2 – GMF synthesis of EtOH-H2O-IL extractive distillation. (Reproduced from Tian and Pistikopoulos [18].)

Two intermediate process solutions using ethylene glycol solvent are also generated during the iterative solution procedure. Fig. 11.5(a) shows the conventional process design (Alternative 1) using an extractive distillation column and a solvent recovery column, featuring a total annualized cost at 2.63×10^6 per year. This solution is identified at the first GBD iteration. Another process solution (Alternative 2) obtained at the 23rd iteration is depicted in Fig. 11.5(b), in which the two distillation columns are thermally integrated via interconnecting vapor and liquid streams. The total annualized cost for this GMF configuration is 1.55×10^6 per year.

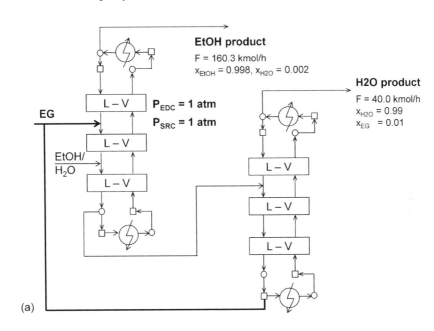

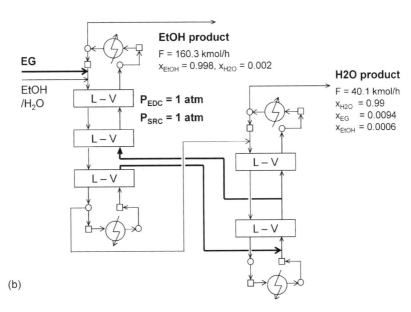

FIGURE 11.5 Case Study 2 – GMF alternative designs. (a) Alternative 1, (b) Alternative 2. (Reproduced from Tian and Pistikopoulos [18].)

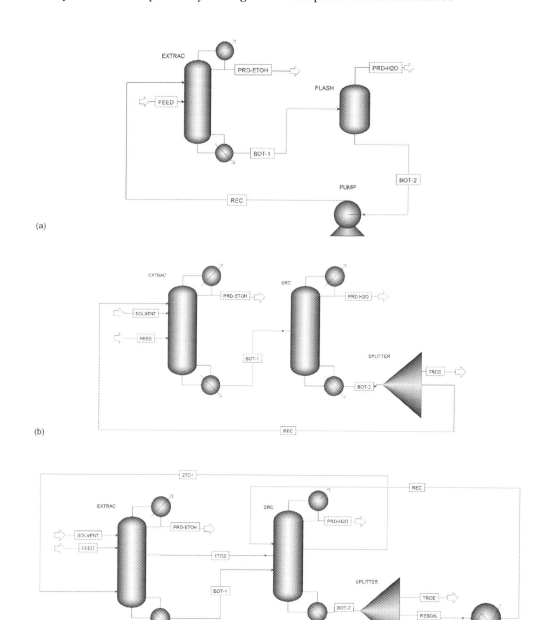

FIGURE 11.6 Case Study 2 – Equipment-based validation. (a) Optimal solution, (b) Alternative 1, (c) Alternative 2. (Reproduced from Tian and Pistikopoulos [18].)

The above process solutions are then translated to equipment-based flowsheet for validation (Fig. 11.6). The Aspen simulation results are summarized in Table 11.6. The process solution with IL solvent, compared to Alternative 1 using ethylene glycol as solvent, features 49% reduction in cooling energy consumption and 53% reduction in heating energy

Chapter 11 • PI synthesis of extractive separation systems with material selection   185

consumption. Assuming that specialized equipment design is not necessitated to accommodate any unique material characteristics of ionic liquids (e.g., viscosity), the capital cost investment can also be expected to be much lower in the IL-assisted process which requires a 15-tray extractive distillation column and a flash column compared to a 18-tray extractive distillation column and a 15-tray solvent recovery column in Alternative 1. For the equipment-based translation of Alternative 2, the design shown in Fig. 11.6c cannot meet the EtOH and H2O product specifications with the same level of energy consumption rates as Alternative 1. This is because the separation target of the second GMF mass/heat exchange module (numbered in descending order from the left), where the heavy components are recycled to the extractive separation section from the solvent recovery section, cannot be met by an aggregation of distillation trays. In this case, GMF can provide the information on separation targets and corresponding module/stream operating conditions, with which the performance promises can be potentially realized. If to construct a new design, these information will serve as a starting point. The stream conditions of this module are: (i) liquid inlet – 355.8 K, 209.3 kmol/h, EtOH 76.44 mol%, H2O 19.11%, EG 4.45%, (ii) liquid outlet – 320.0 K, 130.8 kmol/h, EtOH 0.02%, H2O 30.58%, EG 69.40%, (iii) vapor inlet – 397.0 K, 90.7 kmol/h, H2O 0.36%, EG 99.64%, (iv) vapor outlet: 500.0 K, 169.2 kmol/h, EtOH 94.53%, H2O 0.19%, EG 5.28%.

Table 11.6 Case Study 2 – Identification of GMF solutions to equipment-based designs (reproduced from Tian and Pistikopoulos [18]).

	Optimal solution		Alternative 1		Alternative 2	
	EDC	FC	EDC	SRC	EDC	SRC
Number of trays	15	1	18	15	18	6
EtOH-H2O feed location	10	1	12	–	4	–
Solvent makeup feed location	–	–	3	–	4	–
Solvent recovery feed location	2	–	3	–	5	–
Pressure (atm)	0.29	0.001	1	1	1	1
Reflux ratio	1.1	–	3.0	0.9	3.0	15.0
Condenser temperature (K)	322.9	–	351.5	370.8	351.3	354.3
Reboiler temperature (K)	432.6	432.6	427.7	468.9	385.5	468.7
Condenser duty (MW)	4.01	–	7.32	0.86	7.00	7.40
Reboiler duty (MW)	3.81	0.53	6.94	1.31	5.21	9.34
Solvent	[EMIM][OAc]		Ethylene glycol		Ethylene glycol	
Solvent recovery purity (mol/mol)	0.996		0.997		0.997	
Solvent recovery flowrate (kmol/h)	100.0		223.0		100.0	
EtOH product purity (mol%)	99.8		99.8		91.7	
EtOH product flowrate (kmol/h)	160.0		160.0		160.0	
H2O product purity (mol%)	99.2		99.2		66.6	
H2O product flowrate (kmol/h)	40.0		40.0		40.0	

Note: *EDC* – extractive distillation column, *FC* – flash column, *SRC* – solvent recovery column.

References

[1] A.S. Alshehri, R. Gani, F. You, Deep learning and knowledge-based methods for computer aided molecular design – toward a unified approach: state-of-the-art and future directions, arXiv preprint, arXiv:2005.08968, 2020.

[2] M. Hostrup, P.M. Harper, R. Gani, Design of environmentally benign processes: integration of solvent design and separation process synthesis, Computers & Chemical Engineering 23 (1999) 1395–1414.

[3] A.I. Papadopoulos, P. Linke, Multiobjective molecular design for integrated process-solvent systems synthesis, AIChE Journal 52 (2006) 1057–1070.

[4] M. Folić, C.S. Adjiman, E.N. Pistikopoulos, Design of solvents for optimal reaction rate constants, AIChE Journal 53 (2007) 1240–1256.

[5] P. Lek-utaiwan, B. Suphanit, P.L. Douglas, N. Mongkolsiri, Design of extractive distillation for the separation of close-boiling mixtures: solvent selection and column optimization, Computers & Chemical Engineering 35 (2011) 1088–1100.

[6] D. Valencia-Marquez, A. Flores-Tlacuahuac, R. Vasquez-Medrano, Simultaneous optimal design of an extractive column and ionic liquid for the separation of bioethanol–water mixtures, Industrial & Engineering Chemistry Research 51 (2012) 5866–5880.

[7] N. Medina-Herrera, I.E. Grossmann, M.S. Mannan, A. Jiménez-Gutiérrez, An approach for solvent selection in extractive distillation systems including safety considerations, Industrial & Engineering Chemistry Research 53 (2014) 12023–12031.

[8] T. Zhou, Z. Song, X. Zhang, R. Gani, K. Sundmacher, Optimal solvent design for extractive distillation processes: a multiobjective optimization-based hierarchical framework, Industrial & Engineering Chemistry Research 58 (2019) 5777–5786.

[9] E. Marcoulaki, A. Kokossis, On the development of novel chemicals using a systematic optimisation approach: Part II. Solvent design, Chemical Engineering Science 55 (2000) 2547–2561.

[10] Y. Chen, R. Gani, G.M. Kontogeorgis, J.M. Woodley, Integrated ionic liquid and process design involving azeotropic separation processes, Chemical Engineering Science 203 (2019) 402–414.

[11] A.I. Papadopoulos, P. Linke, Efficient integration of optimal solvent and process design using molecular clustering, Chemical Engineering Science 61 (2006) 6316–6336.

[12] S.R. Ismail, E.N. Pistikopoulos, K.P. Papalexandri, Modular representation synthesis framework for homogeneous azeotropic separation, AIChE Journal 45 (1999) 1701–1720.

[13] T. Waltermann, T. Grueters, D. Muenchrath, M. Skiborowski, Efficient optimization-based design of energy-integrated azeotropic distillation processes, Computers & Chemical Engineering 133 (2020) 106676.

[14] Y. Ge, L. Zhang, X. Yuan, W. Geng, J. Ji, Selection of ionic liquids as entrainers for separation of (water+ ethanol), Journal of Chemical Thermodynamics 40 (2008) 1248–1252.

[15] C. Dai, Z. Lei, X. Xi, J. Zhu, B. Chen, Extractive distillation with a mixture of organic solvent and ionic liquid as entrainer, Industrial & Engineering Chemistry Research 53 (2014) 15786–15791.

[16] H.-H. Chen, M.-K. Chen, B.-C. Chen, I.-L. Chien, Critical assessment of using an ionic liquid as entrainer via extractive distillation, Industrial & Engineering Chemistry Research 56 (2017) 7768–7782.

[17] H.-H. Chen, M.-K. Chen, I.-L. Chien, Using [EMIM][OAC] as entrainer for isopropyl alcohol dehydration via extractive distillation, in: 2017 6th International Symposium on Advanced Control of Industrial Processes (AdCONIP), IEEE, 2017, pp. 257–262.

[18] Y. Tian, E.N. Pistikopoulos, A process intensification synthesis framework for the design of extractive separation systems with material selection, Journal of Advanced Manufacturing and Processing 3 (2021) e10097.

12

Process intensification synthesis of dividing wall column systems

In this chapter, we explore the process design and intensification of multi-component separation systems, with particular interest in the use of dividing wall columns (DWCs). As introduced in Chapter 1, DWC features a fully thermally coupled and single-shell distillation column. The task-integrated design scheme, with improved thermodynamic efficiency, can lead to approximately 30% savings in capital expenditure, space, and energy [1]. Process design approaches for DWC (or thermally coupled columns) have also been investigated by the Process Systems Engineering community. The pioneering process synthesis strategies in references [2–4] used "column sections" as the elementary components to construct a superstructure representation, which enabled the systematic generation of conventional distillation sequences, thermally coupled columns, and dividing wall columns by activating or deactivating the stream connections. However, the columns were generally modeled using short-cut methods (e.g., the Fenske-Underwood-Gilliland method) assuming sharp splits, constant volatility, etc. The pre-postulated superstructure of the column section interconnections also significantly affected the solution space as highlighted by Agrawal [5]. With the recent advancements in deterministic and data-driven optimization algorithms, high-fidelity tray-by-tray models started to be applied for DWC design optimization [6–8]. While the column representation accuracy can be improved in this way, the resulting numerical complexity and computational load normally did not allow for simultaneous considerations of other structural variants beyond DWC systems.

Generalized Modular Representation Framework (GMF) has also been applied to synthesize complex separation systems as presented in the book of Georgiadis and Pistikopoulos [9]. GMF was able to systematically navigate the design space and to generate distillation sequences, heat-integrated distillation, and dividing wall column, while providing a more efficient approach to balance representation accuracy and computational load. However, a simplified superstructure network was adapted which restricted the discovery of non-intuitive structural variants.

Herein, we demonstrate the GMF synthesis approach with full superstructure representation capability to design and intensify a heterogeneous separation system for methyl methacrylate (MMA) purification [20]. MMA is the foundational monomer for large-scale poly-methyl methacrylate production and the co-monomer widely used in plastics, paints, and coatings industry [10]. Such applications also set a very high purity specification for the commercial MMA monomers (typically at 99.8%). A number of process alternatives have been proposed to purify the MMA reactor product at the industrial scale, including

---

Synthesis and Operability Strategies for Computer-Aided Modular Process Intensification
https://doi.org/10.1016/B978-0-32-385587-7.00023-3
Copyright © 2022 Elsevier Inc. All rights reserved.

188 Synthesis and Operability Strategies for Computer-Aided Modular PI

the use of two distillation columns with water decanter [11], two distillation columns with extraction [12], membrane-assisted separation [13], etc. In a recent patent by Jewell et al. [14] from Dow Global Technologies, a dividing wall column design, integrated with a water decanter, was proposed to improve MMA separation efficiency. It has been reported that, with the same number of column trays and the same energy consumption, the recovery of MMA product from the invented DWC design can be 12.2 kmol/h (or 7.6%) higher than a two-column design. This patented MMA purification process will be revisited hereafter as the case study to explore more energy- and cost-efficient process solutions using GMF.

12.1 Case study: methyl methacrylate purification

12.1.1 Process description

For the MMA purification case study defined in Jewell et al. [14], the raw material is considered as a reaction product mixture from MMA preparation which consists of MMA, Water (H2O), Methanol (MeOH), and MMA oligomers (MMAOLG). The oligomers of MMA include the dimer of MMA and smaller amounts of higher oligomers. The component feed flowrates are summarized in Table 12.1.

Table 12.1 Summary of component flowrates in feed mixture. (Reproduced from Tian et al. [20].)

Component	MMA	H2O	MeOH	MMAOLG
Flowrate (kmol/h)	175.259	15.7643	0.332871	9.60129

The quaternary mixture can exhibit liquid-vapor and liquid-liquid phase behaviors. Multiple methanol-MMA and water-MMA azeotropes can be formed in the system under different pressures and temperatures [15]. To accurately capture the complex phase behaviors, the UNIQUAC model and the associated parameters are adapted from Wu et al. [11] as given in Table 12.2, which have been compared and validated with experimental data and other activity coefficient models. The extended Antoine equation is applied for vapor pressure calculation, the coefficients of which are summarized in Table 12.3. Note that the MMA oligomers are treated as a pseudo component, with the major physical properties identical with MMA. The extended Antoine equation coefficients for MMA oligomers are estimated based on a C7 compound (i.e., C7H14O3), which can best approximate its volatility compared to the patent result data.

UNIQUAC model:

$$\ln \gamma_i = \ln \frac{\Phi_i}{x_i} + \frac{z}{2} q_i \ln \frac{\theta_i}{\Phi_i} - q_i' \sum_j \frac{\theta_j' \tau_{ij}}{t_j'} + l_i + q_i' - \frac{\Phi_i}{x_i} \sum_j x_j l_j$$

$$\theta_i = \frac{q_i x_i}{q_T}, \quad q_T = \sum_k q_k x_k, \quad \theta_i' = \frac{q_i' x_i}{q_T'}, \quad q_T' = \sum_k q_k' x_k,$$

Chapter 12 • Process intensification synthesis of dividing wall column systems   189

$$\Phi_i = \frac{r_i x_i}{r_T}, \quad r_T = \sum_k r_k x_k, \quad l_i = \frac{z}{2}(r_i - q_i) + 1 - r_i'$$

$$t_i' = \sum_k \theta_k' \tau_{ki}, \quad \tau_{ij} = exp(a_{ij} + \frac{b_{ij}}{T}), \quad z = 10$$

Table 12.2 UNIQUAC model parameters.
(Reproduced from Tian et al. [20].)

Binary interaction parameters					
component i	component j	a_{ij}	a_{ji}	b_{ij}	b_{ji}
MMA	MMA	0	0	0	0
MMA	H2O	0	0	−474.33	−194
MMA	MeOH	0	0	−411.619	44.6284
MMA	MMAOLG	0	0	0	0
H2O	H2O	0	0	0	0
H2O	MeOH	0.6437	−1.0662	−322.131	432.879
H2O	MMAOLG	0	0	−194	−474.33
MeOH	MeOH	0	0	0	0
MeOH	MMAOLG	0	0	44.6284	−411.619
MMAOLG	MMAOLG	0	0	0	0

Relative molecular volume and surface area				
Parameters	MMA	H2O	MeOH	MMAOLG
r_i	3.92156	0.92	1.43111	3.923
q_i	3.564	1.4	1.432	3.68

Extended Antoine equation:

$$\ln P_i = C_{1i} + \frac{C_{2i}}{T + C_{3i}} + C_{4i} T + C_{5i} \ln T + C_{6i} T^{C_{7i}} \quad (bar, K)$$

Table 12.3 Antoine equation coefficients for vapor pressure calculation.
(Reproduced from Tian et al. [20].)

Component	C_{1i}	C_{2i}	C_{3i}	C_{4i}	C_{5i}	C_{6i}	C_{7i}
MMA	95.8471	−8085.3	0	0	−12.72	8.3307e-6	2
H2O	62.1361	−7258.2	0	0	−7.3037	4.1653e-6	2
MeOH	71.2051	−6904.5	0	0	−8.8622	7.4664e-6	2
MMAOLG	65.8111	−8481.4	0	0	−7.6565	6.4118e-18	6

190   Synthesis and Operability Strategies for Computer-Aided Modular PI

The product specifications are set to obtain:

- MMA product with a MMA purity of at least 99.80 wt% and a maximum of 0.05 wt% water as per The Dow Chemical Company Sales Specification [16]. To ensure the product purity particularly to restrict the existence of low molecular weight components, the MMA product purity is also required to reach at least 99.80 mol%;
- MMA product with a flowrate of at least 172 kmol/h, which is the best product recovery rate reported in the patent [14] with the integrated DWC and water decanter process;
- Water recovery with a purity of at least 99.0 mol%.

12.1.2 Synthesis objective

The *objective* is to synthesize the optimal process solution(s) for the above MMA purification process by minimizing total annualized cost. The DWC-decanter design invented in the patent [14] is used as a base case and new process solutions are to be developed with improved energy efficiency and cost efficiency. The use of dividing wall columns is of particular interest, while the applied GMF synthesis strategy, as detailed in Chapters 3–5, can systematically generate process options without equipment pre-postulation, thus not restricting the design solutions.

The solution of this problem will identify: (i) the unit operation selection and integrated process scheme (for multiple promising process design alternatives), (ii) the optimal process design and operating parameters, and (iii) the estimated total annualized cost.

12.2 Base case design and simulation analysis

Fig. 12.1 depicts the integrated DWC-decanter design for MMA purification proposed by Jewell et al. [14]. The key design characteristics are summarized as follow:

- The dividing wall column consists of 20 column stages (note that theoretical stages are assumed with 100% efficiency)
- The dividing wall extends vertically within the column
- The column section above the dividing wall can have 2 to 6 stages, the section below the dividing wall can have 2 to 6 stages, and the divided section can have 6 to 15 stages
- The column is operated at reduced pressure from 1 to 50 mmHg (note that the process temperature is preferably below 373.5 K to prevent MMA polymerization [17])
- Feed tray is located from the column bottom at a distance of 35% to 65% of the height of the dividing wall
- Besides the top and bottom streams, two side draws are removed from the column:
  - Upper side draw: removed from one stage above the divided section and sent to the water decanter at a flowrate of 376 kmol/h
  - Middle side draw: i.e. the MMA product stream, removed from 35% to 65% height of the dividing wall at a flowrate of 172 kmol/h

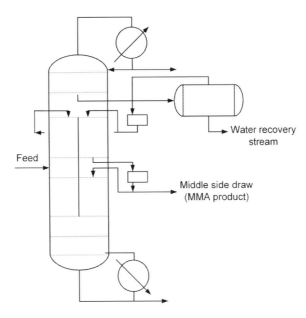

FIGURE 12.1 MMA purification using dividing wall column integrated with water decanter [14]). (Reproduced from Tian et al. [20].)

- The water decanter separates the upper side draw to a dewatered organic stream and an aqueous stream
- The aqueous stream gives the water recovery stream
- The dewatered organic stream is returned to the column one stage below the stage from which the upper side draw is removed
- The dewatered stream is split between the divided sections preferably no more than 52% to each section

The integrated dividing wall column and water decanter flowsheet simulation is set up in Aspen Plus as shown in Fig. 12.2. The Aspen PETLYUK module is used to model the dividing wall column as a Petlyuk column, assuming no heat transfer across the dividing wall. The water decanter is modeled via the DECANTER module in which the liquid-liquid equilibrium calculations are employed for phase separation. Based on the patent design parameters, sensitivity analysis-based optimization is performed to minimize reboiler duty, condenser duty, and number of trays. The degrees of freedom for optimization include: (i) number of trays, (ii) feed tray location, (iii) distillate rate, (iv) reflux ratio, (v) vapor connect stream flowrates, and (vi) liquid connect stream flowrates. The resulting Petlyuk column design results are summarized in Table 12.4. The water decanter is designed at 1 atm and 323.15 K, which requires a heating duty of 596.8 kW. This design configuration, after preliminary optimization in Aspen Plus, will be used as the base case and benchmark the energy and cost improvements resulted by the new design alternatives.

192 Synthesis and Operability Strategies for Computer-Aided Modular PI

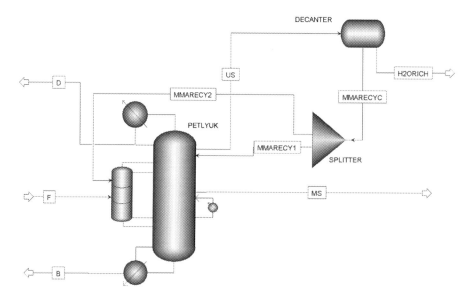

FIGURE 12.2 Base case design and simulation in Aspen Plus (reproduced from Tian et al. [20]).

Table 12.4 Base case – Results summary for Aspen optimization. (Reproduced from Tian et al. [20].)

		Main column	Prefractionator
Column design	Number of stages	17	5
	Feed stage location	6 (MMARECY)	1 (MMARECY)
			3 (F)
	Reflux ratio	69	–
	Distillate rate (kmol/h)	7	–
	Pressure (atm)	0.06	0.06
Connect streams	Stream 1, Liquid	Source, Stage 6	Destination, Stage 1
	Stream 2, Vapor	Source, Stage 15	Destination, Stage 5
	Stream 3, Liquid	Destination, Stage 15	Source, Stage 5
	Stream 4, Vapor	Destination, Stage 6	Source, Stage 1
Energy consumption	**Heating duty (kW)**	5035.5	–
	Cooling duty (kW)	−5560.4	–
MMA product	Purity (wt%)	99.84	
	Purity (mol%)	99.80	
	Flowrate (kmol/h)	172.0	
Water recovery	Purity (mol%)	99.45	
	Flowrate (kmol/h)	12.1	

Note: MMARECY – Dewatered MMA-rich outflow from Decanter, F – Feed stream
Heat duty – Positive numbers for heating, Negative numbers for cooling
Stage numbering – Condenser counted as the 1st stage, Reboiler counted as the last stage
Stage efficiency – Theoretical stages are assumed for design with 100% efficiency

12.3 Process intensification synthesis via GMF

In what follows, we leverage the GMF synthesis to investigate the industrial MMA purification problem, targeting for new process solutions with improved cost performance.

12.3.1 GMF representation for base case design

To ensure the representation accuracy for the MMA purification process, we first validate the GMF representation for the base case design configuration (Fig. 12.2). A GMF modular structure is set up as depicted in Fig. 12.3. The binary variables in the synthesis model are fixed according to the modular structure, e.g. if the interconnecting stream exists, the binary variable takes the value of 1; otherwise, the binary variable is assigned as 0. The feed stream and interconnecting streams are placed in consistency with the Aspen simulation results given in Table 12.4. Each mass/heat exchange module ("M/H") stands for a certain column section as per the base case design, numbered in a descending order from right to left. The decanter ("DE") is modeled as a separate module integrated with the GMF modular building blocks, in which only liquid-liquid equilibrium calculations are performed.

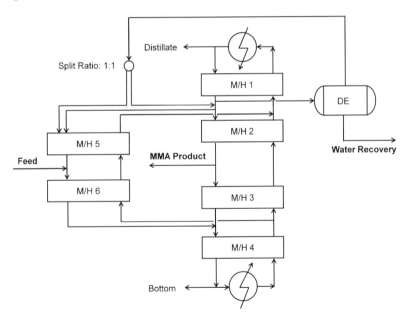

FIGURE 12.3 Base case – GMF representation (adapted from Tian et al. [20]).

In this context, M/H 1 represents the column section above the divided column section which consists of 5 stages. The upper side draw is removed as part of the M/H 1 liquid outlet stream and sent to the decanter module for liquid-liquid separation. The aqueous outlet stream from the decanter is the water recovery stream, while the dewatered organic outlet stream is returned and split equally to enter M/H 2 and M/H 5. M/H 2 and M/H 3 represent the divided section in the main column, between which the middle side draw (i.e., the

MMA product stream) is obtained. As per the base case design, M/H 2 consists of 4 stages and M/H 3 consists of 6 stages. M/H 4 describes the column section below the divided section, comprising 2 stages. Two pure heat exchange modules are respectively placed at the top and the bottom, acting as a total condenser and a total reboiler. For the prefractionator section comprising 5 stages, M/H 5 and M/H 6 are used and the feed stream enters M/H 6. As can be noted, the number of stages captured by each mass/heat exchange module may vary from one to another. Each mass/heat exchange module in general characterizes a mass transfer pattern (e.g., component A and B transfer from the liquid phase to the vapor phase, component C from vapor to liquid). The number of stages necessitated to achieve the separation target by a mass/heat exchange module can be determined via: (i) a heuristic-based trial-and-error approach to translate each module to a minimum number of essential stages with an implicit objective function to minimize equipment size, and (ii) the GMF orthogonal collocation approach to rigorously obtain the optimal number of stages as discussed in Chapter 11 [18].

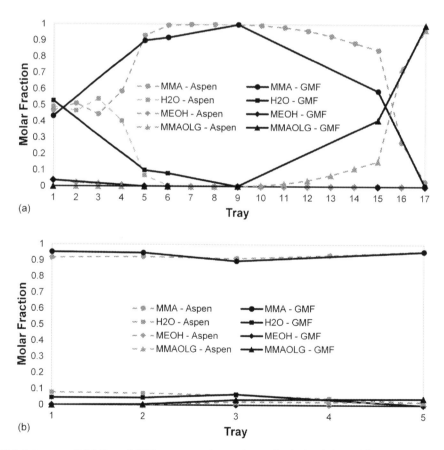

FIGURE 12.4 Base case – Validation of GMF representation vs. Aspen simulation (adpated from Tian et al. [20]). (a) Liquid molar fraction profile in main column, (b) Liquid molar fraction profile in prefractionator.

Chapter 12 • Process intensification synthesis of dividing wall column systems   195

By "simulating" the above GMF structure for the base case design, we can test if GMF can capture the major process characteristics for this heterogeneous separation problem. As shown in Fig. 12.4, the results for GMF representation and Aspen simulation are compared on the liquid composition profiles for the main column and the prefractionator. It can be noted that GMF provides an overall good estimate, although certain tray-wise details are missed, on describing the physical process taking place in the integrated process with the Petlyuk column and the decanter.

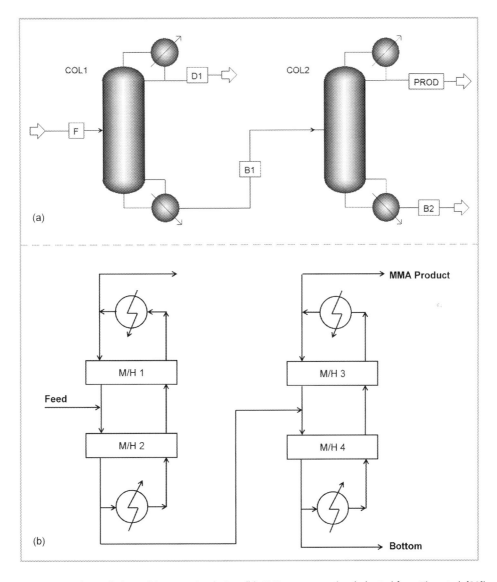

FIGURE 12.5 Two column design – (a) Aspen simulation, (b) GMF representation (adapted from Tian et al. [20]).

The conventional two-column configuration which was used in the patent [14] as a comparative design is also simulated using Aspen Plus RADFRAC module and then represented via GMF as shown in Fig. 12.5. Each of the column has 10 stages, with the feed stream enters the second stage in the first column. By comparing the liquid and vapor molar fraction profiles in Fig. 12.6, it can be concluded that GMF can accurately capture the complex physical behaviors in this MMA purification process.

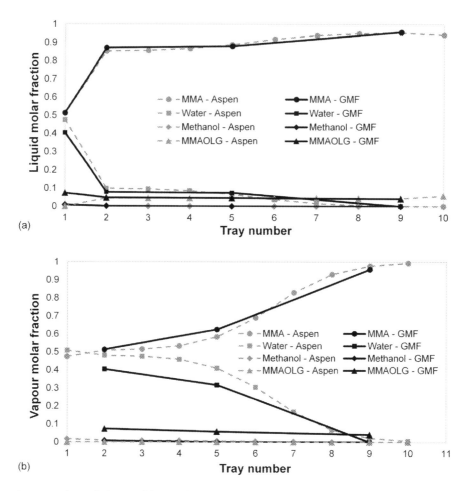

FIGURE 12.6 Two column design – Validation of GMF representation vs. Aspen simulation. (a) Liquid molar fraction profile in the first column, (b) Vapor molar fraction profile in the first column. (Adapted from Tian et al. [20].)

Chapter 12 • Process intensification synthesis of dividing wall column systems   197

12.3.2 GMF synthesis optimization

In this section, we optimize the MMA purification process using GMF synthesis.

12.3.2.1 Retrofit design

We first investigate to retrofit the base case design with minimum structural changes. To this purpose, all the pure heat exchange modules and mass/heat exchange modules in Fig. 12.3 are required to be active. Namely, no addition or reduction of the modules. The existing interconnecting streams and feed/product streams also remain active as that in Fig. 12.3. However, new interconnecting streams are allowed to be added to the structure for optimization.

The resulting GBD Primal Problem comprises 8448 modeling constraints and 4493 continuous variables while the Master Problem with 510 modeling constraints and 783 binary variables. The initial solution structure is the base case design. The GBD solution procedure converges in 28 iterations and the optimal solution is obtained at the 17th iteration as shown in Fig. 12.7 and Table 12.5. The notable differences between the new solution (referred as "Design 1") and the base case design include: (i) the upper side draw flowrate – instead of removing 376 kmol/h upper side draw as indicated by the patent, the upper side draw is removed at a much smaller flowrate of 40.3 kmol/h which reliefs the column for large amount vaporization, (ii) the split ratio of the decanter dewatered organic stream returning to the divided sections – instead of splitting equally to the prefractionator and the main column, the split ratio is suggested as 1:3.7 to return a larger portion of the organic stream to the main column to minimize remixing. The dewatered organic stream in this case actually has a larger MMA molar fraction (~ 92 mol%) than the liquid stream outlet from the prefractionator section, since the prefractionator streams mix with the MMA feed stream. Due to the previous specification of equal distribution, the adjustment in the split ratio is achieved by the GMF structural model via activating another splitting stream to the main column.

The base case design using GMF representation features a total annualized cost of $\$1.56 \times 10^6$, and the optimal retrofit design gives a total annualized cost of $\$1.06 \times 10^6$. Due to the abstract GMF representation which tends to intensify the process design to the utmost efficiency and the approximations on pseudo-capital cost estimation, the actual number on cost improvements will be finalized using equipment-based rigorous simulation. However, the relative cost optimality for different GMF solutions and the corresponding design changes, as will be demonstrated in the following sections, can provide critical instructions in generating better design solutions.

12.3.2.2 Grassroots design

The optimal grassroots design is generated by enabling the full superstructure representation without pre-specifying any existing design components. A maximum of 10 mass/heat

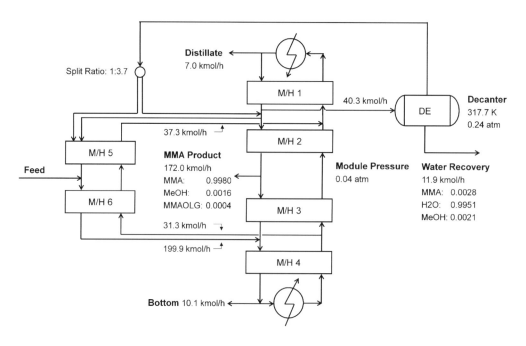

FIGURE 12.7 Design 1 – GMF synthesis solution for retrofit design (reproduced from Tian et al. [20]).

Table 12.5 Design 1 – GBD convergence statistics (adapted from Tian et al. [20]).

Iteration	1	2	3	4	5	6	7	8	9	10
Primal	155.7	109.7	137.8	171.4	171.4	145.5	117.2	108.6	171.4	171.4
Master	−30687.1	−8038.7	−8038.7	−5111.2	−5111.2	−879.5	−879.5	−879.5	−879.5	−879.5
Iteration	11	12	13	14	15	16	17	18	19	20
Primal	168.2	171.4	171.4	124.7	124.7	108.5	106.1	154.7	146.9	infes
Master	−879.5	−879.5	−879.5	−879.5	−879.5	−879.5	−879.5	−879.5	−879.5	−879.5
Iteration	21	22	23	24	25	26	27	28		
Primal	infes	infes	166.1	151.7	171.4	122.3	142.5	120.6		
Master	−879.5	−879.5	−879.5	−879.5	−879.5	−876.4	−271.7	120.6		

exchange modules and 20 heat exchange modules are available for use in the GMF structural model. Only the activated modules are considered in the synthesis model using the GAMS dynamic sets to reduce computational load. The initial design structure is set as the 6-module base case design, but with all the liquid and vapor connect streams activated to avoid pre-postulation. The GBD solution procedure converges in 25 iterations and the optimal solution is obtained at the 19th iteration with a total annualized cost of 8.8×10^5. The design structure of this Design 2 and the GBD convergence statistics are presented respectively in Fig. 12.8 and Table 12.6. The overall structure still features a Petlyuk column type of design, to be further verified with equipment-based simulation. The key design changes are summarized as follow:

- Only four mass/heat exchange modules are selected, indicating a smaller equipment than the base case and Design 1
- No upper side draw is removed from the main column. Instead, the distillate stream is directed to the decanter
- The decanter dewatered organic stream is split into three streams: (i) a stream with 6.8 kmol/h flowrate returned to M/H 4 (i.e., the prefractionator section), (ii) a stream with 54.6 kmol/h flowrate returned to M/H 1 (in main column section), and (iii) another stream with 54.3 kmol/h flowrate returned to M/H 2 (in main column section)
- The MMA feed stream enters M/H 4
- The MMA product stream is obtained as the liquid outlet stream from M/H 1
- M/H 4 is coupled with the two pure heat exchange modules for vapor inlet and outlet streams
- Decanter is operated at 0.30 atm, 300.7 K

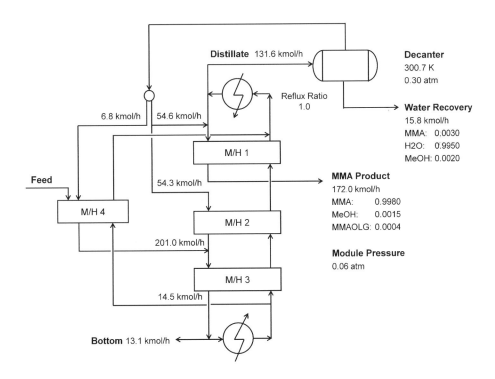

FIGURE 12.8 Design 2 – GMF synthesis solution for grassroots design (reproduced from Tian et al. [20]).

200 Synthesis and Operability Strategies for Computer-Aided Modular PI

Table 12.6 Design 2 – GBD convergence statistics (adapted from Tian et al. [20]).

Iteration	1	2	3	4	5	6	7	8	9	10
Primal	107.5	infes	infes	infes	171.4	171.4	171.4	171.4	171.4	171.4
Master	−22624.4	−22624.4	−22624.4	−22624.4	−22400.8	−22400.8	−10996.1	−5265.0	−694.5	−688.5
Iteration	**11**	**12**	**13**	**14**	**15**	**16**	**17**	**18**	**19**	**20**
Primal	171.4	infes	infes	171.4	171.4	171.4	171.4	infes	88.2	171.4
Master	−435.0	−422.1	−422.1	−189.0	−187.8	−163.2	−159.0	−159.0	−159.0	−23.9
Iteration	**21**	**22**	**23**	**24**	**25**					
Primal	171.4	171.4	171.4	171.4	161.7					
Master	−23.9	−20.8	−20.8	−20.8	94.7					

12.3.2.3 Two-column design

To compare the two-column process with the above derived Petlyuk column designs, GMF synthesis is performed starting from the structure in Fig. 12.5b and enforces the use of four pure heat exchange modules. With the MMA product specifications of 99.8 wt% and 99.8 mol%, the maximum product flowrate is found to be 166.8 kmol/h which is 5.2 kmol/h less than Designs 1 and 2. By setting the MMA product flowrate specification as 165.0 kmol/h for this case, the optimal GMF solution is obtained in 7 iterations as illustrated in Fig. 12.9. The M/H 1 and the pure heat exchanger for heating make a stripping section, the vapor outlet stream from which is sent to the decanter module. As can be noted, the water recovery purity also requires further improvement. The liquid outlet stream, together with the decanter dewatered organic stream, enters M/H 3. M/H 2, M/H 3, and two pure heat exchangers compose a typical distillation column setup. Despite the insufficiency in product specifications, this process solution features a total annualized cost of 8.3×10^5 which makes it another potential design alternative. Therefore, this two-column process solution is referred as Design 3 to proceed with the next-step equipment-based process validation.

12.3.3 Steady-state validation and Aspen simulation

In this section, the Designs 1, 2, and 3 generated by GMF synthesis are translated as equipment-based process alternatives and simulated in Aspen Plus to validate the economic optimality. The Aspen simulation flowsheets are presented in Fig. 12.10 and the results are detailed in Table 12.7.

For Design 1 based on retrofit optimization, the Petlyuk column structure remains identical with that for the base case design (e.g., number of stages, feed stage locations, connect stream source and destination stages). The decanter dewatered organic stream split ratio is specified as 1:3 (prefractionator : main column) for Aspen simulation, adjusted based on the GMF suggested split ratio as 1:3.7. If Design 1 is translated to a dividing wall column,

Chapter 12 • Process intensification synthesis of dividing wall column systems 201

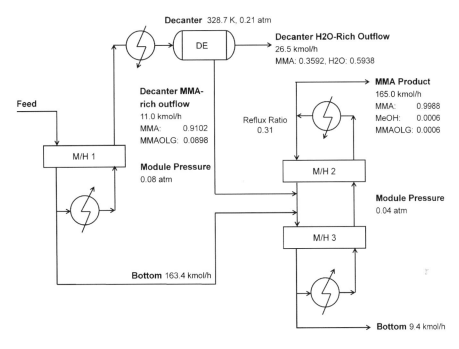

FIGURE 12.9 Design 3 – GMF synthesis solution for two-column process (reproduced from Tian et al. [20]).

the resulting dividing wall position should allow for a larger space for the main column due to the asymmetrical distribution of vapor and liquid streams. GMF also suggests to reduce the upper side draw flowrate from 376 kmol/h in base case to around 40 kmol/h. In the Aspen simulation, 69 kmol/h is found to be the minimum achievable flowrate for the upper side draw. Additional synthesis results for the connect stream flowrates, main column reflux ratio, decanter temperature and pressure are also used as the initial guess in the Aspen simulation setup and further adjusted to reach product specifications and minimize condenser/reboiler duties. As a result, Design 1 achieves 18.4% reduction in the heating duty and 18.7% reduction in the cooling duty.

Design 2 is also translated to a Petlyuk column integrated with water decanter. In consistency with the base case Aspen optimization, the target is to identify the corresponding column design with minimized reboiler duty, condenser duty, and number of stages. The main column, with three mass/heat exchange modules and two pure exchange modules (Fig. 12.8), is translated to a 10-stage column (including condenser and reboiler). More specifically, M/H 1 is translated to 2 stages, M/H 2 to 3 stages, and M/H 3 to another 3 stages. The prefractionator section, with one mass/heat exchange module, is identified as a 3-stage column section. As suggested by GMF synthesis structure, MMA feed stream enters the 1st stage in prefractionator. Around 10% of the decanter dewatered organic stream, the rest of the dewatered stream is split equally and fed respectively to Stages 3 and 5 in the main column. The connect stream stages in the main column are the 2nd and the 10th stage. If this Petlyuk column is further translated to a dividing wall column, the divided section should be extended vertically from the 3rd stage to the 9th stage. The operating specifications in Aspen Plus for reflux ratio and distillate rate are respectively specified as 1.3 and 130.0 kmol/h, based on the GMF reflux ratio 1.0 and distillate rate 131.6 kmol/h. Note that, in this design, the distillate stream is directed to the decanter for water recovery and MMA recycle. Therefore, around 2 mol% methanol exists in the aqueous stream from the decanter. By adding a flash column, the water recovery stream can be purified to 99.5 mol%. In all, Design 2 results in 37.4% reduction in the heating duty and 39.0% reduction in the cooling duty.

The equipment-based flowsheet for Design 3 comprises two conventional distillation columns, a water decanter, and a flash column to purify the aqueous stream from decanter. Column 1 consists of 10 stages, with the MMA feed stream entering on Stage 2. Column 2 consists of 7 stages. The bottom stream from Column 1 is fed onto the 4th stage in Column 2 and the decanter dewatered organic stream is all returned to the 5th stage in Column 2. As mentioned earlier, the GMF two-column synthesis provides a 5.6% lower total annualized cost than Design 2 at the trade-off of 7 kmol/h less MMA recovery flowrate. With the GMF suggested design structure, another trade-off solution is found in the Aspen simulation which can reduce 56.5% heating duty and 56.1% cooling duty. However, the MMA product is off-specification with MMA 99.8 wt% and 99.0 mol% while water is at 0.16 wt% exceeding the maximum 0.05 wt% threshold [16]. The process bottleneck for a higher MMA product purity is the insufficient separation of methanol and water from the decanter dewatered organic stream, which are carried then into the MMA product stream from the top of Column 2. An efficient selective separation method to remove the small amount methanol (0.64 mol%) and water (6.21 mol%) from the decanter dewatered organic stream (23 kmol/h), such as membrane-assisted separation [13,19], may help to realize the potential energy savings using the two-column design.

To summarize, two Petlyuk column-based process alternatives, i.e. Design 1 and Design 2, have been identified by process intensification synthesis using GMF and then validated by rigorous steady-state simulation using Aspen Plus to achieve energy savings and equipment size reduction for the MMA purification task. Assuming no heat transfer across the dividing wall, the Petlyuk columns can be converted to dividing wall column designs as shown in Fig. 12.11. The key design and operating considerations in the DWCs have been discussed in the previous section.

Chapter 12 • Process intensification synthesis of dividing wall column systems   203

Table 12.7 Designs 1, 2 and 3 – Results summary for Aspen simulation. (Adapted from Tian et al. [20].)

		Design 1	Design 2	Design 3
Column 1	Number of stages	17	10	10
	Feed stage location	6 (MMARECY)	3 (MMARECY1)	2 (F)
			5 (MMARECY2)	
	Reflux ratio	56	1.3	0.001
	Pressure (atm)	0.06	0.06	0.06
	Vapor split ratio	22.8	1.43	–
	Liquid split ratio	51.7	7.66	–
Column 2	Number of stages	5	3	7
(or Prefractionator)	Feed stage location	1 (MMARECY)	1 (MMARECY)	4 (from COL1)
		3 (F)	1 (F)	5 (MMARECY)
	Reflux ratio	–	–	0.05
	Pressure (atm)	0.06	0.06	0.04
Connect stream 1	Source	COL2, Stage 1	COL2, Stage 1	–
(Vapor)	Destination	COL1, Stage 6	COL1, Stage 2	–
Connect stream 2	Source	COL1, Stage 6	COL1, Stage 2	–
(Liquid)	Destination	COL2, Stage 1	COL2, Stage 1	–
Connect stream 3	Source	COL1, Stage 15	COL1, Stage 10	–
(Vapor)	Destination	COL2, Stage 5	COL2, Stage 3	–
Connect stream 4	Source	COL2, Stage 5	COL2, Stage 3	–
(Liquid)	Destination	COL1, Stage 15	COL1, Stage 10	–
Decanter	Temperature (k)	317.8	323.2	290
	Pressure (atm)	0.23	0.40	0.40
	MMA recovery split ratio	1:3	1:4.5:4.5	–
	(COL2:COL1)			
Energy consumption	Total reboiler duty (kW)	4517.3	3248.0	2387.1
	Total condenser duty (kW)	−4521.4	−3391.8	−2431.9
	Decanter duty (kW)	78.3	193.5	−10.3
	Flash column duty (kW)	–	82.2	59.8
Energy savings	**Heating duty**	**18.4%**	**37.4%**	**56.5%**
(compared to base case)	**Cooling duty**	**18.7%**	**39.0%**	**56.1%**
MMA Product	MMA wt%	99.96	99.80	99.80
	MMA mol%	99.80	99.80	99.04
	Water wt%	0.04	0.02	0.16
	Flowrate (kmol/h)	172.0	172.0	172.0
Water recovery	Purity (mol%)	99.24	99.48	99.52
	Flowrate (kmol/h)	12.2	9.6	10.9

Note: MMARECY – Dewatered MMA-rich outflow from Decanter, F – Feed stream
COL1 – Column 1, COL2 – Column 2 (or Prefractionator)
Heat duty – Positive numbers for heating, Negative numbers for cooling
Tray numbering – Condenser counted as the 1st tray, Reboiler counted as the last tray
Stage efficiency – Theoretical stages are assumed for design with 100% efficiency

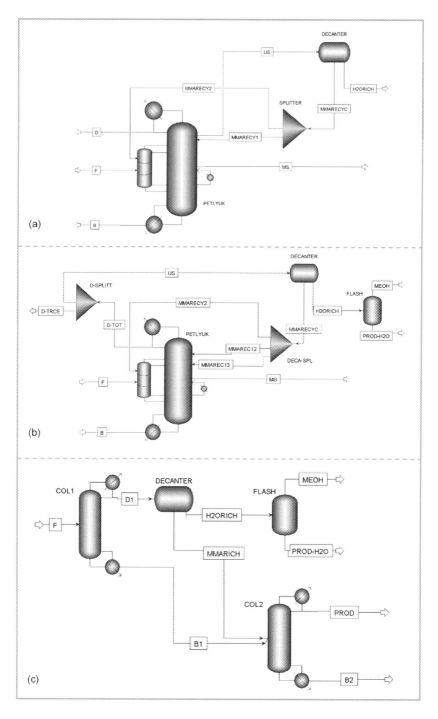

FIGURE 12.10 Aspen simulation flowsheets for Designs 1, 2 and 3 (reproduced from Tian et al. [20]).

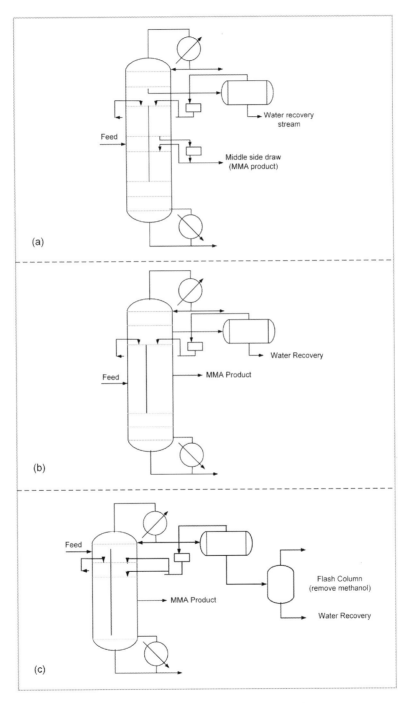

FIGURE 12.11 Dividing wall column designs for the MMA purification case study. (a) Patent design, (b) Design 1, (c) Design 2 (adapted from Tian et al. [20]).

References

[1] I. Dejanović, L. Matijašević, Ž. Olujić, Dividing wall column – a breakthrough towards sustainable distilling, Chemical Engineering and Processing: Process Intensification 49 (2010) 559–580.

[2] R. Sargent, K. Gaminibandara, Optimum design of plate distillation columns, 1976.

[3] J.A. Caballero, I.E. Grossmann, Design of distillation sequences: from conventional to fully thermally coupled distillation systems, Computers & Chemical Engineering 28 (2004) 2307–2329.

[4] G. Dünnebier, C.C. Pantelides, Optimal design of thermally coupled distillation columns, Industrial & Engineering Chemistry Research 38 (1999) 162–176.

[5] R. Agrawal, Synthesis of distillation column configurations for a multicomponent separation, Industrial & Engineering Chemistry Research 35 (1996) 1059–1071.

[6] C.O. Okoli, T.A. Adams, Design of dividing wall columns for butanol recovery in a thermochemical biomass to butanol process, Chemical Engineering and Processing: Process Intensification 95 (2015) 302–316.

[7] E.S. Rawlings, Q. Chen, I.E. Grossmann, J.A. Caballero, Kaibel column: modeling, optimization, and conceptual design of multi-product dividing wall columns, Computers & Chemical Engineering 125 (2019) 31–39.

[8] T. Waltermann, S. Sibbing, M. Skiborowski, Optimization-based design of dividing wall columns with extended and multiple dividing walls for three- and four-product separations, Chemical Engineering and Processing: Process Intensification 146 (2019) 107688.

[9] M.C. Georgiadis, E.N. Pistikopoulos, Energy and Process Integration, Begell House, 2006.

[10] R. Wilczynski, J. Jerrick Juliette, Methacrylic acid and derivatives, Kirk-Othmer Encyclopedia of Chemical Technology (2000).

[11] Y.C. Wu, C. Hsu, H.-P. Huang, I.-L. Chien, Design and control of a methyl methacrylate separation process with a middle decanter, Industrial & Engineering Chemistry Research 50 (2011) 4595–4607.

[12] Y. Bernardin, R. Billon, X. Marcarian, F. Vallet, Unit and process for purification of crude methyl methacrylate, US Patent 10,793,505, 2020.

[13] S. Li, S. Zhou, M. Yu, Process for purification of methyl methacrylate using molecular sieve membranes, US Patent 9,487,469, 2016.

[14] D.W. Jewell, J.G. Pendergast, W.G. Worley, et al., Process for purification of methyl methacrylate, US Patent 10,392,337, 2019.

[15] J. Gmehling, J. Menke, J. Krafczyk, K. Fischer, Azeotropic Data, Wiley, 1994.

[16] The Dow Chemical Company, The Dow Chemical Company sales specification on methyl methacrylate, https://www.dow.com/en-us/pdp.methyl-methacrylate-mma-10-ppm-mehq.154299z.html, 2018. (Accessed 10 April 2021).

[17] W.G. Worley, S.W. Hoy IV, Process for purification of methyl methacrylate, US Patent 10,487,038, 2019.

[18] P. Proios, E.N. Pistikopoulos, Hybrid generalized modular/collocation framework for distillation column synthesis, AIChE Journal 52 (2006) 1038–1056.

[19] I. Kumakiri, K. Hashimoto, Y. Nakagawa, Y. Inoue, Y. Kanehiro, K. Tanaka, H. Kita, Application of fau zeolite membranes to alcohol/acrylate mixture systems, Catalysis Today 236 (2014) 86–91.

[20] Y. Tian, V. Meduri, R. Bindlish, E.N. Pistikopoulos, A Process Intensification Synthesis Framework for the Design of Dividing Wall Column Systems, Computers & Chemical Engineering (2022) 107679.

13

Operability and control analysis in modular process intensification systems

Process intensification (PI) systems can have very different process physics and dynamics from those in well-established conventional processes, normally posing more demanding requirements on process control [1,2]. For example, a key difficulty with process intensification is that it can result in tight integration of tasks with less degrees of freedom (DOFs), narrower operating windows, and typically faster process dynamics. In this chapter, we aim to develop a fundamental understanding of the unique operability and control characteristics in modular and intensified process systems. We will discuss the following key open questions using rigorous model-based analyses:

i. How does the loss of DOFs affect the operation and control of an intensified process compared to its conventional process counterpart?
ii. Operability concerns result from the violation of inequality process constraints during operation (under uncertainty and disturbances). For the role of process constraints, what is the difference between intensified versus conventional processes?
iii. While PI is affected by the loss of DOFs, the numbering up of modular designs can contribute to additional DOFs. What is the trade-off between economics and operability/control in an intensified modular production process?

13.1 Loss of degrees of freedom

The loss of DOFs occurs mostly in PI systems which combine multiple process steps into a single unit or with tight mass/energy integration between units. Note that the basis of comparison is at the process level to design an intensified process or its conventional counterpart process to complete a given production task, instead of comparing a piece of intensified equipment versus a conventional one. To showcase how the DOFs are impacted by process intensification, we give an illustrative example comparing an intensified reactive distillation (RD) process against a conventional reactor-distillation-recycle process as shown in Fig. 13.1.

Synthesis and Operability Strategies for Computer-Aided Modular Process Intensification
https://doi.org/10.1016/B978-0-32-385587-7.00024-5
Copyright © 2022 Elsevier Inc. All rights reserved.

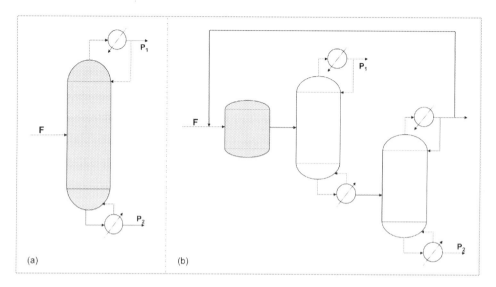

FIGURE 13.1 DOF comparison – (a) Reactive distillation, (b) Reactor-distillation-recycle. (Reproduced from Pistikopoulos et al. [10].)

13.1.1 DOF analysis

To understand how the DOFs change in these two processes at different process design stages, we first perform a general DOF analysis using three types of models: (i) high fidelity dynamic modeling, (ii) steady-state modeling, and (iii) superstructure-based synthesis modeling. The detailed DOF analysis results are presented in Appendix E. Regardless of model types, RD consistently has less degrees of freedom compared to its conventional counterpart. Some general rules can also be summarized which result in the loss of DOFs in such a task-integrated system as reactive distillation:

- The coupling of reaction and separation in a single unit requires reaction and separation to take place at the same temperature and pressure conditions. This mostly affects the thermodynamic DOFs and results in the concern of a reduced operating window.
- The full integration of the reactor and the distillation in a single RD unit, at the same time, converts the interconnecting streams between units (e.g., reactor outlet stream, recycle stream) to internal flows within the unit. Thus, the potential measurement and manipulation of these steams (for example, through valve) are no longer available.
- The reduction in the number of units in RD results in the loss of thermodynamic, design, and operational DOFs.

In addition to these empirical observations, it is worth highlighting a theoretical development from Baldea [3]. The author rigorously proved that, an integrated process, which consists of n units in series with identical material holdup and an infinitely large material recycle stream connecting the last and the first units, is statically equivalent to an

Chapter 13 • Operability and control analysis in modular PI systems   209

intensified process consisting of a single unit with the same material holdup as each integrated unit. However, the dynamics of the intensified process is n times faster than the integrated one, while losing manipulated variables for control in the streams between each unit.

13.1.2 A numerical case study: olefin metathesis

To compare the open-loop and closed-loop dynamic behaviors, in what follows we present a comparative case study for olefin metathesis to produce 2-butene and 3-hexene from 2-pentene. The task is to produce 50 kmol/h of 0.98 mol/mol butene and 50 kmol/h of 0.98 mol/mol hexene at 1 atm, given as raw material a saturated liquid stream of 100 kmol/h pure pentene. The RD column and the reactor-distillation-recycle flowsheet depicted in Fig. 13.1 are modeled using high fidelity dynamic models built in PSE gPROMS ModelBuilder. The generalized distillation model, reactive distillation model, and cost functions can be found in Appendix C. The design and operation parameters, determined via steady-state optimization to minimize total annualized cost, are presented in Table 13.1. All the parameters associated with controller design are summarized in Appendix E.

Table 13.1 Design and operation parameters for olefin metathesis. (Adapted from Pistikopoulos et al. [10].)

	Reactive distillation	Reactor-distillation-recycle		
		Reactor	Distillation 1	Distillation 2
Number of stages	17	/	6	9
Feed tray locations	7 & 12	/	5	3
Reactive volume (m^3)	23.2	51.9	/	/
Diameter (m)	5.0	5.7	6.0	7.7
Pressure (atm)	1	1	1	1
Reflux ratio	3.0	/	5.0	0.46
Capital cost ($\times 10^5$ $)	5.51	6.66	2.56	5.40
Operating cost ($\times 10^5$ $)	2.11	5.39	3.96	5.46
Total annualized cost ($\times 10^5$ $)	3.94	2.76	3.88	7.25

Considering a fixed design configuration, the reactive distillation column has two available operational degrees of freedom, i.e., reflux ratio and distillate flowrate. On the other hand, six operational DOFs are available for the control of the reactor-distillation-recycle process: (i) reactor outlet flowrate controlled via valve stem position, (ii) distillate flowrate and reflux ratio for distillation column 1, (iii) reflux ratio and boilup ratio for column 2, and (iv) recycle ratio. We first perform a series of open-loop analyses on each of the above DOFs (or manipulated variables). The dynamic responses in RD and reactor-distillation-recycle are illustrated in Fig. 13.2.

210 Synthesis and Operability Strategies for Computer-Aided Modular PI

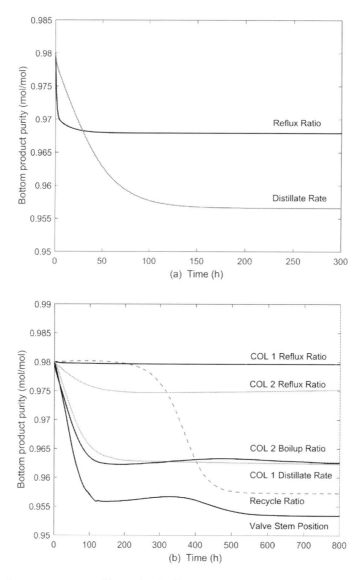

FIGURE 13.2 Open-loop step response – (a) Reactive distillation, (b) Reactor-distillation-recycle. (Adapted from Pistikopoulos et al. [10].)

Two differences can be summarized based on Fig. 13.2: (i) the dynamic response of RD is much faster than that of the reactor-distillation-recycle, and (ii) the reactor-distillation, with significant material recycling, shows a typical two-time-scale behavior in such integrated systems. For example, the response of bottom product purity to a step change in the valve stem position (Fig. 13.2b) clearly exhibits an initial fast transient followed by slow dynamics. However, this dynamic behavior is not observed in the RD process.

For RD closed-loop control, two PI controllers are designed using reflux ratio and distillate rate as manipulated variables to respectively control the top and bottom product purity. Thus, this results in two single-input single-output (SISO) systems. As shown in Fig. 13.3, PI control can meet the requirement of set point tracking to RD top and bottom product purity. Among the six DOFs for reactor-distillation-recycle, two pairing schemes are selected and tested to control the top and bottom product purity with SISO PI controllers. The results are shown in Fig. 13.4. It can be noted that PI control can also perform satisfactorily for set point tracking in this conventional process but only when a good pairing scheme is selected.

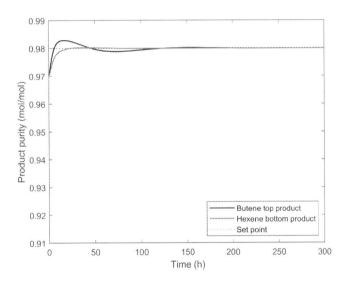

FIGURE 13.3 PI control for set point tracking in RD column (adapted from Pistikopoulos et al. [10]).

13.2 Role of process constraints

From a model-based point of view, any operability, safety, or control concerns can be viewed as the violation of process modeling constraints – typically the inequality ones which define the product specifications, safe operation region, design capacity, etc. These violations can be caused by the existence of disturbances and uncertainties, changes in operating conditions, product specifications, etc. In this section, by analyzing the role of constraints, we compare the operability performance of the reactive distillation process and reactor-distillation-recycle process as a motivating example.

13.2.1 Temperature and pressure bounds

The combination of reaction and separation into a single unit restricts these two tasks taking place under the same temperature and pressure conditions. In this context, the

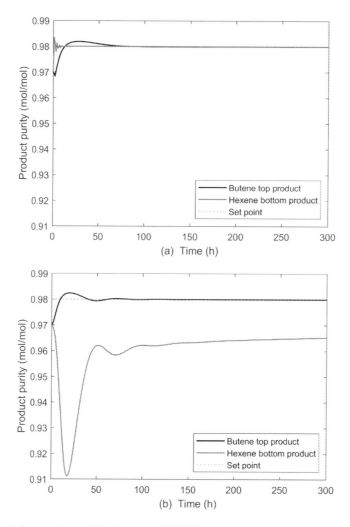

FIGURE 13.4 PI control for set point tracking in reactor-distillation-recycle process: (a) Manipulated variables – COL1 distillate flowrate and COL2 boilup ratio, (b) Manipulated variables – COL1 distillate flowrate and flowsheet recycle ratio (adapted from Pistikopoulos et al. [10]).

operating window of RD is conceptually the *intersection* of the reaction operating window and the distillation operating window. However, the operating window of reactor-distillation-recycle is ideally the *union* of the operating windows of reaction and distillation.

For the olefin metathesis case study – the reaction occurs at atmospheric pressure between 270.15 K to 340.15 K which results in a rather limited operating window in the space of temperature-pressure (T-P). Fig. 13.5 depicts the T-P window for reaction, distillation, and reactive distillation by sampling through the design and operation space for the specific production task. It can be observed that, in this example, RD is faced with a significant

Chapter 13 • Operability and control analysis in modular PI systems 213

reduction of operating window as well as much stricter design and operation constraints, compared to the much more flexible operating window enabled by distillation process.

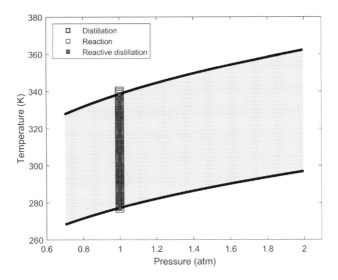

FIGURE 13.5 Temperature-pressure operating window for olefin metathesis. (Adapted from Pistikopoulos et al. [10].)

13.2.2 Flowrate bounds

For the stream flows connecting different units in a process, flowrate bounds are normally given either to avoid unreasonably large flows (e.g., for recycle stream) or to ensure feasible equipment design (e.g., pipe). A notable difference of reactive distillation, compared to the flowsheet of a reactor followed by a train of distillation columns, is the conversion of external connecting streams (e.g., reactor outlet stream, recycle stream) to internal flows within the unit – which emphasizes the impact of design parameters on flowrate bounds. Column diameter has been found to be the most limiting design parameter for distillation operation under uncertainty. A given column diameter can only accommodate a certain range of flowrate uncertainties, beyond which will cause flooding issues. Using the example of olefin metathesis, we compare the impact of column diameter on an allowable feed flowrate range in the RD column and the reactor-distillation-recycle process.

For a given reactive distillation column diameter (or the diameter of Column 1 or 2 in the reactor-distillation-recycle process), we first characterize the "steady-state feasible region" in which the process is feasible for operation and can satisfy the top and bottom product purity specifications under uncertainties in feed flowrate. Then, the "dynamic feasible region" is determined where the process is required to meet process specifications with PI control under time-variant feed flowrate uncertainties. Conceptually, the dynamic feasible region should be a subset of the steady-state region since the latter assumes perfect control. Several observations can be made based on the results depicted in Fig. 13.6:

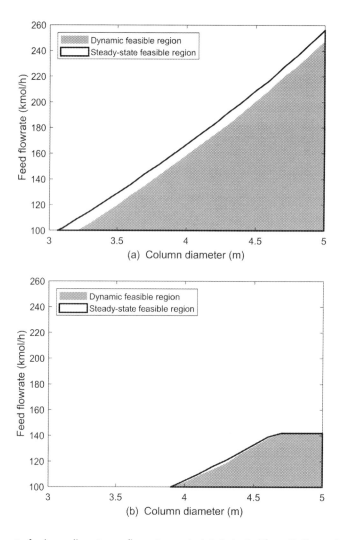

FIGURE 13.6 The impact of column diameter on flowrate constraints (adapted from Pistikopoulos et al. [10]).

- RD can accommodate a larger set of feed flowrate uncertainties than its conventional counterpart at the same over-design factor (i.e., over-design compared to the minimum required column diameter at the nominal feed flowrate). This is because the conventional process has around twice larger external and internal flowrates than RD.
- The reactor-distillation-recycle process reaches a maximum flowrate at 142 kmol/h regardless of the increase of column diameter – because a pre-specified upper bound is met for recycle flowrate at 425 kmol/h compared to the nominal at 300 kmol/h.

- The dynamic feasible region of RD has a notable mismatch with its steady-state feasible region, which can be up to 11 kmol/h. Due to the more complex interaction between process variables in RD, the system is more vulnerable to constraint violation under uncertainties and/or control actions. This emphasizes the need of simultaneous dynamic design and control under uncertainties in such intensified systems.

13.2.3 Process specifications

Given a step change in feed flowrate (+1 kmol/h), Fig. 13.7 shows that RD has a larger deviation from the purity specifications than that in the reactor-distillation-recycle process.

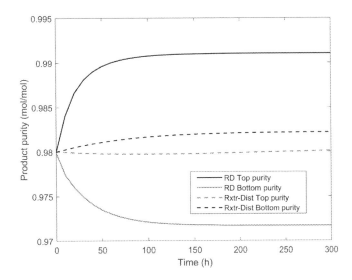

FIGURE 13.7 Open-loop response to feed flowrate step change (adapted from Pistikopoulos et al. [10]).

To test the closed-loop performance of these two processes, a sinusoidal disturbance is considered in the feed flowrate with a period of 24 hours and an amplitude of 8 kmol/h. With the PI controllers on, the control output profiles for RD and reactor-distillation-recycle are given in Fig. 13.8. While the conventional process is well controlled, the RD process has notable off-specifications.

Given the insufficiency of the SISO PI controllers for RD disturbance rejection, another explicit/multi-parametric model predictive controller (mp-MPC) is designed following the PAROC framework (Chapter 8), treating the RD column as a multi-input multi-output (MIMO) process system. The closed-loop performance of both control schemes are shown in Fig. 13.9, indicating better control performance with the mp-MPC controller in terms of meeting product purity specifications under disturbance.

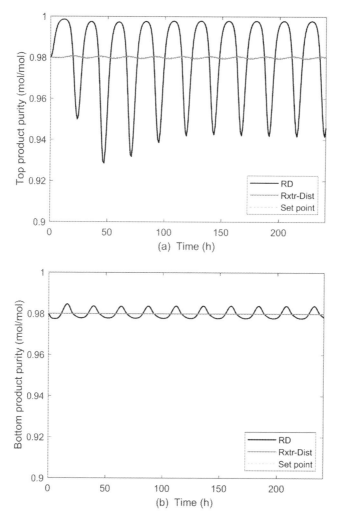

FIGURE 13.8 PI control for disturbance rejection in RD and reactor-distillation-recycle. – (a) Top butene product purity profile, (b) Bottom hexene product purity profile (adapted from Pistikopoulos et al. [10]).

13.3 Numbering up vs. scaling up

In this section, we investigate the impact of modularization on operability and control. We focus on the question if/how the previously mentioned operability and control concerns of an intensified unit can be addressed by introducing extra modular and parallel units (at the expense of more investment cost). Another potential benefit of using modular and parallel units is to enhance process reliability which indicates the probability of a system to perform its designated function over a specified time interval. While this topic is beyond the scope of this chapter, we refer the readers to [4,5] for more information.

Chapter 13 • Operability and control analysis in modular PI systems 217

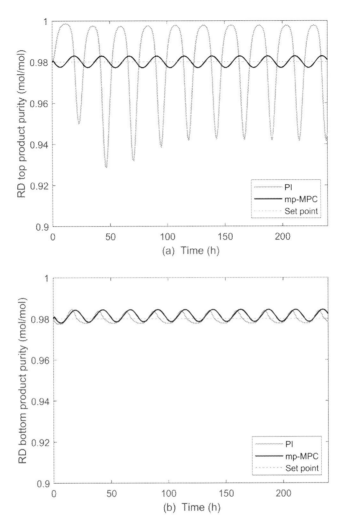

FIGURE 13.9 PI and mp-MPC control for RD disturbance rejection – (a) Top butene product purity profile, (b) Bottom hexene product purity profile (adapted from Pistikopoulos et al. [10]).

Continuing with the example on olefin metathesis, an alternative flowsheet is constructed against the single RD unit which consists of two modular RD units operated in parallel as shown in Fig. 13.10. From economics point of view, the modular flowsheet features a total annualized cost *18.8% higher* than that of the single RD unit.

Recall the discussion on loss of DOFs due to process intensification, modularization on the other hand provides the opportunity to increase DOFs because of the increased number of process units (including also auxiliary units such as mixers and splitters) and interconnecting streams. To test the control performance for rejecting the sinusoidal feed

218 Synthesis and Operability Strategies for Computer-Aided Modular PI

flowrate disturbance in the previous section, PI controllers are designed to control each modular RD unit as per the pairing scheme 1 given in Table 13.2.

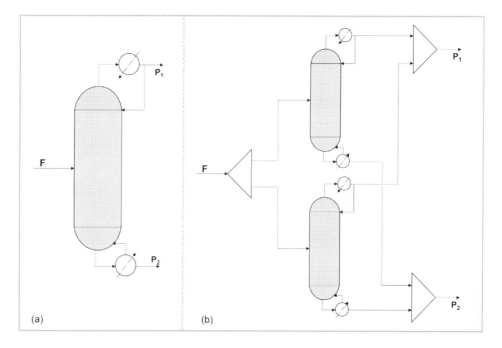

FIGURE 13.10 Modular reactive distillation for olefin metathesis – (a) A single unit, (b) Two modular units. (Adapted from Pistikopoulos et al. [10].)

Table 13.2 PI control pairing schemes for modular RD units. (Reproduced from Pistikopoulos et al. [10].)

	Pairing Scheme 1	
	Input	Output
Controller 1	Feed split ratio	RD1 top product purity
Controller 2	RD1 distillate rate	RD1 bottom product purity
Controller 3	RD2 distillate rate	RD2 top product purity
Controller 4	RD2 reflux ratio	RD2 bottom product purity
	Pairing Scheme 2	
	Input	Output
Controller 1	Feed split ratio	Top product purity after mixing
Controller 2	RD2 reflux ratio	Bottom product purity after mixing
Controller 3	RD2 distillate rate	RD2 top product purity

Chapter 13 • Operability and control analysis in modular PI systems   219

If the two modular RD units adopt the same controller designs, they will function synchronously showing a closed-loop performance similar to the original single RD unit. However, the overall process closed-loop performance may be improved by selecting different pairs of control inputs and outputs as depicted in Fig. 13.11a. A second pairing scheme, as detailed in Table 13.2, can be designed to control units at a process level to ensure that the final products are on specification without monitoring product specifications from individual units. The additional DOFs resulted by modularization can again contribute to a better disturbance rejection performance in terms of the final product purity as shown in Fig. 13.11b. This pairing scheme also brings the benefit of one remainder DOF which can be used to achieve other operational objectives of interest, such as to minimize energy consumption. Note that under both pairing schemes the two modular RD units are operated along different trajectories. This observation indicates further opportunities to explore the design and control optimization of this modular flowsheet by simultaneously considering modular unit design, control structure, and closed-loop performance.

13.4 Remarks

Based on the above dynamic and operating characteristics in PI and modular designs, the following research needs can be identified and highlighted for intensified and/or modular systems (although also important for conventional processes):

- Theoretical developments are necessitated to understand the process dynamics, operability, and control for a wider range of modular and intensified systems (e.g., task-integrated systems, micro-reaction systems, rotating equipment).
- Conceptual design and operational optimization approaches for modular processing systems are needed to systematically analyze the trade-off between module sizing, profitability, and operational flexibility.
- Advanced model-based control will be key enabling tools to ensure the actual operational performances of an integrated process with multiple modular and/or intensified process equipment. Advanced model reduction techniques and dynamic optimization algorithms should also be developed in support of this goal.
- As indicated above, the operational flexibility can be improved with a number of modular designs by enabling synchronized operation or tailed operation for individual units. Thus it opens up questions on how to optimize the operation/control strategies for each of the units and how to perform optimal decision making when the units need to alter operation states in response to changes in production plan.
- The trade-off between the number of control variables vs control efficiency needs to be addressed when the units are numbering up. For example, how to ensure a certain temperature profile across the parallel modular units? In addition to the unit-by-unit control analogous to controlling a single unit operation, indirect control systems can be a good option to reduce the control variables as shown in Hasebe [6] for microreactors.

220 Synthesis and Operability Strategies for Computer-Aided Modular PI

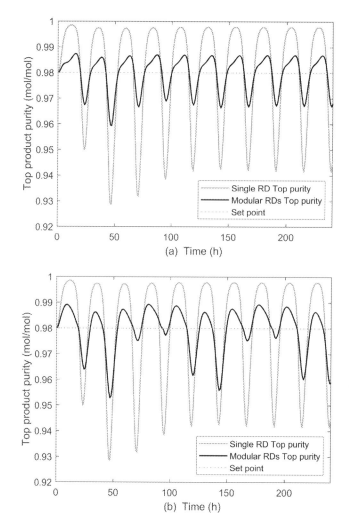

FIGURE 13.11 PI control for modular RD units vs. a single RD. – (a) Pairing scheme 1, (b) Pairing scheme 2 (adapted from Pistikopoulos et al. [10]).

- Distributed decision making is also necessitated which decomposes the large-scale modular intensified process into constituent subsystems with corresponding localized decision making agents [7,8]. The control decisions are then coordinated for the overall process following approaches such as distributed model predictive control [9].
- The integration of operability and control metrics at an early design stage will be beneficial, due to the tight interaction between design and operating parameters in such systems. A holistic framework for computer-aided design, operability analysis, and control optimization of modular PI systems can be highly beneficial.

References

[1] A.I. Stankiewicz, J.A. Moulijn, Process intensification: transforming chemical engineering, Chemical Engineering Progress 96 (2000) 22–34.

[2] N.M. Nikačević, A.E. Huesman, P.M. Van den Hof, A.I. Stankiewicz, Opportunities and challenges for process control in process intensification, Chemical Engineering and Processing: Process Intensification 52 (2012) 1–15.

[3] M. Baldea, From process integration to process intensification, Computers & Chemical Engineering 81 (2015) 104–114.

[4] Y. Ye, I.E. Grossmann, J.M. Pinto, S. Ramaswamy, Modeling for reliability optimization of system design and maintenance based on Markov chain theory, Computers & Chemical Engineering 124 (2019) 381–404.

[5] H.H. Chin, P.S. Varbanov, J.J. Klemeš, M.F.D. Benjamin, R.R. Tan, Asset maintenance optimisation approaches in the chemical and process industries – a review, Chemical Engineering Research and Design 164 (2020) 162–194.

[6] S. Hasebe, Design and operation of micro-chemical plants – bridging the gap between nano, micro and macro technologies, Computers & Chemical Engineering 29 (2004) 57–64.

[7] P. Daoutidis, A. Allman, S. Khatib, M.A. Moharir, M.J. Palys, D.B. Pourkargar, W. Tang, Distributed decision making for intensified process systems, Current Opinion in Chemical Engineering 25 (2019) 75–81.

[8] Y. Shao, V.M. Zavala, Modularity measures: concepts, computation, and applications to manufacturing systems, AIChE Journal (2020) e16965.

[9] P.D. Christofides, R. Scattolini, D.M. de la Pena, J. Liu, Distributed model predictive control: a tutorial review and future research directions, Computers & Chemical Engineering 51 (2013) 21–41.

[10] E.N. Pistikopoulos, Y. Tian, R. Bindlish, Operability and control in process intensification and modular design: Challenges and opportunities, AIChE Journal 67 (2021) e17204.

14

A framework for synthesis of operable and intensified reactive separation systems

In this chapter, we provide a detailed and step-wise tutorial on applying the SYNOPSIS framework (Chapter 9) to synthesize intensified, optimal, and operable methyl tert-butyl ether (MTBE) production systems [10,11]. This chapter is structured as follows:

- Section 14.1 states the process synthesis and design problem to clarify the necessary input information and the expected process outputs delivered by the framework, with particular focus on this MTBE production case study.

- Section 14.2 presents step by step how to synthesize MTBE production process alternatives with considerations of economics, flexibility, inherent safety, and/or control. For each step as summarized below, details are provided on process model setup, results interpretation, cross-validation between steps to ensure framework consistency, etc.
  - Step 1: Process intensification synthesis representation
  - Step 2: Superstructure optimization
  - Step 3: Steady-state design with operability and safety
  - Step 4: Optimal intensified steady-state designs
  - Step 5: Simultaneous design and control optimization
  - Step 6: Verifiable and operable process intensification designs

14.1 Process description

MTBE can be made by the catalytic reaction of isobutylene (IB4) and methanol (MeOH) in the liquid phase with a suitable catalyst as shown by Eq. (14.1):

$$MeOH + IB4 \rightleftharpoons MTBE, \quad \Delta_r H^o_{298K} = -37.7 \, kJ/mol \tag{14.1}$$

Recall the generalized problem definition depicted in Fig. 9.1, the required input information for synthesizing intensified and operable MTBE production systems include:

- **Feed conditions**

  Two feed streams are available, including a liquid feed with pure methanol (MeOH, reactant) and a saturated vapor feed with isobutylene (IB4, reactant) and 1-butene (NB4,

224   Synthesis and Operability Strategies for Computer-Aided Modular PI

inert). The temperatures, pressures, flowrates, and molar compositions of these feed streams are all fixed, as summarized in Table 14.1.

Table 14.1 Summary of feed data (nominal conditions). (Reproduced from Tian and Pistikopoulos [10].)

	Methanol Feed (Liquid)	Butenes Feed (Vapor)	Unit
Temperature	320	350	K
Flowrate	215.5	545	mol/s
Pressure	11	11	atm
x_{MeOH}	1	0	mol/mol
x_{IB4}	0	0.3578	mol/mol
x_{NB4}	0	0.6422	mol/mol
x_{MTBE}	0	0	mol/mol

- **Product specifications**
  The desired product is liquid MTBE with a purity of at least 0.98 mol/mol and a flowrate of at least 197 mol/s.

- **Reaction kinetics**
  An ion-exchange resin, Amberlyst 15, is selected. No side reaction is considered (e.g., the formation of diisobutylene from isobutylene) as this can be eliminated by the selection of proper Amberlyst 15 particles [1]. The intrinsic reaction rate is given by Eqs. (14.2)–(14.4) [1,2]:

$$r = k\left[\frac{a_{IB4}}{a_{MeOH}} - \frac{1}{K_a}\frac{a_{MTBE}}{a_{MeOH}^2}\right] \quad \text{kmol}/(\text{h}\cdot\text{kg cat}) \tag{14.2}$$

$$k = 8.5132 \times 10^{13}\exp\left[\frac{-11,113.78}{T}\right] \quad \text{kmol}/(\text{h}\cdot\text{kg cat}) \tag{14.3}$$

$$\ln K_a = -10.0982 + \frac{4254.05}{T} + 0.2667\ln T \tag{14.4}$$

where r gives the molar reaction rate per unit mass of dry catalyst resin, a denotes the activity of each component, k is the rate constant, and K_a represents the reaction equilibrium constant.

- **Physical properties**
  The liquid mixtures of MeOH, IB4, NB4, and MTBE are highly nonideal. Thus, UNIQUAC equations are employed to calculate the liquid activity coefficients, as shown in Eq. (14.5). The specific UNIQUAC parameters for this mixture system are given in Table 14.2 [1]. The saturated vapor pressures are calculated via the Antoine equation (Eq. 14.6). The component-specific parameters are given in Table 14.3 [3].

Chapter 14 • A framework for operable PI synthesis   225

Table 14.2   UNIQUAC parameters.

Binary Interaction Parameters Int_{ij}, kcal/kmol				
	MeOH	**IB4**	**NB4**	**MTBE**
MeOH	0.0	−70.003	−70.003	−174.94
IB4	1403.5	0.0	0.0	103.73
NB4	1403.5	0.0	0.0	103.73
MTBE	931.43	−48.931	−48.931	0.0
Relative Molecular Volume and Surface Areas				
	MeOH	**IB4**	**NB4**	**MTBE**
r_i	1.4311	2.9195	2.9209	4.0693
q_i	1.432	2.864	2.564	3.556

Table 14.3   Parameters for Antoine equation.

	C1	C2	C3
MeOH	23.49989	3643.31362	−33.434
IB4	20.64556	2125.74886	−33.16
NB4	20.64917	2132.42000	−33.15
MTBE	20.71616	2571.5846	−48.406

UNIQUAC Equation

$$\ln\gamma_i = \ln\gamma_i^c + \ln\gamma_i^r$$

$$\ln\gamma_i^r = q_i\left[1 - \ln\left(\sum_j \theta_j\tau_{ji}\right) - \sum_k\left(\frac{\theta_k\tau_{ik}}{\sum_j\theta_j\tau_{jk}}\right)\right]$$

$$\ln\gamma_i^c = 1 - J_i + \ln J_i - 5q_i\left[1 - \frac{J_i}{L_i} + \ln\left(\frac{J_i}{L_i}\right)\right] \tag{14.5}$$

$$J_i = \frac{r_i}{\sum_j r_j x_j} \qquad L_i = \frac{q_i}{\sum_j q_j x_j}$$

$$\tau_{ij} = \exp\left(\frac{-Int_{ij}}{RT}\right) \qquad \tau_{ii} = 1$$

Antoine Equation

$$\ln P^{sat} = C1 - \frac{C2}{C3+T} \qquad \text{(Pa. K)} \tag{14.6}$$

226   Synthesis and Operability Strategies for Computer-Aided Modular PI

- **Cost data**
  Steam and cooling water are available as utilities, respectively at the price of 137.27 US\$/(kW·yr) and 26.19 US\$/(kW·yr).
  Rigorous equipment cost data will be provided later depending on the identified process solutions via GMF synthesis. The pseudo-capital cost estimation used in GMF is detailed in Section 4.3.

- **Uncertainty in operating conditions**
  Methanol feed flowrate varies within the range of 215.5 ± 10 mol/s.

- **Disturbances**
  A random disturbance exists during operation in IB4 inlet composition within a range of 0.3578 ± 0.05 mol/mol.

- **Hazardous properties**
  The toxicity, flammability, and explosiveness property data for each component are summarized in Table 14.4.

Table 14.4 Hazardous property data. (Adapted from Tian and Pistikopoulos [10].)

	LC_{50} (rat,1h,inh)/mg	Flam	TNT equivalence/kg
MeOH	65.6	Yes	4.62
IB4	155	Yes	2.05
NB4	164.5	Yes	2.03
MTBE	21.3	Yes	2.62

- **Equipment failure frequency**
  The equipment failure frequency data are given in Table 14.5, classified by each type of processing units as documented in the *Handbook of Failure Frequencies* [4].

Table 14.5 Failure frequencies for different types of processing units. (Reproduced from Tian and Pistikopoulos [10].)

	Heat exchanger	Process vessel	Reactor Vessel
$freq/(\text{module} \cdot \text{yr})^{-1}$	5×10^{-5}	5×10^{-6}	5×10^{-6}

The synthesis objectives are to:

- Determine design solution(s) for the above MTBE production process with the optimal economic performance with respect to total annualized cost,

- as well as with feasible and inherently safe operation despite the influence of disturbances and uncertainty.

14.2 Synthesis of intensified and operable MTBE production systems

14.2.1 Step 1: process intensification synthesis representation

This step aims to first validate the physical representation of this MTBE production process using Generalized Modular Representation Framework. Available designs in open literature and/or in industrial practice are simulated with GMF to provide preliminary insights into this process. Then, an enriched and generalized superstructure representation is formulated in preparation for the next optimization step.

The reference design is adapted herein as a reactive distillation (RD) column from Jacobs and Krishna [5] undertaking the same MTBE production task (Fig. 14.1a). The GMF modular representation for this RD column, given in Fig. 14.1b, comprises a total of seven mass/heat exchange modules and two pure heat exchange modules. The rectification section consists of one mass/heat exchange module for pure separation and the stripping section consists of two separation modules, while four reactive separation modules make the reaction zone. The pure heat exchange modules serve as the column's reboiler and condenser, respectively. Under the same operating conditions, the GMF modular configuration fulfills the product requirements (i.e. product flowrate and purity specification).

This GMF simulation study also provides an estimate on how many modules should be used for the next step superstructure optimization (Fig. 14.2). As the number of available modules increases, the quality of the representation via GMF modules is improved while

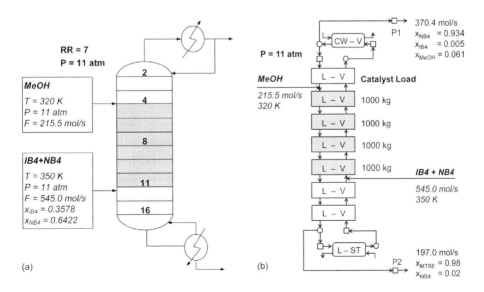

FIGURE 14.1 Simulation of reactive distillation in reference design. (a) Reference design [5], (b) GMF representation. L: Liquid, V: Vapor, CW: Cooling water, ST: Steam, RR: Reflux ratio, P: Pressure.
Shade modules: Reactive separation, Blank modules: Separation. (Adapted from Tian and Pistikopoulos [10].)

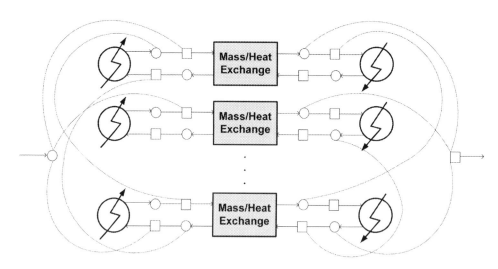

FIGURE 14.2 GMF modular superstructure for process intensification synthesis. (Interconnecting streams simplified for illustration purpose.)

the numerical complexity of the model increases as well. In this context, we first assign a maximum of ten mass/heat exchange modules, with their associated pure heat exchange modules, to be available during the next optimization step. Once this bound is active, the value is relaxed to allow more modules to be used.

In this enriched superstructure representation, it is worth highlighting that – each mass/heat exchange module is not pre-postulated with any underlying reaction and/or separation tasks. Instead, these tasks will be determined through the optimization procedure aiming at the "optimal" performance while their thermodynamic feasibilities are ensured by the driving force constraints (Chapter 3).

14.2.2 Step 2: superstructure optimization

This step aims to synthesize the optimal GMF modular process solution for MTBE production by minimizing total annualized cost (i.e., cost-optimal without operability considerations). Note that the optimization is performed using the full superstructure representation (Fig. 14.2), and not constrained by the initial simulation structure (Fig. 14.1b). The derived GMF process solution is then translated to equipment-based process alternative and validated with rigorous steady-state simulation, e.g. using Aspen Plus.

In addition to the general GMF synthesis model formulation detailed in Chapter 4, some problem-specific constraints are also considered for this case study:

- Vapor flowrate around mass/heat exchange modules are enforced to be less than 3000 mol/s to limit the equipment size [6].
- No more than 2000 kg catalysts can be used in each mass/heat exchange module [7].

Chapter 14 • A framework for operable PI synthesis   229

- Assume constant system pressure with an upper bound of 11 atm to ensure safe operation [8].
- For clarity, mass/heat exchange modules are numbered in a descending order from left to right.

The objective function is formulated as Eq. (14.7):

$$\text{Total Annualized Cost ($/yr)} = \sum_e C_{cw} \times Qc_e + \sum_e C_{steam} \times Qh_e + \sum_e C_{cap,e} \qquad (14.7)$$

where e is the set of GMF mass/heat exchange modules, $C_{cap,e}$ denotes the pseudo-capital cost for each mass/heat exchange module, Qc_e and Qh_e are respectively the heating and cooling duty in each pure heat exchange module. C_{cw} and C_{steam} are the utility cost data respectively for cooling water and steam.

The resulting synthesis model involves 14,594 modeling constraints, 8,098 continuous variables, and 734 binary variables. The mixed-integer nonlinear programming (MINLP) problem is solved with the Generalized Bender Decomposition (GBD) strategy implemented in General Algebraic Modeling System (GAMS), with optimality gap set as 0.01%. More details on GBD solution strategy can be found in Section 4.4.

The initial design structure for the 1st GBD iteration, as suggested by the reference design, comprises seven mass/heat exchange modules and two pure heat exchange modules. However, the reaction/separation phenomena taking place in each module are not pre-postulated. All the interconnection flows between existing modules are also available as optimization variables.

The solution of this MINLP problem by minimizing total annualized cost results in a modular structure as shown in Fig. 14.3. It features the same reactive distillation type of configuration as shown in the previous reference design simulation study. The "column section" includes a separation module in the rectification section, two reactive separation modules in the reaction zone, and another separation module in the stripping section. Note that the reflux ratio, labeled in Fig. 14.3, is not explicitly considered as a variable but calculated posteriorly according to the flowrates of the top product and the rectifying module liquid inlet. This design alternative features a total cost of 8.70×10^5/yr, including a capital cost of 6.10×10^4/yr and a utility cost of 8.09×10^5/yr.

For verification with equipment-based design, the GMF modular layout and operating conditions (i.e., pressure, reflux ratio) are then utilized as design parameters for steady-state simulation of reactive distillation in Aspen Plus using the rigorous distillation module RADFRAC. Simultaneous chemical and physical equilibrium are assumed. The chemical equilibrium is applied in the form of Eq. (14.4). Thermodynamic properties of this system are described via the property set SYSOP11 (UNIQUAC/Redlich-Kwong). Each GMF mass/heat exchange module is translated to a number of column trays, the determination of which is carried out through trial-and-error with the goal of minimizing the total number of column stages in Aspen simulation. However, for this GMF optimal structure (Fig. 14.3), the resulting Aspen reactive distillation column cannot be operated with a reflux ratio of 0.66 since the stripping section will dry up with hardly any vapor flows. A larger

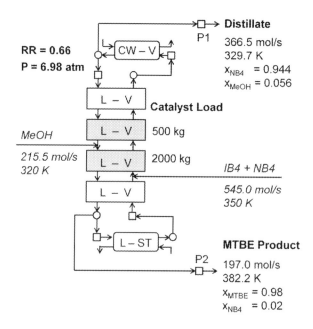

FIGURE 14.3 Optimal GMF design configuration (Adapted from Tian and Pistikopoulos [10]).

reflux ratio, with a rough estimate of 2 obtained from Aspen analysis, is required to ensure a feasible reactive distillation operation and to satisfy the desired process specifications.

For this sake, the reflux ratio is explicitly considered in the GMF synthesis model. A minimum value of 1.7 is set to the reflux ratio, as an illustrative bound to show its impact on the optimal design solution. Starting from the same initial structure, a new optimal solution is generated and the optimization results for this design are shown in Fig. 14.4a. The major design change, resulted by the increase of reflux ratio, lies in an extra separation module in the stripping section. For a comparison purpose, the previous optimal structure is recovered as a feasible solution with constrained reflux ratio (Fig. 14.4b). Although the new optimal configuration has a larger capital cost, it outperforms in terms of hot utility cost. This is due to a smaller liquid inlet flowrate into the heating module, as there are more stripping trays to drive the unreacted MeOH and IB4 back to the reaction zone.

This new optimal modular solution (Fig. 14.4a) is then translated to a reactive distillation column in Aspen Plus: each reactive separation module and the bottom separation module are translated to three column trays while the other separation modules to two column trays respectively, thus featuring a total of 13 column trays (not including reboiler or condenser). The column operating conditions, at this equipment design level, are determined using Aspen optimization, with an objective to minimize operating cost. Column pressure is selected as a manipulated variable due to its effect on the column temperature profile, which affects the reaction rate, physical equilibrium, and condenser/reboiler duty. Reflux ratio is also manipulated to meet process specifications and to adjust condenser/re-

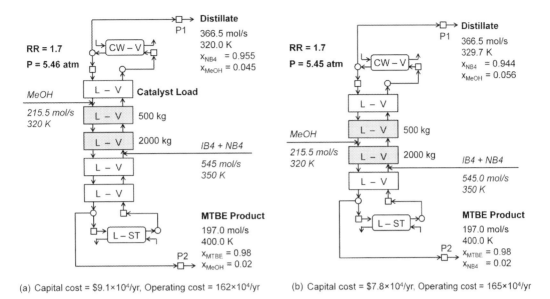

FIGURE 14.4 GMF design alternatives with consideration of reflux ratio. (a) optimal solution, (b) feasible solution. (Adapted from Tian and Pistikopoulos [10].)

boiler duty. More rigorous optimization of the resulting reactive distillation column, to account for more design variables (e.g., column tray numbers, feed/reactive tray location, column diameter), will be addressed in a later step of this framework via dynamic optimization using high fidelity process models.

Fig. 14.5 presents the optimal operating conditions for the resulting reactive distillation. Fig. 14.6 shows the comparison of column temperature and liquid compositions between GMF synthesis and Aspen simulation. Although tray-wise details are lost within the mass/heat exchange modules, the trends of mass and heat exchange in the 15-tray reactive distillation column are well captured using only five GMF modules.

Several notes can be made here regarding this process intensification synthesis step:

i. The observed "intensification" of reflux ratio suggested by GMF synthesis aligns with the findings in Luyben and Hendershot [9] that process intensification tends to drive the minimization of liquid holdups in reflux drum and column base, through the reduction of reflux ratio, for savings in investment and utility costs but simultaneously may cause problems from operational perspective. This necessitates the integrated framework to ensure the consistency between phenomenological synthesis, rigorous unit operation design, and operational analysis for the delivery of verifiable intensified designs.

ii. Despite this, the GMF representation is able to capture the combined reaction/separation phenomena and to suggest optimal intensified process configuration for MTBE production without a pre-postulation of plausible units/tasks. Actually, the first optimization problem, without any explicit consideration of reflux ratio, contributes to

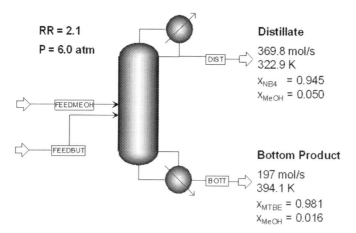

FIGURE 14.5 Aspen validation of optimal MTBE reactive distillation (adapted from Tian and Pistikopoulos [10]).

identify a set of tasks for this process (i.e., separation and reactive separation). Then more detailed optimization, incorporating operational perspectives, is carried out to bridge the gap from task to unit operation.

iii. Although GMF is built on the exploitation of thermodynamic space, we do not assume chemical or physical equilibrium. Table 14.6 presents the driving force values of $G2$ for individual component in each module. As introduced in Section 3.3, $G2$ denotes how far the process is from the equilibrium. If it equals to 0, the process is approximately at its thermodynamic equilibrium. In Table 14.6, the magnitude of $G2$ indicates that reaction takes a more significant role in driving this process compared to separation.

Table 14.6 $G2$ values for each mass/heat exchange module. (Adapted from Tian and Pistikopoulos [10].)

M/H Module	Task	MeOH	IB4	NB4	MTBE
1	Separation	−0.12	0	0.07	0
2	Reactive Separation	0	−17.2	0.004	0
3	Reactive Separation	30.3	−20	−0.53	0
4	Separation	0	0.002	0	−2.97
5	Separation	−0.038	0.002	0.002	−3.93

iv. Although capital cost can be readily incorporated into the cost objective in Aspen optimization, the current objective function formulation in Aspen Plus only considers operating cost, for the reasons that: (a) operating cost constitutes the dominant driving force in the conceptual design of distillation columns; (b) the major goal is to ensure the feasibility from modular representation to equipment-based flowsheet; and (c) the optimal design parameters will be fully investigated in a later step of simultaneous design and control.

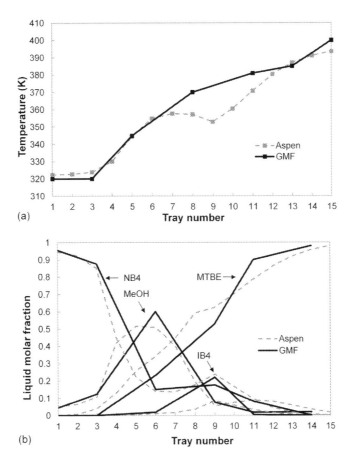

FIGURE 14.6 Synthesis results validation – (a) GMF modular results vs. (b) equilibrium simulation. (Adapted from Tian and Pistikopoulos [10].)

14.2.3 Step 3: integrated design with operability and safety

After deriving the cost-optimal GMF modular design from Step 2 (Fig. 14.4a), this step aims to verify, or improve if necessary, its operability and inherent safety performances via the integration of model-based metrics such as flexibility analysis and risk analysis.

14.2.3.1 Risk analysis for inherent safety performance evaluation

Risk analysis, as detailed in Section 7.4, evaluates the process inherent safety performance based on equipment failure frequency and consequence severity. For this MTBE production process, three types of risks are considered (i.e., toxicity risk, flammability risk, and explosiveness risk). The corresponding hazardous properties of MeOH, IB4, NB4, and MTBE are given in Table 14.4 as part of the process input data. Following the risk calculation procedure, the risk values for each mass/heat exchange module in the above cost-optimal design (hereafter referred as "nominal design") are summarized in Table 14.7. In what follows, these risk values will be used as base case to be improved.

Table 14.7 Toxicity, flammability, and explosiveness risk values for nominal design.

Module	Task	$Risk_{tox}$	$Risk_{flam}$	$Risk_{expl}$
1	Separation	0.089	3.7×10^{-4}	0.002
2	Reaction/Separation	0.083	3.8×10^{-4}	0.002
3	Reaction/Separation	0.097	4.5×10^{-4}	0.002
4	Separation	0.036	1.7×10^{-4}	7.3×10^{-4}
5	Separation	0.033	1.5×10^{-4}	6.3×10^{-4}
Sum		0.338	1.5×10^{-3}	0.007

The first target towards an inherently safer design is set to reduce the overall process risk, for each type of risk considered, by at least 20% compared to that of the nominal design (i.e. $R_{risk}^{overall} \leq 20\% \cdot R_{risk}^{overall,norm}$). The overall process risk is considered as a sum of the individual module risks. Incorporating this constraint into the GMF synthesis model, the cost-optimal and inherently safer design is illustrated in Fig. 14.7a, which actually recovers the modular structure of the cost-optimal design without reflux ratio constraint (Fig. 14.3). **The overall process risk is reduced in an intuitive way by reducing the number of modules in the process, which implies that a smaller process unit can be inherently safer.**

However, it is noted that the 2nd reactive separation module, numbered from top to bottom in a descending order, takes up more than 1/3 of the overall process risk, which

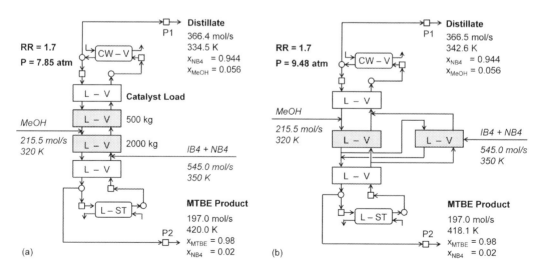

FIGURE 14.7 Inherently safer GMF design alternatives. (a) reduced process risk, (b) reduced process risk & constrained module risk (adapted from Tian and Pistikopoulos [10].)

Chapter 14 • A framework for operable PI synthesis   235

makes it a comparatively more risky component in this design. To address this issue, the risk of an individual mass/heat exchange module is further constrained to be no more than 30% of the overall process risk (i.e. $R_{risk}^{mod} \leq 30\% \cdot R_{risk}^{overall}$). **This results in the solution illustrated in Fig. 14.7b, where a bypass stream is added to introduce more degrees of freedom for design, thus to release the mass/heat exchange burden of the 2nd reactive separation module.**

14.2.3.2 Flexibility analysis for operation under uncertainty

Considering the uncertain methanol feed flowrate in the range of 215.5 ± 10 kmol/s, flexibility test (Chapter 6) identifies the critical point as $f_{MeOH}^l = 225.5$ mol/s. The multiperiod MINLP model for integrated design with flexibility and inherent safety results in the same optimal structure configurations as shown in Fig. 14.7. However, **an increase of module diameters is required to accommodate the uncertainty in flowrates.**

A summary of the resulting GMF designs is presented in Table 14.8, with different cost, safety, and operability performances. **The results indicate that operability and safety considerations can result in significant structure changes in optimal design solutions.**

Table 14.8 Summary of different GMF design options. (Adapted from Tian and Pistikopoulos [10].)

	Flexibility	Reduced overall process risk	Constrained individual module risk	Total annualized cost ($\times 10^6$ $/yr)
Nominal Design (Fig. 14.4a)				1.71
Operable Design 1 (Fig. 14.7a)	✓	✓		1.82
Operable Design 2 (Fig. 14.7b)	✓	✓	✓	1.90

14.2.4 Step 4: optimal and operable intensified steady-state designs

This step concludes the steady-state synthesis and design of MTBE production systems. The above derived GMF nominal design, together with two operable designs, are translated to equipment-based process alternatives and validated with rigorous steady-state simulation.

In Step 2, we have translated and validated the GMF nominal design (i.e., cost-optimal without operability/safety considerations) as a reactive distillation column with design and operating conditions summarized in Fig. 14.5. In what follows, we discuss how to identify the two operable GMF modular solutions as equipment-based process alternatives.

Operable Design 1

This GMF design (Fig. 14.7a) is translated, in a similar manner as that for Fig. 14.4a, as a reactive distillation column. A total of 11 column trays (not including reboiler or condenser) are used, with the simulation results illustrated in Fig. 14.8a.

236 Synthesis and Operability Strategies for Computer-Aided Modular PI

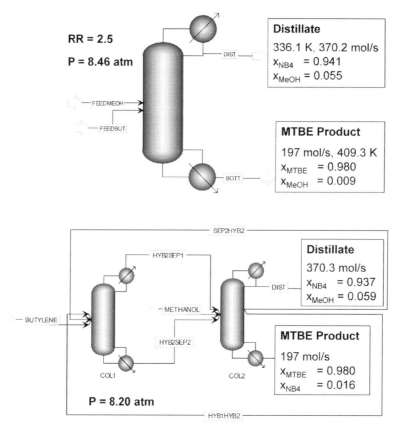

FIGURE 14.8 Aspen validation of steady-state optimal operable designs.
(a) Operable Design 1, (b) Operable Design 2 (adapted from Tian and Pistikopoulos [10].)

Operable Design 2

This design solution (Fig. 14.7b) can be identified as a reactive Petlyuk column: (i) the major part of the design, excluding the second reactive separation module, can be translated as a reactive distillation column; (ii) the second reactive separation module can be translated as a side reactive column which is fully thermally coupled with the main column.

As for the equipment-based validation, existing distillation column modules in Aspen Plus cannot perform reactive Petlyuk column simulation. Thus two RADFRAC columns, i.e. the side column (COL1) and the main column (COL2), are used and integrated via interconnecting streams (Fig. 14.8b). The main column comprises 8 trays in the column section with the 3rd tray to the 5th tray constituting the reactive zone, while the side column has 3 trays, all reactive. The interconnecting streams in the GMF modular design are placed at the corresponding tray (for example, a liquid connecting stream leaves COL2 at the 6th tray to the 1st tray in COL1). Note that the condenser and reboiler are not used in the side column design. The detailed simulation results are presented in Fig. 14.8b.

A detailed comparison of GMF synthesis results and the corresponding Aspen simulation results are summarized in Table 14.9. The resulting three reactive distillation-based designs obtained from steady-state design are depicted in Fig. 14.9.

Table 14.9 Comparison of GMF synthesis and Aspen validation. (Adapted from Tian et al. [11].)

	Nominal Design		Operable Design 1		Operable Design 2	
	GMF	Aspen	GMF	Aspen	GMF	Aspen
Pressure (atm)	5.46	6.00	7.85	7.95	9.48	8.20
Reflux ratio	1.70	2.10	1.70	2.50	1.70	3.30
Reboiler duty (MW)	7.5	6.6	8.4	9.6	8.9	20
Condenser duty (MW)	23	22	23	24	23	34
Module/Tray number	7	15	6	13	6	10+3[a]
Product flowrate (mol/s)	197.0	197.0	197.0	197.0	197.0	197.0
Product purity (mol/mol)	0.98	0.98	0.98	0.98	0.98	0.98

[a] Note: Main column – 10 trays, Side column – 3 trays. Reboiler and condenser are included in the tray numbering.

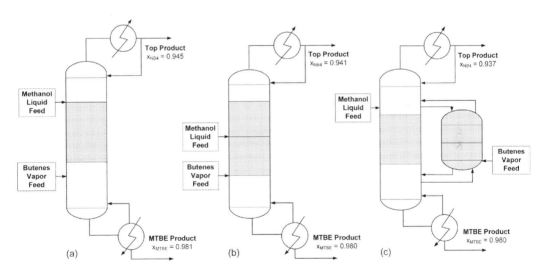

FIGURE 14.9 Equipment-based flowsheet alternatives for MTBE production. (Shaded section: Reactive zone.) (Reproduced from Tian et al. [11].)

14.2.5 Step 5: simultaneous design and control optimization

This step aims to take the steady-state process solutions to dynamic analysis, design, and control optimization. Specifically, the following tasks are included: (i) high fidelity modeling and simulation, (ii) open-loop analysis and validation with flexibility and safety considerations; (iii) explicit/multi-parametric model predictive control, and (iv) simultaneous design

238 Synthesis and Operability Strategies for Computer-Aided Modular PI

and control optimization to close the loop for delivering optimal design solutions with optimal dynamic operation strategies.

14.2.5.1 Dynamic high fidelity modeling and simulation

Dynamic high fidelity models for the above three MTBE production systems (Fig. 14.9) are developed based on a general reactive distillation model developed by Schenk et al. [8]. The key features of the model, mathematical formulation, and simulation setup are detailed in Appendix C. This task takes place in PSE gPROMS ModelBuilder. Physical properties (e.g., activity, fugacity, enthalpy, saturated pressure) are calculated using the MultiFlash thermodynamic package.

On converting the steady-state design in Aspen Plus to dynamic simulation, a major difference in the modeling assumption is that – the steady-state model assumes physical and chemical equilibrium (Table 14.9) while the high fidelity dynamic model utilizes rate-based reaction kinetic calculation to reflect the actual kinetic-controlled characteristics in MTBE reactive distillation. This results in more column trays in the dynamic reactive distillation column design as shown in Table 14.10. The design and operating parameters of Nominal Design, Operable Design 1, and Operable Design 2 are summarized in Table 14.10. Model statistics can be found in Table 14.11.

Table 14.10 Design and operating parameters for dynamic simulation. (Adapted from Tian et al. [11].)

	Nominal Design	Operable Design 1	Operable Design 2
Number of stages	17	15	11 (main)
			3 (side)
Reactive stages	4-11	4-11	2-6 (main)
			1-3 (side)
Pressure (atm)	6.0	7.9	8.2
Reflux ratio	2.11	2.75	4.00
Reboiler duty (MW)	6.84	11.4	20.0
Condenser duty (MW)	21.8	25.4	33.9
MTBE purity (mol/mol)	0.98	0.98	0.98
Product flowrate (mol/s)	197	197	197

Note: *main* – Main column, *side* – Side column or prefractionator.
Reboiler and condenser are included in the tray numbering.

Table 14.11 Model statistics for dynamic simulation.

	Nominal Design	Operable Design 1	Operable Design 2
Modeling equations	1098	962	900
Initial conditions	60	52	48
Algebraic variables	1038	910	852
Differential variables	60	52	48

14.2.5.2 Open-loop analysis with flexibility and safety considerations

The open-loop analysis aims to verify if the derived dynamic MTBE production systems are sustaining their desired level of operability and inherent safety as promised by steady-state design with flexibility and risk considerations (Step 3).

Risk analysis for inherent safety assessment

The model-based risk analysis is incorporated in the dynamic models to calculate process risk values as an "inherent safety indicator". Since the reduction of process risk is specified on a comparative basis, we define a metric, i.e. "Risk Ratio", with a scale of 0-1. As a base case, the Nominal Design has a Risk Ratio of 1. Thereby the Risk Ratios of Operable Designs 1 and 2 are expected to be around 0.8 for consistency with the steady-state synthesis with safety considerations (i.e., reducing at least 20% of process risk). The Risk Ratios calculated from dynamic simulation are 0.91 and 0.81 respectively for Operable Designs 1 and 2. Both of the designs are inherent safer than Nominal Design, although Design 1 not fully achieving the desired inherent safety level. However, since the identification and translation of the equipment-based process designs from phenomena-based synthesis are based on "trial-and-error" attempts, we will later incorporate risk calculation in dynamic optimization to maintain process risks at the desired level.

Flexibility analysis for operation under uncertainty

Regarding flexibility considerations at steady-state, Nominal Design is not flexible over the uncertainty range of methanol feed flowrate (i.e. 205.5 mol/s – 225.5 mol/s), while Design 1 and 2 are derived to be flexible using the multiperiod GMF synthesis approach with flexibility test. To verify the operation under uncertainty at this dynamic stage, the feasible regions of the MTBE production systems are depicted in Fig. 14.10. We also analyze these designs with respect to the disturbance in the IB4 inlet composition, which will be introduced later for control investigations.

As can be noted from Fig. 14.10, Nominal Design is not feasible over the entire range of uncertainty or disturbance while Operable Designs 1 and 2 are, which is consistent with the results obtained via steady-state design with operability. In this regard, only Operable Designs 1 and 2 will be considered in the following steps for controller design and dynamic optimization.

14.2.5.3 Explicit/multi-parametric model predictive control

In this task, we develop explicit/multi-parametric model predictive control (mp-MPC) strategies for Operable Designs 1 and 2 following the PAROC (PARametric Optimization and Control) framework, to maintain stable and operable conditions in the presence of process disturbances. More details on the PAROC framework can be found in Chapter 8.

To set up the control problem, two sets of disturbances exist during system operation: (i) a disturbance in the methanol liquid feed flowrate, and (ii) a disturbance in the IB4 inlet composition. The MTBE molar composition in the bottom product is treated as the control variable with a desired set point of 0.98 mol/mol. The vapor molar flowrate from the

240 Synthesis and Operability Strategies for Computer-Aided Modular PI

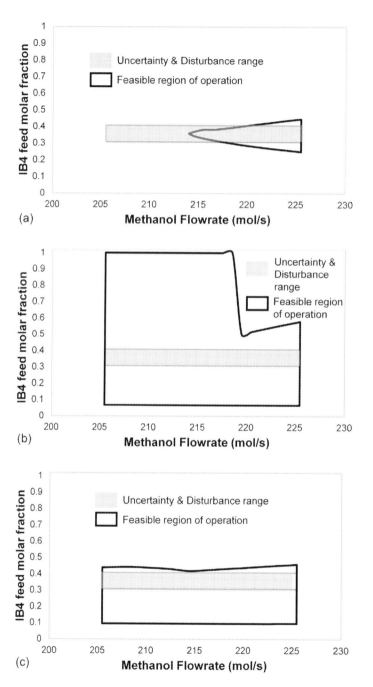

FIGURE 14.10 Process operating window vs. uncertainty range. (a) Nominal Design, (b) Operable Design 1, (c) Operable Design 2. (Adapted from Tian et al. [11].)

Chapter 14 • A framework for operable PI synthesis 241

first tray is also selected as a measured output to monitor the system operation. For the selection of manipulated variable, input step changes are imposed and the vapor butenes flowrate shows the most impact on the control output. It should be noted that the relationships between the manipulated and the controlled variables can be complex with possible gain inversion due to the nonlinearities that exist in the model formulation. More detailed analyses can be found in Appendix F. In this context, both Operable Designs 1 and 2 are single input single output (SISO) systems. The steps for mp-MPC controller design are introduced in what follows:

(i) Model approximation

Although the high fidelity model of the MTBE production systems developed in Step 5.1, although it accurately captures the process dynamics, it is challenging to be integrated within a dynamic optimization formulation due to its model complexity. Moreover, it has been shown that in an explicit MPC framework, model approximation can be applied to reduce the numerical complexity while maintaining the desired level of prediction accuracy. The MATLAB® System Identification Toolbox is utilized to develop the reduced order models for the two operable process solutions (i.e. Operable Designs 1 & 2), using random input-output data based on the original high fidelity model. To recap the general structure of the approximated linear state-space model (Eq. (14.8)):

$$\overline{x}_{k+1} = A\overline{x}_k + Bu_k + Cd_k \tag{14.8a}$$

$$y_k = D\overline{x}_k \tag{14.8b}$$

where the matrices A, B, C, D define the state-space model, the index k denotes the current time instant, $\overline{x}$ is the vector of state variables (pseudo-states which do not provide physical meanings due to the model reduction), u is the vector of manipulated variables, y represents the vector of control variables, and d is the vector of disturbances. Y and De will also be included respectively as binary structural variables and design variables if for design-aware controller design. The physical meanings for each type of variables are explained in Table 14.12 which are applicable to Operable Designs 1 and 2.

The resulting discrete linear state-space models for Operable Designs 1 and 2 are summarized in what follows:

Operable Design 1

A discretization step, $T_s = 10$ s, is selected. The discrete reduced order model consists of 6 state variables. The parameter matrices are:

$$A = \begin{bmatrix} 0.9898 & 0.0084 & -0.0499 & 0.0024 & 0.0362 & -0.1017 \\ -0.0668 & 0.8165 & 0.0670 & 0.0848 & 0.0179 & -0.0244 \\ 0.1875 & -0.0313 & 0.6780 & -0.1176 & 0.1388 & -1.0977 \\ 0.0229 & 0.4052 & -0.1391 & 0.4649 & -0.4364 & 1.9624 \\ 0.0159 & -0.0877 & 0.1333 & -0.1057 & 0.3208 & 0.4096 \\ -0.0001 & 0.0219 & 0.0457 & -0.0426 & -0.4609 & -0.6961 \end{bmatrix}$$

242　Synthesis and Operability Strategies for Computer-Aided Modular PI

Table 14.12 Types of variables for controller design.

Symbol	Definition	Variables	Physical description
$x(t)$	State variables	$M_{i,k}, i = 1, ..., NC$ $k = 1, ..., N$	Molar holdups
$y(t)$	Control variables	$x_{MTBE}^{product}$ V_1	MTBE composition in product Vapor flowrate from Tray 1 (for monitoring)
$u(t)$	Manipulated variable	F_{feed}^{vap}	Vapor feed flowrate
$d(t)$	Disturbances	F_{feed}^{liq} $x_{feed,IB4}$	Liquid feed flowrate IB4 composition in liquid feed
Y	Binary variables	$yf_{feed,k}, k = 1, ..., N$ $yr_k, k = 1, ..., N$	Feed tray structure Reflux tray structure
De	Design variables	D_{col} $Catalyst_k, k = 1, ..., N$	Column diameter Catalyst load

$$B = \begin{bmatrix} 0.0005 \\ 0.0058 \\ -0.0023 \\ -0.0336 \\ -0.0098 \\ -0.0025 \end{bmatrix} \quad C = \begin{bmatrix} 0.0006 & 0.3367 \\ -0.0044 & 4.4020 \\ 0.0096 & -4.4309 \\ 0.0233 & -27.3449 \\ 0.0123 & -11.7250 \\ 0.0114 & -5.7239 \end{bmatrix} \tag{14.9}$$

$$D = \begin{bmatrix} 0.3025 & -0.0015 & -0.0079 & 0.0005 & 0.0007 & -0.0002 \\ -83.0519 & 585.6174 & 100.1365 & 57.3670 & 6.1853 & -0.4021 \end{bmatrix}$$

Operable Design 2

The discretization step is also selected as $T_s = 10$ s. The discrete reduced order model consists of 2 state variables with the following parameter matrices:

$$A = \begin{bmatrix} 0.9587 & -0.0001 \\ 0.0009 & 0.9581 \end{bmatrix} \quad B = \begin{bmatrix} 37.8345 \\ 233.4201 \end{bmatrix}$$

$$C = \begin{bmatrix} 158.6232 & -85.3915 \\ 0.4062 & -0.0658 \end{bmatrix} \quad D = \begin{bmatrix} 0.4062 & -0.0658 \end{bmatrix} \tag{14.10}$$

(ii) Explicit model predictive controller design

In deriving mp-MPC control strategies, the objective is to express the optimal control actions as explicit functions of the parameters of the system. In the case of Operable Designs 1 and 2, the different types of variables involved in controller design are shown in Table 14.12. The explicit model predictive control strategy is constructed based on the approximated model presented in Eq. (14.8). The control objective is to achieve a bottom MTBE molar fraction of 0.98 mol/mol by manipulating the vapor butenes feed flowrate. In addition, to account for the restriction of the boil up flowrate V_b by the column diameter

Chapter 14 • A framework for operable PI synthesis   243

D_{col}, the following constraint is imposed in the control problem formulation:

$$0.04V_b \leq D_{col}^2 \tag{14.11}$$

Table 14.13 gives the controller tuning parameters for Operable Designs 1 and 2. The resulting mp-MPC problem is solved in the MATLAB/Parametric OPtimization (POP) Toolbox. The optimal map of solutions for Operable Design 1 includes 17 critical regions described by the corresponding active sets, and 126 critical regions for Operable Design 2.

Table 14.13 Design-aware mp-MPC controller tuning parameters (adapted from Tian et al. [11]).

MPC Parameters	Operable Design 1	Operable Design 2
OH	2	4
CH	1	3
QR	$5E6$	$1E5$
R	$1E5$	$1E5$
u_{min}	490	490
u_{max}	750	750
y_{min}	$[0 \quad 0]^{\mathsf{T}}$	$[0 \quad 0]^{\mathsf{T}}$
y_{max}	$[1 \quad 1600]^{\mathsf{T}}$	$[1 \quad 1600]^{\mathsf{T}}$
d_{min}	$[205.5 \quad 0.3]^{\mathsf{T}}$	$[205.5 \quad 0.3]^{\mathsf{T}}$
d_{max}	$[235.5 \quad 0.4078]^{\mathsf{T}}$	$[235.5 \quad 0.4078]^{\mathsf{T}}$

(iii) Closed-loop validation of mp-MPC control

The closed-loop performance of the above derived mp-MPC controllers against the original high fidelity model is validated as shown in Fig. 14.11. In the case that the controller performances are not satisfying operational expectations, the model approximation step needs to be re-visited to generate better reduced order models and to re-design mp-MPC controllers.

14.2.5.4 Simultaneous design and control optimization

As discussed in Section 8.3, a mixed-integer dynamic optimization problem can be formulated to integrate design and control, the solution of which allows us to derive explicit closed loop strategies that maintain economic, stable, and operable conditions in the presence of process disturbances. The mixed-integer dynamic optimization formulation is set up in gPROMS as shown in Eq. (14.12). The optimization objective is given by Eq. (14.12a) to minimize total annual cost as calculated in Appendix C. Eqs. (14.12b) and (14.12c) describe the high fidelity model presented in Appendix C. Eq. (14.12d) integrates the derived explicit control actions. Eq. (14.12e) defines the parameter space from multi-parametric programming point of view. The variable bounds are given in Eqs. (14.12f)–(14.12h). Eq. (14.12i) defines Y as binary variables.

$$\min_{Y, De} F = \int_0^\tau Annualized\ Cost(x, y, u, Y, d, De)dt \tag{14.12a}$$

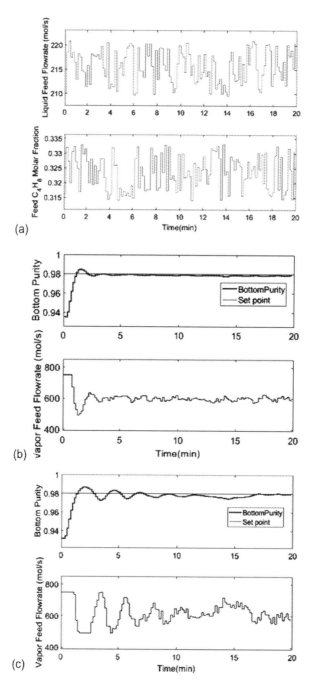

FIGURE 14.11 Closed-loop validation of mp-MPC control. (a) Disturbance profile, (b) Operable Design 1 output-input profile, (c) Operable Design 2 output-input profile. (Adapted from Tian et al. [11].)

$$\text{s.t. } \frac{dx}{dt} = f(x, y, u, Y, d, De) \tag{14.12b}$$

$$y = h(x, u_c, Y, d, De) \tag{14.12c}$$

$$u_T = K_i \theta_T + r_i, \ \theta_T \in CR^i = \{CR_i^A \theta \le CR_i^b\} \tag{14.12d}$$

$$\theta_T = [x_T, y_T, y_T^{sp}, Y, d_T, De, u_{T-1}] \tag{14.12e}$$

$$\underline{y} \le y \le \overline{y} \tag{14.12f}$$

$$\left[\underline{x}^T \ \underline{d}^T\right]^T \le \left[x^T \ d^T \ \right]^T \le \left[\overline{x}^T \ \overline{d}^T\right]^T \tag{14.12g}$$

$$\underline{De} \le De \le \overline{De} \tag{14.12h}$$

$$Y \in \{0, 1\}^q \tag{14.12i}$$

The mp-MPC integrated dynamic optimization results for the MTBE production systems are summarized in Table 14.14. Also note that the Risk Ratio of both designs are kept under 80% of the Nominal Design, thus consistent with the steady-state synthesis operability and safety promises.

14.2.6 Step 6: verifiable and operable process intensification designs

Up to this stage, two designs (Fig. 14.12) have been generated for the MTBE production task of interest. The trade-offs between their cost, operability, safety, and control performances have been thoroughly investigated as shown in Table 14.14, which can be used to assist further decision making.

Table 14.14 Simultaneous design and control optimization. (Adapted from Tian et al. [11].)

	Operable Design 1	Operable Design 2
Column Diameter (m)	2.0	2.3
Number of Trays	15	11 (main)
		3 (side)
Feed Tray Location 1	10	3 (side)
Feed Tray Location 2	7	3 (main)
Catalyst Mass (ton)	4.4	6.5
Column Cost ($/yr)	0.042×10^6	0.090×10^6
Catalyst Cost ($/yr)	0.076×10^6	0.113×10^6
Operating Cost ($/yr)	2.290×10^6	3.397×10^6
Total Annual Cost ($/yr)	2.408×10^6	3.601×10^6
Risk Ratio	0.79	0.78

Note: *main* – Main column, *side* – Side column or prefractionator.
Reboiler and condenser are included in the tray numbering.

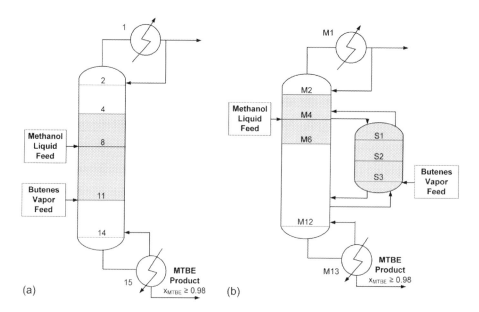

FIGURE 14.12 Operable and intensified process design alternatives. (a) Operable Design 1, (b) Operable Design 2. Shaded trays: reactive zone, M: main column, S: side column. (Reproduced from Tian et al. [11].)

References

[1] A. Rehfinger, U. Hoffmann, Kinetics of methyl tertiary butyl ether liquid phase synthesis catalyzed by ion exchange resin – I. Intrinsic rate expression in liquid phase activities, Chemical Engineering Science 45 (1990) 1605–1617.
[2] F. Colombo, L. Cori, L. Dalloro, P. Delogu, Equilibrium constant for the methyl tert-butyl ether liquid-phase synthesis using UNIFAC, Industrial & Engineering Chemistry Fundamentals 22 (1983) 219–223.
[3] S.R. Ismail, P. Proios, E.N. Pistikopoulos, Modular synthesis framework for combined separation/reaction systems, AIChE Journal 47 (2001) 629–649.
[4] Flemish Government, Handbook failure frequencies 2009 for drawing up a safety report, 2009.
[5] R. Jacobs, R. Krishna, Multiple solutions in reactive distillation for methyl tert-butyl ether synthesis, Industrial & Engineering Chemistry Research 32 (1993) 1706–1709.
[6] H. Eldarsi, P. Douglas, Methyl-tert-butyl-ester catalytic distillation column: part II: optimization, Chemical Engineering Research and Design 76 (1998) 517–524.
[7] S. Hauan, T. Hertzberg, K.M. Lien, Why methyl tert-butyl ether production by reactive distillation may yield multiple solutions, Industrial & Engineering Chemistry Research 34 (1995) 987–991.
[8] M. Schenk, R. Gani, D. Bogle, E. Pistikopoulos, A hybrid modelling approach for separation systems involving distillation, Chemical Engineering Research and Design 77 (1999) 519–534.
[9] W.L. Luyben, D.C. Hendershot, Dynamic disadvantages of intensification in inherently safer process design, Industrial & Engineering Chemistry Research 43 (2004) 384–396.
[10] Y. Tian, E.N. Pistikopoulos, Synthesis of operable process intensification systems – Steady-state design with safety and operability considerations, Industrial & Engineering Chemistry Research 58 (2018) 6049–6068.
[11] Y. Tian, I. Pappas, B. Burnak, J. Katz, E.N. Pistikopoulos, A Systematic Framework for the synthesis of operable process intensification systems – Reactive separation systems, Computers & Chemical Engineering 134 (2020) 106675.

15

A software prototype for synthesis of operable process intensification systems

In this chapter, we present the SYNOPSIS software prototype for SYNthesis of Operable ProcesS Intensification Systems [1–3], featuring the phenomena-based modular process intensification synthesis and model-based operability and control analysis. In what follows, we will first introduce the software prototype structure and detail each functional suite. Then, we will demonstrate the prototype with a tutorial case study on the process design and intensification of a reactive separation for olefin metathesis.

15.1 The SYNOPSIS software prototype

The prototype platform stands on three major suites: (i) Process Intensification Synthesis Suite, (ii) Process Intensification Model Library, and (iii) Process Operability & Control Suite. As illustrated by Fig. 15.1, the user interface built in Python integrates these suites together in a seamless manner to provide a consolidated environment while allowing user input and interaction. The different suites are connected with multiple commercial software to take advantage of their expertized functionalities, such as GAMS for mixed-integer nonlinear optimization, Aspen Plus for steady-state process simulation, gPROMS for dynamic modeling, and MATLAB® for control analysis.

15.1.1 User interface

The user interface serves as a central node for the information flow as shown in Fig. 15.1. From the interface, users can select individual suite for targeted equipment or flowsheet intensification, design, or control analysis. Moreover, the interface can navigate the users to perform integrated design with operability and control in a step-by-step manner leveraging different suites. It can provide key input and output information to the users at every stage and automatically transfer process data from suite to suite.

15.1.2 Process intensification synthesis suite

The synthesis suite provides the tool to systematically generate optimal and intensified process solutions using the Generalized Modular Representative Framework (GMF). To briefly recap the GMF basics detailed in Chapters 3–5: GMF utilizes two phenomena-based

---

Synthesis and Operability Strategies for Computer-Aided Modular Process Intensification
https://doi.org/10.1016/B978-0-32-385587-7.00026-9
Copyright © 2022 Elsevier Inc. All rights reserved.

248 Synthesis and Operability Strategies for Computer-Aided Modular PI

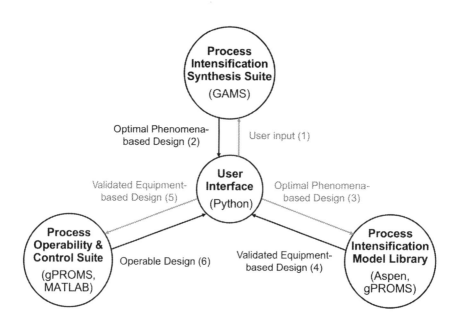

FIGURE 15.1 Information flow chart for the software prototype platform.

modular building blocks, i.e. a pure heat exchange module and a mass/heat exchange module. A superstructure network is built to interlink these modules and to encapsulate the plausible process design structures, no matter conventional or intensified. A single mixed-integer nonlinear optimization problem is formulated to identify the optimal process solutions, solved by the Generalized Benders Decomposition algorithm coded in GAMS. The parameters of the model are assumed to be deterministic and are provided by the users through the user interface. The Python Application Programming Interface (API) is employed to enable the data change between Python and GAMS.

15.1.3 Process intensification model library

The model library stores a number of steady-state and dynamic models for intensified process systems (e.g., reactive distillation, dividing wall column). The models, built with different levels of complexity, can be hosted in Python, Aspen Plus, gPROMS, etc. The model library supports the synthesis suite by validating the GMF module-based design with rigorous equipment-based models. It also provides dynamic high fidelity models to be used in the operability and control suite.

15.1.4 Process operability and control suite

The following model-based analyses are currently enabled to account for PI operability and control considerations: (i) flexibility analysis which examines the operational feasibility under process uncertainty (Chapter 6), (ii) risk analysis which evaluates inherent

Chapter 15 • A software prototype for synthesis of operable PI systems   249

safety performance at the conceptual design stage (Chapter 7), and (iii) explicit/multi-parametric model predictive control which delivers optimal dynamic operation strategies under disturbances [4] (Chapter 8). These operability and control analysis approaches can also be integrated with the above synthesis suite and model library to identify optimal and intensified process designs with guaranteed operability, safety, and/or control criterion [5].

15.2 Case study: pentene metathesis reaction

15.2.1 Problem statement

In this section, we will introduce the distinct functionalities of the prototype via a case study for 2-Pentene (C_5H_{10}) metathesis to form 2-Butene (C_4H_8) and 2-Hexene (C_6H_{12}). We will also highlight the integration between the different prototype functions following the SYNOPSIS framework introduced in Chapter 9. The metathesis reaction takes place in the liquid phase at atmospheric pressure, and can be described by ideal vapor-liquid equilibrium. The production target is to obtain 50 kmol/h of 98 mol% butene and 50 kmol/h of 98 mol% hexene from a saturated liquid feed stream of 100 kmol/h pure pentene at atmospheric pressure. The design objectives are: (i) to synthesize a process design with minimum operating cost, and (ii) to achieve product specifications under a feed flowrate disturbances of $\pm$ 10 kmol/h with control considerations.

15.2.2 Software prototype home page

The home page of the integrated prototype platform is shown in Fig. 15.2. It aims to provide an overview of the prototype functionalities and to navigate the users through distinct toolkits.

15.2.3 Process intensification synthesis suite

Fig. 15.3 presents the introduction page of the synthesis suite, which provides a brief overview on GMF for users to learn about this modular process intensification synthesis strategy. By clicking on the heading of "Generalized Modular Representation Framework (GMF)", users will be directed to our research group website and relevant journal publications (https://parametric.tamu.edu/research/#Intensification).

The current version of Process Intensification Synthesis Suite supports the simulation and synthesis of reactive separation systems using GMF. Take the pentene metathesis problem as an example, we can start investigating this process by simulating the plausible process design configurations. As shown in Fig. 15.4, the "Reaction/Separation Simulation" function provided in the Suite allows the users to simulate conventional reactor-separator systems and intensified reactive separation systems using GMF modules. Users can input different feed conditions (e.g., flowrate, compositions) and the number of pre-specified GMF modules to construct the conventional or intensified process. By clicking on the bottom left button to "Run GMF Simulation Results", the users will be updated on the numerical results regarding product purity, flowrate, and the graphical layouts of GMF modules (Fig. 15.5). Note that the GMF codes are implemented in GAMS and connected

250 Synthesis and Operability Strategies for Computer-Aided Modular PI

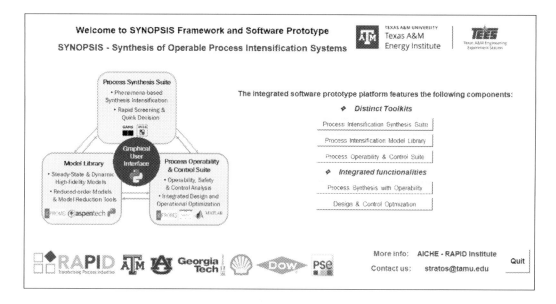

FIGURE 15.2 SYNOPSIS software prototype – Home page.

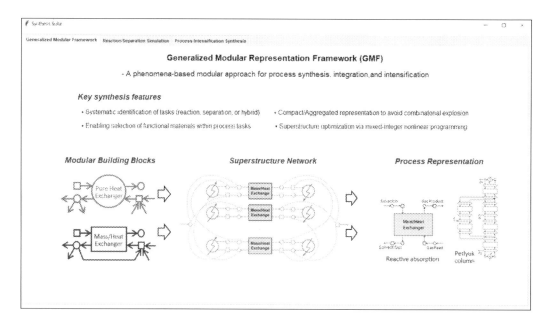

FIGURE 15.3 Process Intensification Synthesis Suite – Introduction page.

Chapter 15 • A software prototype for synthesis of operable PI systems   251

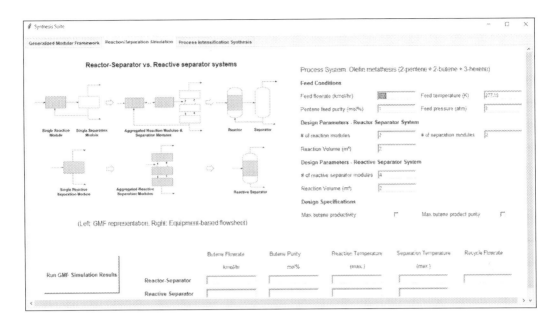

FIGURE 15.4 Process Intensification Synthesis Suite – Reaction/Separation Simulation.

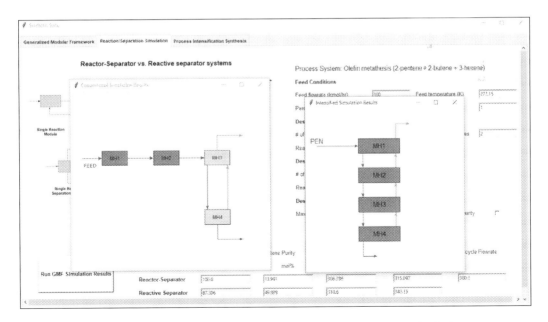

FIGURE 15.5 Process Intensification Synthesis Suite – Reaction/Separation Simulation results display. Left: Conventional system; Right: Intensified system.

with Python via the designated Application Programming Interface. However, all the data input and results output are communicated with the users via the python-based user interface. The users do not need to navigate to the GAMS software environment.

The "Reaction/Separation Simulation" function aims to help the users to gain an initial insight on if intensified designs, compared to their conventional counterparts, will outperform by enhancing productivity and/or product purity. Based on this, the next "Process Intensification Synthesis" function aims to expand from fixed design simulation to superstructure-based optimization and to identify the optimal process design solutions without pre-postulation of plausible process structures. As illustrated in Fig. 15.6, the user interface window is designed with different panels to show input/output data, run time data, superstructure layout, and comparison between alternatives of the generated flowsheet. The Python *tkinter* package is used for this purpose. To allow for the dynamic updates of user interface panels, they are created as *objects* of python defined *classes*.

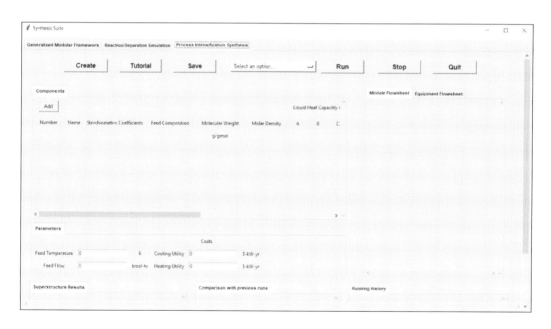

FIGURE 15.6 Process Intensification Synthesis Suite – Process Intensification Synthesis user interface.

To start, users should add the components involved in the synthesis problem. The AIChE DIPPR databank (https://www.aiche.org/dippr) has been linked with the Suite to provide component search capability and to automatically load related physical properties such as molecular weight, heat capacity coefficients, standard Gibbs energy of formation, etc. – as demonstrated in Fig. 15.7. Users can also modify the physical data based on their own databank. The other input information required from the users include reaction stoichiometry coefficients, kinetic parameters, feed conditions, utility costs, etc.

Chapter 15 • A software prototype for synthesis of operable PI systems   253

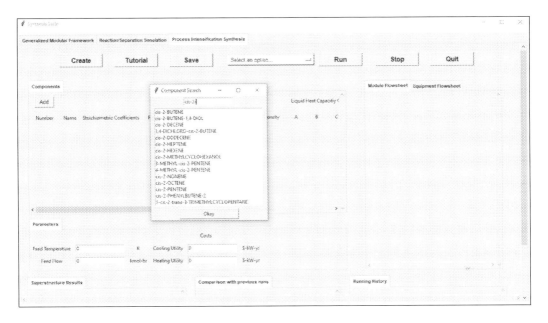

FIGURE 15.7 Process Intensification Synthesis Suite – Process Intensification Synthesis component search.

To synthesize a GMF module-based process solution, the default maximum number of modules is set to 8. This number can be increased to acquire more information about the identified unit operations or be decreased to reduce the computational time. The optimal process solution determined via mixed-integer nonlinear programming includes values for the number of pure heat exchange modules and mass/heat exchanger modules (integer variables), the presence of connection between stream and modules (binary variables), and the operating conditions (continuous variables). The GMF optimization results obtained from GAMS are exported to a database file to be used for results display in the python-based user interface.

The resulting "Optimal Process Solution" with the information on the constituent phenomena is shown in Fig. 15.8. To visualize the process solution under the "Module Flowsheet" panel, a specialized python script is created using conditional statements to read the binary values in the exported database file. The existence of module and stream can be automatically handled, i.e. a module is added to the process if the associated binary variable is assigned as 1, or 0 otherwise. The python script accounts for an optimal layout of the module-based solution by automatically: (i) sizing the modules and text in the flowsheet based on the number of modules versus the size of the allocated window, (ii) displaying process streams without overlapping with the modules, (iii) using the red and blue (gray and dark gray in print version) lines to delegate vapor and liquid streams respectively, and (iv) using white and gray blocks to respectively illustrate pure heat exchange modules and mass/heat exchange modules (light gray for separation task, dark gray for reaction/reactive separation task).

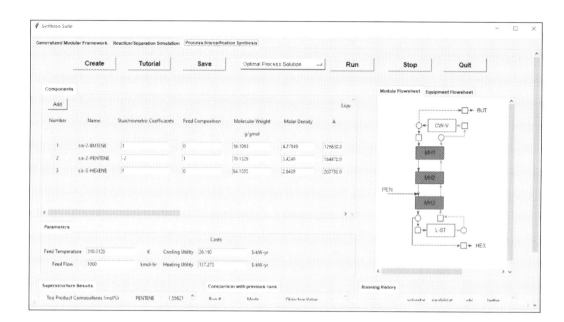

FIGURE 15.8 Process Intensification Synthesis Suite – Process Intensification Synthesis optimal process solution.

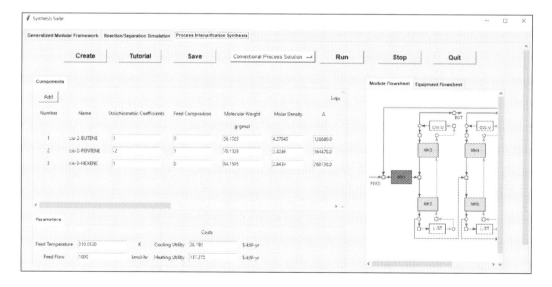

FIGURE 15.9 Process Intensification Synthesis Suite – Process Intensification Synthesis conventional process solution.

Chapter 15 • A software prototype for synthesis of operable PI systems   255

Specifically to the solution obtained in Fig. 15.8, the pure heat exchange modules represented by white blocks at the top and the bottom can be translated into condenser and reboiler respectively. The mass/heat exchange modules can be translated into trays of a reactive distillation column – for which some suggested parameters based on GMF synthesis results are provided under the tab of "Equipment-based Flowsheet". With these design parameters, users can directly click on the button of "Go to Model Library" to load the steady-state reactive distillation model in Aspen Plus for next-step validation and rigorous design optimization. The synthesis function also supports the generation of "Conventional Process Solution". As shown in Fig. 15.9, a process design with one reactor followed by two distillation columns is generated for the pentene metathesis problem. The comparisons of optimal design and conventional design are also recorded by the user interface under the panel of "Comparison with Previous Runs".

15.2.4 Process intensification model library

The resulting GMF module-based solutions are then identified and translated to equipment-based process alternatives with the help of a specialized PI model library. The model library consists of validated rigorous models for various intensified reaction and/or separation systems. For example, Fig. 15.10 shows the steady-state and dynamic high-fidelity models available for reactive distillation and dividing wall columns. Moreover, the model library integrates different commercial software platforms (e.g., Aspen, gPROMS, Python) to leverage the existing unit operation models and to enable the flexible use of different types of models for different computational purposes (e.g., simulation, optimization, control). Figs. 15.11 and 15.12 illustrate the excel-based interfaces for the simulation of reactive distillation, which connect the python-based model library interface respectively with Aspen and gPROMS. In this way, the users can simulate the process models by inputting the design and operating parameters without looking into the Aspen or gPROMS environment and associated models or codes.

15.2.5 Process operability and control suite

After generating an intensified process design with minimized operating cost, we proceed to design explicit/multi-parametric model predictive controller (mp-MPC) to ensure operation under disturbances, using the Process Operability and Control Suite. The introduction page of this suite is presented in Fig. 15.13, which provides an overview on the PAROC framework and guides the users to our research group website and journal publications for more information. The suite is designed following the steps of the PAROC framework – starting with high fidelity dynamic modeling interacted with the PI model library. Users can input different process design parameters directly via the python-based user interface to test on the operability and control performances of different design configurations (Fig. 15.14).

Based on the high fidelity model, simulation data can be generated by sampling through the users specified disturbance range, input variables, and output variables (Fig. 15.15) – in order to identify a linear state-space model which is suitable for mp-MPC

256 Synthesis and Operability Strategies for Computer-Aided Modular PI

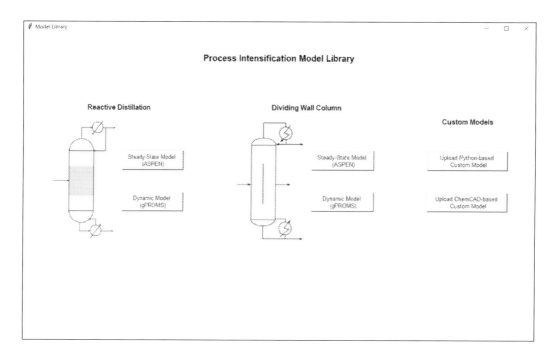

FIGURE 15.10 Process Intensification Model Library – Introduction page.

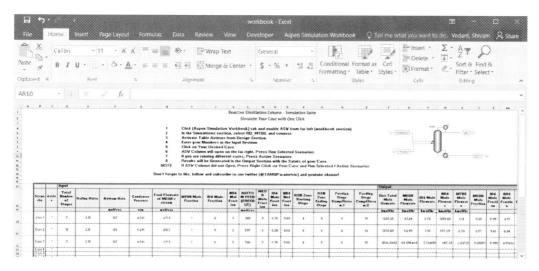

FIGURE 15.11 Process Intensification Model Library – Reactive distillation steady-state model. (Linked with Aspen Plus.)

Chapter 15 • A software prototype for synthesis of operable PI systems 257

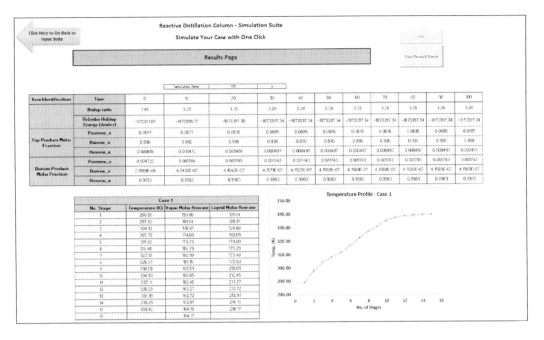

FIGURE 15.12 Process Intensification Model Library – Reactive distillation dynamic model (linked with gPROMS).

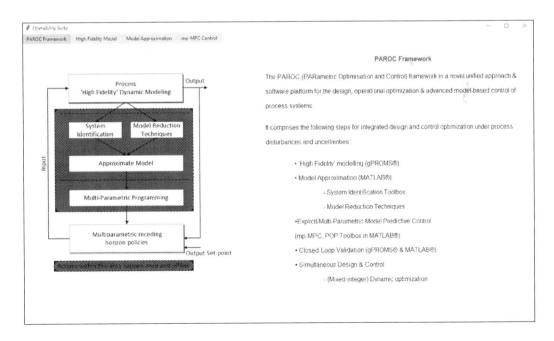

FIGURE 15.13 Process Operability and Control Suite – Introduction page.

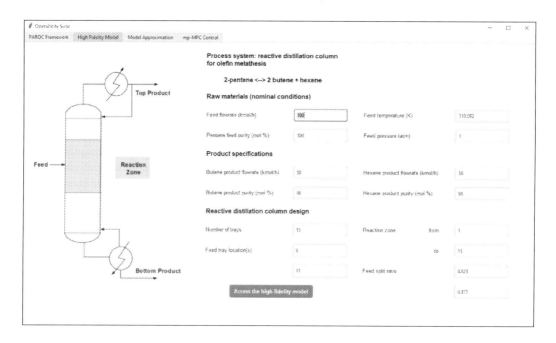

FIGURE 15.14 Process Operability and Control Suite – High fidelity model.

FIGURE 15.15 Process Operability and Control Suite – Model approximation.

Chapter 15 • A software prototype for synthesis of operable PI systems   259

control. A number of approximated models can be automatically generated with different numbers of identified state variables (Fig. 15.16). The two major criteria for approximated model selection are sufficient model output accuracy and correct step responses – the results of which are all provided to users to check.

After selecting the approximated model, we proceed to design the mp-MPC controller by assigning tuning parameters as depicted in Fig. 15.17. The mp-MPC controller can be generated via the link with MATLAB POP toolbox by clicking on the corresponding button. Users can also test the resulting controller for closed-loop validation to check the control performance for disturbance rejection (Fig. 15.18).

Up to this stage, we have utilized this software prototype to synthesize an optimal process design for pentene metathesis, with guaranteed dynamic control performance under disturbances.

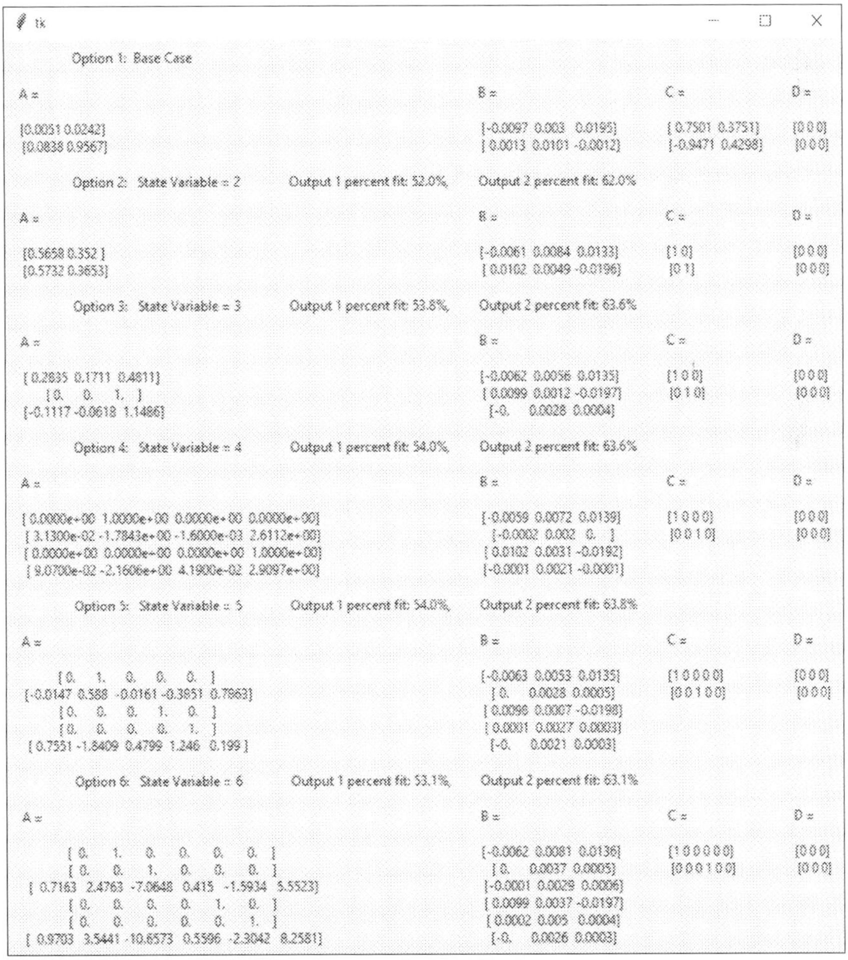

FIGURE 15.16 Process Operability and Control Suite – Selection of approximated models.

260 Synthesis and Operability Strategies for Computer-Aided Modular PI

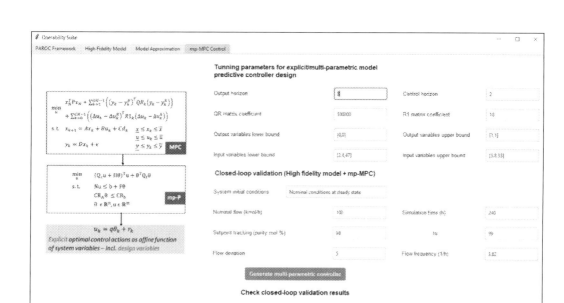

FIGURE 15.17 Process Operability and Control Suite – mp-MPC controller design.

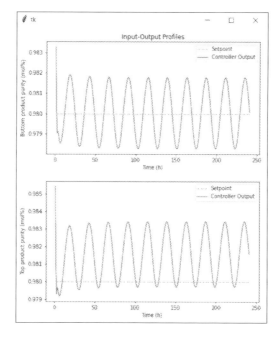

FIGURE 15.18 Process Operability and Control Suite – Closed-loop validation.

References

[1] E.N. Pistikopoulos, M.M.F. Hasan, J.S. Kwon, M.J. Realff, F. Boukouvala, M.R. Eden, S. Cremaschi, B.J. Tatarchuk, J.B. Powell, L. Spanu, R. Bindlish, S. Leyland, SYNOPSIS – Synthesis of Operable Process Intensification Systems, RAPID Institute Project 9.3, DE-EE0007888-09-03, 2020.

[2] Y. Tian, I. Pappas, B. Burnak, J. Katz, E.N. Pistikopoulos, A systematic framework for the synthesis of operable process intensification systems – reactive separation systems, Computers & Chemical Engineering 134 (2020) 106675.

[3] S. Vedant, M.R. Atencio, Y. Tian, V. Meduri, E.N. Pistikopoulos, Towards a Software Prototype for Synthesis of Operable Process Intensification Systems, Computer Aided Chemical Engineering, vol. 50, Elsevier, 2021, pp. 767–772.

[4] E.N. Pistikopoulos, N.A. Diangelakis, R. Oberdieck, M.M. Papathanasiou, I. Nascu, M. Sun, PAROC – an integrated framework and software platform for the optimisation and advanced model-based control of process systems, Chemical Engineering Science 136 (2015) 115–138.

[5] Y. Tian, E.N. Pistikopoulos, Synthesis of operable process intensification systems – steady-state design with safety and operability considerations, Industrial & Engineering Chemistry Research 58 (2018) 6049–6068.

A

Process modeling, synthesis, and control of reactive distillation systems

In this appendix as the supplementary information for Chapter 2, we provide a brief overview on the modeling, synthesis, and control of reactive distillation systems, which is one of the pioneering process intensification technologies in chemical engineering research and industrial practice.

A.1 Modeling of reactive distillation systems

Available models for reactive distillation (RD) can be clustered into short-cut models, equilibrium-stage models, rate-based models, and computational fluid dynamic (CFD) models in the ascending order of complexity. The rigorousness of these models depends on the details used in describing mass transfer, reaction, and hydrodynamics.

Mass transfer models can be built on the assumption of thermodynamic equilibrium between separate phases, as in equilibrium-stage models, or based on detailed diffusion calculations (e.g., Maxwell-Stefan diffusion equations), as in the rate-based models [1]. In rate-based models, fluid dynamics at the interface can be described as one-dimensional flow or stagnant elements by using the two-film theory [2], penetration theory [3], or surface renewal theory [4]. However, some model parameters (e.g., film thicknesses in the two-film theory, residence time in the penetration and surface renewal theories) need to be experimentally determined as functions of the physicochemical and operational conditions and column geometry [5].

Reaction is taken into account either via kinetic rate expressions or equilibrium models. Depending on the nature of the reaction (i.e., homogeneous, heterogeneous, acid-base catalyzed, or auto-catalyzed), reaction models may also differ in detail [6]. For instance, for heterogeneous systems, mass transport effects in the solid phase become important. A common practice in this case is to use pseudo-homogeneous models with lumped internal and external mass transfer resistances. This assumption is valid when the catalyst surface is completely exposed to the liquid bulk phase [7]. Reaction rate is also an important factor. The relative rate between intrinsic reaction rate and mass transfer rate, quantified by the Ha number, is often used to define the model. For instance, if the reaction is very fast, then the reaction is assumed to be in equilibrium and non-reactive equilibrium model can be used in combination with reaction equilibrium equations. If the reaction is relatively slow, however, kinetic rate expressions are necessary to describe the operation of the reactive separation column.

Crucial operational parameters to describe the hydrodynamics in RD columns include axial dispersion, liquid hold-up, and pressure drop. These can be captured either via experimental correlations or detailed CFD simulations [8]. Here, CFD methods are highly useful in predicting the performance of the reactive separators with different column internals and geometries. Furthermore, these simulations might serve as virtual experiments to obtain some model parameters required for rate-based models without the need for physical experimentation.

Similar to the steady-state modeling approaches, dynamic modeling of RD processes can also be performed via equilibrium or rate-based models [9]. A full high fidelity dynamic model for general reactive distillation systems is documented in Appendix C, which considers: (i) dynamic material and energy balances for each column tray, (ii) nonideal phase equilibrium or non-equilibrium calculations, (iii) liquid and vapor, material and energy holdups, (iv) liquid hydraulics and liquid level on each tray using modified Francis weir formulation, (v) equations for the pressure drop from tray to tray, (vi) flooding and entrainment correlations and evaluation of minimum allowable column diameter. The accuracy of these models is critical for analyzing start-up and shut-down processes and for determining appropriate control structures. More details on steady-state and dynamic modeling of RD columns can be found in Taylor and Krishna [10].

A.2 Short-cut design of reactive distillation

Design of RD columns is a more challenging task than the design of ordinary distillation columns as some of the commonly used assumptions are no longer valid for RD designs. For instance, constant molar overflow assumption is not applicable for RDs and liquid holdup becomes an important design variable as reaction takes place in the liquid phase. Several short-cut methods for RD design have been proposed: boundary value methods [11], pinch based methods [12,13], and shortest stripping line method [14]. In boundary value method, tray-by-tray calculations are performed starting from the two ends of the column and a feasible configuration is obtained when two profiles intersect each other [11]. Locating the intersection point, however, is challenging as an appropriate value of reflux ratio should be guessed beforehand and this requires an iterative procedure. The visual identification of the profile intersection also hinders its wider applicability. Pinch-based methods provide remedy for the problems associated with boundary value methods. For instance, Avami et al. [13] used pinch analysis to design RD columns for multi-component and multi-reaction systems, mixtures with liquid–liquid phase split, and double-feed columns with minimal energy consumption. Shortest stripping line method is also a systematic procedure for determination of the minimum energy requirements for the RD columns. It relies on nonlinear programming (NLP) to find the shortest distance along the tray-to-tray stripping lines, in which boil-up ratio appears as the only optimiza-

Appendix A • Process modeling, synthesis, and control of RD systems   265

tion variable [14]. A more detailed discussion on these short-cut methods can be found in Avami et al. [13].

A.3 Synthesis design of reactive distillation

Under given input and output specifications, synthesis design can be performed to provide the equipment design and operating parameters required by detailed simulations and to test the feasibility of the proposed reactive distillation system. Synthesis methods for RD columns can be categorized as heuristic, graphical, and optimization-based approaches. An example of heuristic methods was proposed by Subawalla and Fair [15], which was a post-design procedure applicable to a known RD structure. A set of rules were defined to determine the design and operational parameters, such as the existence of a pre-reactor, operating pressure, reaction zone location, feed location, number of theoretical stages, reflux ratio, reboil ratio, reactive section height, etc. More recently, Shah et al. [16] proposed a systematic framework to characterize the feasibility of RD columns based on a set of heuristics. If a design successfully fulfilled these feasibility requirements, it would be further evaluated according to the economic viability.

Graphical techniques for RD design include residue curve mapping technique, attainable region technique, reactive cascades, thermodynamic-based approach, phenomena-based approach, and scalar/vectorial difference points technique [17]. In general, these methods are with limited scope as they rely on thermodynamic and chemical equilibrium assumption. A more detailed discussion on these techniques with their advantages and limitations can be found in Almeida-Rivera et al. [17]. Optimization-based approaches, however, provide a more systematic procedure for the synthesis of RD columns. These methods are based on either NLP or mixed-integer nonlinear programming (MINLP) formulations with an objective to minimize total annual cost in the most cases. The first optimization method for the synthesis of RD columns was proposed by Ciric and Gu [18]. The proposed MINLP model used discrete decisions to decide the existence of each column tray. This model was based on MESH constraints (i.e., mass balances, vapor-liquid equilibria, mole balance summations, enthalpy balances) and its solution via Generalized Benders Decomposition (GBD) gave optimum number of trays, feed flow rates and locations, liquid holdup, reflux ratio as well as temperature and composition profiles. However, the bilinear terms introduced by the multiplication of the tray binary variables and their related constraints increased the difficulty for solution. For the RD design with two immiscible liquid phases, Gumus and Ciric [19] formulated a bi-level programming problem in which Gibbs free energy minimization at each tray was embedded in an upper level problem to minimize the total annual cost. A Generalized Disjunctive Programming (GDP) model for RD synthesis was proposed by Jackson and Grossmann [20], which accounted for the addition (or removal) of heat from each tray. An initialization procedure was also provided to solve the resulting MINLP problem using logic-based outer approximation method [21]. Stichlmair and Frey [22] formulated a MINLP problem based on superstructures derived from RD thermodynamic analysis and demonstrated this method on the production of methyl tert-butyl ether (MTBE) and methyl acetate.

Simulated annealing (SA) method [23] and differential evolution algorithm [24] are two examples of stochastic optimization methods used in RD synthesis. Segovia-Hernández et al. [25] provided an extensive review on the use of deterministic and stochastic optimization techniques for the optimization and synthesis of reactive distillation columns. Stochastic optimization methods, however, may induce high computational load as they require a large number of iterations. The deterministic optimization solution methods are also challenged by the large number of discrete variables and by the strong nonconvexity inherently in the RD optimization model, which result in poor local solutions and necessitate good starting points. Hence, hybrid methods are investigated to combine the strengths of both solution strategies. For instance, Gómez et al. [26] used rate-based models to synthesize and design a RD column for ethyl tert-butyl ether (ETBE) production. They proposed a two-stage solution strategy in which a NLP subproblem was solved with rSQP (i.e., reduced hessian successive quadratic programming) to determine the catalyst load at each stage and the values of some other design variables. A master problem in the form of MINLP was then solved with SA to decide on the values of integer variables and the remaining continuous variables. Urselmann et al. [27] used a memetic algorithm for the solution of a RD model using MINLP formulation, in which the evolutionary algorithm (EA) was applied to globally optimize the design decisions including all discrete variables and some continuous variables, whereas the deterministic local NLP method was used to solve a subsystem of equations to simulate the behavior of the design proposed by the EA.

A.4 Process control of reactive distillation

Sharma and Singh [28] reviewed the available types of controllers and control structures for several representative RD systems (e.g., MTBE, ETBE, tertiary-amyl methyl ether (TAME)). However, most of these studies applied conventional decentralized proportional-integral-derivative (PID) controllers, which has shown inadequacy in handling the high directionality of the process gain in reactive distillation [29]. More advanced control strategies, such as Model Predictive Control (MPC), have been proposed for RD columns in several works. Kawathekar and Riggs [30] applied nonlinear MPC (NMPC) to an ethyl acetate reactive distillation column where they showed a factor of 2–3 smaller variability in MPC controller performance comparing to PI controllers. In their work, the original differential algebraic equations (DAE) system was approximated with orthogonal collocation on finite elements resulting in a NLP problem. To address the trade-off between model accuracy and computational tractability, Balasubramhanya and Doyle [31] utilized travel waving phenomena to derive lower order model used for nonlinear model-based control of a batch reactive distillation. Other MPC formulations have also been developed for RD columns based on neural networks, genetic algorithm, simulated annealing, etc. [32,33].

However, for a conventional sequential design and control approach, there is no guarantee that the steady-state economically optimal process system will remain optimal

Appendix A • Process modeling, synthesis, and control of RD systems   267

and/or has good dynamic performance when met with external disturbances and parametric/model uncertainties [34]. Thus, an integrated approach to simultaneously address process design, operability, and control is necessitated to ensure process economical performance while avoiding undesirable inherent process dynamic characteristics. Specifically for reactive distillation systems, Mansouri et al. [35] proposed a step-wise computer-aided framework to design RD columns at maximum driving force which also had the best inherent "controllability". Georgiadis et al. [36] compared optimization-based sequential versus simultaneous design and control of a RD column with fixed column structure and PI control scheme. It was shown that a simultaneous consideration of RD design and control could lead to more economically beneficial and better controlled system than a sequential approach. Panjwani et al. [37] further extended the simultaneous design and PI control optimization problem to include column structural optimization variables and control structure/design variables resulting in a mixed-integer dynamic optimization (MIDO) problem. Bernal et al. [38] presented the simultaneous optimal design and control of an ETBE catalytic distillation column using an economic-oriented nonlinear model predictive controller (EO-NMPC). The resulting design and control optimization problem was fully discretized using orthogonal collocation giving a nonlinear programming (NLP) formulation rather than dynamic optimization due to the demanding computational requirement of online MPC. In Chapter 14, we present a case study for simultaneous design and control of reactive distillation systems leveraging explicit/multi-parameter model predictive control (mp-MPC) strategies and mixed-integer dynamic optimization to ensure the operability and optimality of a reactive distillation system while delivering its expected functionality in a computationally effective manner.

A.5 Software tools for modeling, simulation, and design of reactive distillation

ASPEN Plus readily provides a column, RATEFRACTM, for simulating the rate-based reactive absorption, distillation, and stripping processes. Another RADFRAC module provided by ASPEN Plus can be used to simulate conventional distillation systems and reactive distillation systems. Attempts have also been made to develop specialized process simulators for RDs, together with other reactive separation processes. An integrated computer-aided tool for the synthesis and design of reactive distillation was developed by Kenig et al. [39]. The framework consisted of: (i) SYNTHESIZER as a predictive tool for rapid evaluation and basic design of RD; (ii) DESIGNER as a process simulator to model, design, size, and understand the complex behavior within a RD column for integration into industrial processes; and (iii) PREDICTOR as a user interface to integrate the above tools. PROFILER, a process simulator developed for reactive separation columns in the context of European research project Intelligent Column Internals for Reactive Separations (INTINT), was proposed by Klöker et al. [40]. PROFILER was modeled via Aspen Custom Modeler (ACM) and embedded with three different types of model: (i) a detailed rate-based model with Maxwell-Stefan diffusivities, (ii) a simpler rate-based model with effective diffusivities instead of binary Maxwell-Stefan diffusivities, and (iii) an extended equilibrium stage model with equilibrium stages and chemical reaction rates.

References

[1] C. Noeres, E. Kenig, A. Górak, Modelling of reactive separation processes: reactive absorption and reactive distillation, Chemical Engineering and Processing: Process Intensification 42 (2003) 157–178.

[2] W.K. Lewis, W.G. Whitman, Principles of gas absorption, Industrial and Engineering Chemistry 16 (1924) 1215–1220.

[3] R. Higbie, The rate of absorption of a pure gas into a still liquid during short periods of exposure, Transactions of the American Institute of Chemical Engineers 31 (1935) 365–389.

[4] P. Danckwerts, Significance of liquid-film coefficients in gas absorption, Industrial and Engineering Chemistry 43 (1951) 1460–1467.

[5] E. Kenig, Complementary modelling of fluid separation processes, Chemical Engineering Research and Design 86 (2008) 1059–1072.

[6] H. Schmidt-Traub, A. Górak, Integrated Reaction and Separation Operations, Springer, 2006.

[7] J. Holtbrügge, A.K. Kunze, A. Niesbach, P. Schmidt, R. Schulz, D. Sudhoff, M. Skiborowski, Reactive and Membrane-Assisted Separations, Walter de Gruyter GmbH & Co KG, 2016.

[8] M. Klöker, E. Kenig, A. Górak, On the development of new column internals for reactive separations via integration of cfd and process simulation, Catalysis Today 79 (2003) 479–485.

[9] M. Schenk, R. Gani, D. Bogle, E. Pistikopoulos, A hybrid modelling approach for separation systems involving distillation, Chemical Engineering Research and Design 77 (1999) 519–534.

[10] R. Taylor, R. Krishna, Modelling reactive distillation, Chemical Engineering Science 55 (2000) 5183–5229.

[11] D. Barbosa, M.F. Doherty, Design and minimum-reflux calculations for single-feed multicomponent reactive distillation columns, Chemical Engineering Science 43 (1988) 1523–1537.

[12] J. Bausa, R.v. Watzdorf, W. Marquardt, Shortcut methods for nonideal multicomponent distillation: I. Simple columns, AIChE Journal 44 (1998) 2181–2198.

[13] A. Avami, W. Marquardt, Y. Saboohi, K. Kraemer, Shortcut design of reactive distillation columns, Chemical Engineering Science 71 (2012) 166–177.

[14] A. Lucia, A. Amale, R. Taylor, Distillation pinch points and more, Computers & Chemical Engineering 32 (2008) 1342–1364.

[15] H. Subawalla, J.R. Fair, Design guidelines for solid-catalyzed reactive distillation systems, Industrial & Engineering Chemistry Research 38 (1999) 3696–3709.

[16] M. Shah, A.A. Kiss, E. Zondervan, A.B. De Haan, A systematic framework for the feasibility and technical evaluation of reactive distillation processes, Chemical Engineering and Processing: Process Intensification 60 (2012) 55–64.

[17] C. Almeida-Rivera, P. Swinkels, J. Grievink, Designing reactive distillation processes: present and future, Computers & Chemical Engineering 28 (2004) 1997–2020.

[18] A.R. Ciric, D. Gu, Synthesis of nonequilibrium reactive distillation processes by minlp optimization, AIChE Journal 40 (1994) 1479–1487.

[19] Z.H. Gumus, A.R. Ciric, Reactive distillation column design with vapor/liquid/liquid equilibria, Computers & Chemical Engineering 21 (1997) S983–S988.

[20] J.R. Jackson, I.E. Grossmann, A disjunctive programming approach for the optimal design of reactive distillation columns, Computers & Chemical Engineering 25 (2001) 1661–1673.

[21] M. Türkay, I.E. Grossmann, Logic-based minlp algorithms for the optimal synthesis of process networks, Computers & Chemical Engineering 20 (1996) 959–978.

[22] J. Stichlmair, T. Frey, Mixed-integer nonlinear programming optimization of reactive distillation processes, Industrial & Engineering Chemistry Research 40 (2001) 5978–5982.

[23] M. Cardoso, R. Salcedo, S.F. De Azevedo, D. Barbosa, Optimization of reactive distillation processes with simulated annealing, Chemical Engineering Science 55 (2000) 5059–5078.

[24] B. Babu, M. Khan, Optimization of reactive distillation processes using differential evolution strategies, Asia-Pacific Journal of Chemical Engineering 2 (2007) 322–335.

Appendix A • Process modeling, synthesis, and control of RD systems   269

[25] J.G. Segovia-Hernández, S. Hernandez, A.B. Petriciolet, Reactive distillation: a review of optimal design using deterministic and stochastic techniques, Chemical Engineering and Processing: Process Intensification 97 (2015) 134–143.

[26] J.M. Gómez, J.-M. Reneaume, M. Roques, M. Meyer, X. Meyer, A mixed integer nonlinear programming formulation for optimal design of a catalytic distillation column based on a generic nonequilibrium model, Industrial & Engineering Chemistry Research 45 (2006) 1373–1388.

[27] M. Urselmann, S. Barkmann, G. Sand, S. Engell, Optimization-based design of reactive distillation columns using a memetic algorithm, Computers & Chemical Engineering 35 (2011) 787–805.

[28] N. Sharma, K. Singh, Control of reactive distillation column: a review, International Journal of Chemical Reactor Engineering 8 (2010).

[29] N.M. Nikačević, A.E. Huesman, P.M. Van den Hof, A.I. Stankiewicz, Opportunities and challenges for process control in process intensification, Chemical Engineering and Processing: Process Intensification 52 (2012) 1–15.

[30] R. Kawathekar, J.B. Riggs, Nonlinear model predictive control of a reactive distillation column, Control Engineering Practice 15 (2007) 231–239.

[31] L.S. Balasubramhanya, F.J. Doyle, Nonlinear model-based control of a batch reactive distillation column, Journal of Process Control 10 (2000) 209–218.

[32] C. Venkateswarlu, A.D. Reddy, Nonlinear model predictive control of reactive distillation based on stochastic optimization, Industrial & Engineering Chemistry Research 47 (2008) 6949–6960.

[33] N. Sharma, K. Singh, Model predictive control and neural network predictive control of tame reactive distillation column, Chemical Engineering and Processing: Process Intensification 59 (2012) 9–21.

[34] Z. Yuan, B. Chen, G. Sin, R. Gani, State-of-the-art and progress in the optimization-based simultaneous design and control for chemical processes, AIChE Journal 58 (2012) 1640–1659.

[35] S.S. Mansouri, J.K. Huusom, R. Gani, M. Sales-Cruz, Systematic integrated process design and control of binary element reactive distillation processes, AIChE Journal 62 (2016) 3137–3154.

[36] M.C. Georgiadis, M. Schenk, E.N. Pistikopoulos, R. Gani, The interactions of design control and operability in reactive distillation systems, Computers & Chemical Engineering 26 (2002) 735–746.

[37] P. Panjwani, M. Schenk, M. Georgiadis, E.N. Pistikopoulos, Optimal design and control of a reactive distillation system, Engineering Optimization 37 (2005) 733–753.

[38] D.E. Bernal, C. Carrillo-Diaz, J.M. Gomez, L.A. Ricardez-Sandoval, Simultaneous design and control of catalytic distillation columns using comprehensive rigorous dynamic models, Industrial & Engineering Chemistry Research 57 (2018) 2587–2608.

[39] E. Kenig, K. Jakobsson, P. Banik, J. Aittamaa, A. Gorak, M. Koskinen, P. Wettmann, An integrated tool for synthesis and design of reactive distillation, Chemical Engineering Science 54 (1999) 1347–1352.

[40] M. Klöker, E.Y. Kenig, A. Hoffmann, P. Kreis, A. Górak, Rate-based modelling and simulation of reactive separations in gas/vapour–liquid systems, Chemical Engineering and Processing: Process Intensification 44 (2005) 617–629.

B

Driving force constraints and physical and/or chemical equilibrium conditions

In this appendix as the supplementary information for Chapter 3, we discuss the relationship between the GMF driving force constraints with physical and/or chemical equilibrium conditions in reaction systems, separation systems, and reactive separation systems. We aim to: (i) elucidate the thermodynamic approximations used in GMF driving force constraints, and (ii) reveal the conjunctive thermodynamic fundamentals of GMF with other equilibrium-based model formulations (e.g., distillation tray-by-tray modeling).

B.1 Pure separation systems

The GMF driving force constraints for pure separation systems can be expressed as:

$$\begin{aligned}
&G1_i \times G2_i \leq 0 \quad \forall i = 1, ..., NC \\
&G1_i = dn_i^L = f^{LO} x_i^{LO} - f^{LI} x_i^{LI} \\
&G2_i = \left[\frac{\partial (nG)^{tot}}{\partial (n_i^L)}\right]_{T,P} = \ln\left[\frac{\gamma_i^L x_i^L P_i^{sat,L}}{\phi_i^V x_i^V P}\right]
\end{aligned} \quad (B.1)$$

Note that in this appendix, $G2_i$ is defined between the liquid and vapor outlet streams to investigate the equilibrium conditions in the mass/heat exchange module, instead of at the two ends of module as defined in Section 3.3. We also assume that, $\sum_i (G1_i)^2 \neq 0$ for the module of interest; otherwise, it gives a non-functioning module with no mass transfer taking place.

To dictate that the module is at physical equilibrium conditions, the $G2_i$ term should be forced as 0 for each component:

$$G2_i = \left[\frac{\partial (nG)^{tot}}{\partial (n_i^L)}\right]_{T,P} = \ln\left[\frac{\gamma_i^L x_i^L P_i^{sat,L}}{\phi_i^V x_i^V P}\right] = 0 \quad i = 1, 2, ..., NC \quad (B.2)$$

Given the expressions of liquid and vapor chemical potentials:

$$\begin{aligned}
\mu_i^L &= \Delta G_i^f + RT \ln(\gamma_i^L x_i^L P_i^{sat,L}) \\
\mu_i^V &= \Delta G_i^f + RT \ln(\phi_i^V x_i^V P)
\end{aligned} \quad (B.3)$$

271

272   Synthesis and Operability Strategies for Computer-Aided Modular PI

It can be obtained that $\mu_i^L = \mu_i^V$ holds true for each component under the context of Eq. (B.2). In other words, for separation systems with $G2_i = 0$, the systems reach actual physical equilibrium. It can also be proved that when a separation system at its physical equilibrium, i.e. $\mu_i^L = \mu_i^V$, its corresponding GMF driving force constraints also have $G2_i = 0$. In other words, **$G2_i = 0$ is the necessary and sufficient conditions for physical equilibrium**.

B.2 Reactive separation systems

Recall the GMF driving force constraints $G2_i$ definition for reactive separation systems:

$$G2_i = \left[\frac{\partial (nG)^{tot}}{\partial (n_i^L)}\right]_{T,P} = \ln\left[\frac{\gamma_i^L x_i^L P_i^{sat,L}}{\phi_i^V x_i^V P}\right] + \sum_k \sum_j \nu_{jk}\left[\frac{\Delta G_j^f}{RT} + \ln(\phi_j^V x_j^V P)\right]\frac{\partial \epsilon_k}{\partial n_i^L} \tag{B.4}$$

However, if postulating $G2_i = 0$, it only gives:

$$\ln\left[\frac{\gamma_i^L x_i^L P_i^{sat,L}}{\phi_i^V x_i^V P}\right] + \sum_k \sum_j \nu_{jk}\left[\frac{\Delta G_j^f}{RT} + \ln(\phi_j^V x_j^V P)\right]\frac{\partial \epsilon_k}{\partial n_i^L} = 0 \tag{B.5}$$

Rewrite Eq. (B.5) in the form of chemical potentials:

$$(\mu_i^L - \mu_i^V) + \sum_k \left(\sum_j \nu_{jk}\mu_j^V\right)\frac{\partial \epsilon_k}{\partial n_i^L} = 0 \tag{B.6}$$

which is a necessary but not sufficient condition for the actual physical and chemical equilibrium defined by Eqs. (B.7) and (B.8). In other words, the approximate equilibrium surface characterized by the driving force constraints $G2_i = 0$ for reactive separation systems is a superset than that of the actual physical and chemical equilibrium.

$$\mu_i^L - \mu_i^V = 0 \qquad i = 1, 2, ..., NC \tag{B.7}$$

$$\sum_i \nu_{ik}\mu_i = 0 \qquad k = 1, 2, ..., NR \tag{B.8}$$

B.3 Pure reaction systems

For pure reaction system assuming with a single liquid phase, the GMF driving force constraints $G2_i$ term is defined by:

$$G2_i = \sum_k \sum_j \nu_{jk}\left[\frac{\Delta G_j^f}{RT} + \ln(\gamma_i^L x_i^L P_i^{sat,L})\right]\frac{\partial \epsilon_k}{\partial n_i^L} = \sum_k \left(\sum_j \nu_{jk}\mu_j^L\right)\frac{\partial \epsilon_k}{\partial n_i^L} \tag{B.9}$$

Appendix B • Driving force constraints and physical and/or chemical Eq. conditions 273

Note that $\frac{\partial \epsilon_k}{\partial n_i^L} = v_{ik} \neq 0$ for components involved in this liquid-phase reaction. By postulating $G2_i = 0$, we have:

$$G2_i = \sum_k (\sum_j v_{jk} \mu_j^L) \frac{\partial \epsilon_k}{\partial n_i^L} = \sum_k (\sum_j v_{jk} \mu_j^L) v_{ik} = 0 \tag{B.10}$$

which requires that

$$\sum_j v_{jk} \mu_j^L = 0 \tag{B.11}$$

i.e. the reaction system is at chemical equilibrium. Reversely, when a liquid-phase reaction system reaches its chemical equilibrium, its driving force constraints also have $G2_i = 0$. Thus, **$G2_i = 0$ *is the necessary and sufficient conditions for chemical equilibrium.***

C

Reactive distillation dynamic modeling

In this appendix, we present the high fidelity dynamic modeling of a generalized reactive distillation (RD) column. Note that it is also applicable to the modeling of conventional distillation systems if the reaction kinetics equations are neglected. The basis of this dynamic model has been validated with experimental data and open literature data in our previous works for different reactive distillation systems, e.g. MTBE production [1], ethyl acetate production [2,3]. Some key features of this generalized RD model, which enables its prediction accuracy and representation capability for design and/or control optimization, are listed as follow:

- A superstructure model formulation to enable the selection of the number of trays and feed tray locations via integer variables
- Dynamic material and energy balances for each tray
- Accurate calculation of nonideal phase equilibrium via the use of physical property models
- The consideration of liquid and vapor, material and energy holdups
- The consideration of liquid hydraulics and liquid level on each tray using modified Francis weir formulation
- The consideration of phase equilibrium or non-equilibrium using Murphree tray efficiencies
- Equations for the pressure drop from tray to tray correlated with the vapor flow through the openings at the bottom of each tray and its hydrostatic pressure
- Detailed calculation of flooding and entrainment correlations and evaluation of minimum allowable column diameter.

Hereafter we present the modeling equations for the reactive distillation column:

C.1 Process structure

The general column superstructure is depicted in Fig. C.1 which enables the number of trays and feed tray locations to be optimally determined via the following modeling formulation:

$$\sum_{k=1}^{N} yf_{feed,k} = 1 \qquad (C.1)$$

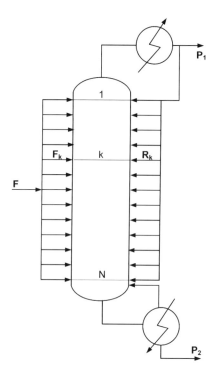

FIGURE C.1 Reactive distillation modeling – Column superstructure.

$$\sum_{k=1}^{N} yr_k = 1 \qquad (C.2)$$

where $feed$ is the index set for feed streams, k is the index set for trays, N is the maximum number of column trays available which provides a reasonable estimate of the upper bound of number of trays, $yf_{feed,k}$ and yr_k are binary variables to denote if tray k is receiving (or not) feed or reflux. If $yf_{feed,k} = 1$ then all feed f enters tray k; similarly $yr_k = 1$ indicates that all reflux enters tray k (otherwise the binary variables take the value of 0). Also note that with Eqs. (C.1) and (C.2), no feed or reflux splitting is considered. The reflux is also constrained via Eq. (C.3) to enter a tray below the feed:

$$\sum_f yf_{feed,k} - \sum_{k'=1}^{k} yr_{k'} \leq 0, \qquad k = 1, ..., N \qquad (C.3)$$

C.2 Tray modeling

The following assumptions are made in the current sieve tray model:
- Liquid and vapor phases are well-mixed;
- Liquid and vapor phases are in thermal and mechanical equilibrium with each other;

Appendix C • Reactive distillation dynamic modeling 277

– Negligible downcomer dynamics;
– Negligible entrainment, weeping, draw-offs, or external heat inputs for the trays.

The tray modeling equations are presented in what follow:

• Component molar balances:

$$(\sum_{k'=k}^{N} yr_{k'}) \cdot \frac{dM_{i,k}}{dt} = \sum_{feed} F_{feed,k} z_{i,feed} + R_k x_{i,d} + L_{k-1} x_{i,k-1} + V_{k+1} y_{i,k+1}$$
$$- L_k x_{i,k} + V_k y_{i,k} + v_i Rate_k \quad i = 1, ...NC, \ k = 1, ..., N$$

(C.4)

where i is the index set for components, $M_{i,k}$ refers to the molar holdup of component i on tray k, $F_{feed,k}$ gives the flowrate of $feed$ to tray k, $z_{i,feed}$ is the molar fraction of component i in inlet stream $feed$, R_k is the reflux flow to tray k, L_k and V_k refer to liquid and vapor flowrates from tray k respectively, x_i, k and y_i, k are the molar fractions of component i in the liquid and vapor outlet streams from tray k, v_i is the reaction stoichiometric coefficient, and $Rate_k$ gives the rate of reaction on tray k determined via specific reaction kinetics (e.g., Eq. (C.5)):

$$Rate_k = r(P, T, \mathbf{x})$$

(C.5)

• Energy balances (note that if constant column pressure is assumed, the energy balances are considered at steady-state):

$$(\sum_{k'=k}^{N} yr_{k'}) \cdot \frac{dU_k}{dt} = \sum_{feed} F_{feed,k} h_{feed} + R_k h_d^l + L_{k-1} h_{k-1}^l + V_{k+1} h_{k+1}^v$$
$$- L_k h_k^l + V_k h_k^v \quad k = 1, ..., N$$

(C.6)

where U_k denotes the internal energy holdup for tray k, h_k^l and h_k^v respectively refer to molar liquid and vapor molar enthalpies. Note that an additional term of heat of reaction is not needed if enthalpies are calculated on element-basis, while required for component-based enthalpy calculation.

• Component molar holdups:

$$M_{i,k} = M_k^l x_{i,k} + M_k^v y_{i,k} \quad i = 1, ...NC, \ k = 1, ..., N$$

(C.7)

where M_k^l (or M_k^v) is the total molar liquid (or vapor) holdup on tray k.

• Energy holdups:

$$U_k = M_k^l h_k^l + M_k^v h_k^v - 0.1 P_k Vol_{tray} \quad k = 1, ..., N$$

(C.8)

where Vol_{tray} stands for tray volume, P_k is the stage pressure.

278　Synthesis and Operability Strategies for Computer-Aided Modular PI

- Volume constraints:

$$\frac{M_k^l}{\rho_k^l} + \frac{M_k^v}{\rho_k^v} = Vol_{tray} \qquad k = 1, ..., N \tag{C.9}$$

where ρ represents molar density.

- Equilibrium vapor phase composition:

$$\Phi_{i,k}^v y_{i,k}^* = \Phi_{i,k}^l x_{i,k} \qquad i = 1, ...NC, \ k = 1, ..., N \tag{C.10}$$

where Φ defines the vapor or liquid fugacity coefficient or activity coefficient.

- Murphree tray efficiency definition:

$$y_{i,k} = y_{i,k+1} + Eff_{i,k}(y_{i,k}^* - y_{i,k+1}) \qquad i = 1, ...NC, \ k = 1, ..., N \tag{C.11}$$

where $Eff_{i,k}$ stands for the Murphree tray efficiency.

- Molar fraction normalization:

$$\sum_{i=1}^{NC} x_{i,k} = \sum_{i=1}^{NC} y_{i,k} = 1 \qquad k = 1, ..., N \tag{C.12}$$

- Liquid levels:

$$Level_k = \frac{M_k^l}{\rho_k^l A_{tray}} \qquad k = 1, ..., N \tag{C.13}$$

where $Level_k$ gives the liquid level on tray k, A_{tray} denotes column tray area.

- Liquid outlet flowrates (modified Francis formula for liquid flow over a rectangular weir):

$$L_k = \begin{cases} 0, \ \text{if } Level_k \leq Height_{weir} \\ 1.84 \cdot \rho_k^l \cdot Length_{weir} \cdot (Level_k - Height_{weir})^{1.5}, \ \text{otherwise} \end{cases} \tag{C.14}$$

- Pressure driving force for vapor inlet:

$$P_{k+1} - P_k = (\sum_{k'=k}^{N} yr_{k'}) \cdot (vel_{k+1}^2 \cdot \tilde{\rho}_{k+1}^v + \tilde{\rho}_k^l \cdot g \cdot Level_k) \tag{C.15}$$

where $\tilde{\rho}$ refer to mass density, g is the gravity constant, vel_k is the velocity of vapor leaving tray k.

- Vapor velocity calculation:

$$vel_k = \frac{V_k}{\rho_k^v A_{holes}} \qquad k = 1, ..., N \tag{C.16}$$

where A_{holes} refers to the total area of all active holes.

Appendix C • Reactive distillation dynamic modeling 279

The following equations are used for tray geometry calculation:

- Free volume between trays:

$$Vol_{tray} = Space \cdot A_{tray} \qquad (C.17)$$

- Cross-sectional area of the column

$$A_{col} = \frac{\pi}{4} D_{col}^2 \qquad (C.18)$$

where $Space$ represents tray spacing, and D_{col} stands for column diameter. The other tray design parameters need to be specified, such as weir length, weir height, active area, etc.

The following modeling equations are used for flooding and entrainment correlations:

- Fractional entrainment (80% flooding factor):

$$ent_k = 0.224exp(-2) + 2.377exp(-9.394FLV_k^{0.314}) \qquad k = 1, ..., N \qquad (C.19)$$

where ent_k is the fractional entrainment for tray k, FLV_k represents for Sherwood flow parameter for tray k.

- Sherwood flow parameter definition:

$$FLV_k = \frac{\tilde{L}_k}{\tilde{V}_k} \cdot \left(\frac{\tilde{\rho}_k^v}{\tilde{\rho}_k^l}\right)^{0.5} \qquad k = 1, ..., N \qquad (C.20)$$

where the superscript ˜ denotes variables in mass basis.

- Mass flowrates:

$$\tilde{L}_k = L_k \cdot \sum_{i=1}^{NC} x_{i,k} MW_i \qquad k = 1, ..., N \qquad (C.21)$$

$$\tilde{V}_k = V_k \cdot \sum_{i=1}^{NC} y_{i,k} MW_i \qquad k = 1, ..., N \qquad (C.22)$$

- Flooding velocity:

$$vel_k^{flood} = \left(\frac{\sigma_k^l}{20}\right)^{0.2} \cdot K1_k \cdot \left(\frac{\tilde{\rho}_k^l - \tilde{\rho}_k^v}{\tilde{\rho}_k^v}\right)^{0.5} \qquad k = 1, ..., N \qquad (C.23)$$

where σ_k^l is surface liquid tension, $K1_k$ is an empirical coefficient given by:

$$K1_k = 0.0105 + 0.1496 \cdot Space^{0.755} \cdot exp(-1.463FLV_k^{0.842}) \qquad k = 1, ..., N \qquad (C.24)$$

- Minimum column diameter and area:

$$D_{col,k}^{min} = \left(\frac{4A_{col,k}^{min}}{\pi}\right)^{0.5} \qquad k = 1, ..., N \qquad (C.25)$$

$$A_{net,k}^{min} = 0.9 \times A_{col,k}^{min} \qquad k = 1, ..., N \qquad (C.26)$$

280 Synthesis and Operability Strategies for Computer-Aided Modular PI

- Minimum net area for vapor-liquid disengagement

$$A_{net,k}^{min} = \frac{V_k}{0.8 \cdot \rho_k^v \cdot vel_k^{flood}} \qquad k = 1, ..., N \tag{C.27}$$

C.3 Reboiler and condenser modeling

The modeling of reboiler and condenser is in an analogous way to that of column trays, but with addition of heat input considerations in energy balances and without the pressure drop equation, flooding or entrainment correlations.

C.4 Physical properties

This column model is independent of the selection of physical property models. Thus, the required physical properties can be generally described as:

$$h^l = h^l(P, T, \mathbf{x}) \tag{C.28}$$
$$h^v = h^v(P, T, \mathbf{y}) \tag{C.29}$$
$$\rho^l = \rho^l(P, T, \mathbf{x}) \tag{C.30}$$
$$\rho^v = \rho^v(P, T, \mathbf{y}) \tag{C.31}$$
$$\tilde{\rho}^l = \tilde{\rho}^l(P, T, \mathbf{x}) \tag{C.32}$$
$$\tilde{\rho}^v = \tilde{\rho}^v(P, T, \mathbf{y}) \tag{C.33}$$
$$\sigma^l = \sigma^l(P, T, \mathbf{x}) \tag{C.34}$$
$$\Phi_i^l = \Phi^l(P, T, \mathbf{x}) \qquad i = 1, ..., NC \tag{C.35}$$
$$\Phi_i^v = \Phi^v(P, T, \mathbf{y}) \qquad i = 1, ..., NC \tag{C.36}$$

C.5 Initial conditions

In the case that the process is initially at steady-state, the initial conditions are:

$$\frac{dM_{i,\sqcup}}{dt}\Big|_{t=0} = 0, \qquad i = 1, ..., NC, \qquad \sqcup = \{k = 1, ..., N\} \tag{C.37}$$

$$\frac{dU_{\sqcup}}{dt}\Big|_{t=0} = 0, \qquad \sqcup = \{k = 1, ..., N\} \tag{C.38}$$

C.6 Equipment cost correlations

The reactive distillation column capital cost is calculated as [4]:

$$Total\ Cost = Operating\ Cost + Column\ Cost + Catalyst\ Cost \tag{C.39}$$

$$Operating\ Cost = 137.27\ \frac{\$}{\text{kW year}} \times Q_{Reboiler} + 26.19\ \frac{\$}{\text{kW year}} \times Q_{Condenser} \tag{C.40}$$

which incorporates the steam and cooling water costs (i.e., 137.27 \$/kW per year and 26.19 \$/kW per year) required for the reboiler and the condenser of the column. $Q_{Reboiler}$ and $Q_{Condenser}$ are respectively the heat duty of reboiler and condenser.

$$Column\ Cost = 1.4 \times 3.18 \times 101.9 \times \frac{M\&S}{280} \times \left(\frac{100D_{col}}{12 \times 2.54}\right)^{1.066} (HS+15)^{0.802}$$
$$+1.4 \times \frac{M\&S}{280} \times 4.7 \times \left(\frac{100D_{col}}{12 \times 2.54}\right)^{1.55} \times HS \tag{C.41}$$

where HS denotes the heighstack of the column, with the following definition

$$HS = Space\left(\sum_{k=1}^{N}(yr_k) - 1\right) \tag{C.42}$$

$$Catalyst\ Cost = \sum_{i=1}^{Reactive\ Trays} 0.2 \times 87\frac{\$}{\text{kg}} \times Catalyst\ Mass \tag{C.43}$$

where the factor 0.2 refers to the fact that the catalyst can be used for five consecutive years.

References

[1] M. Schenk, R. Gani, D. Bogle, E. Pistikopoulos, A hybrid modelling approach for separation systems involving distillation, Chemical Engineering Research and Design 77 (1999) 519–534.

[2] M.C. Georgiadis, M. Schenk, E.N. Pistikopoulos, R. Gani, The interactions of design control and operability in reactive distillation systems, Computers & Chemical Engineering 26 (2002) 735–746.

[3] P. Panjwani, M. Schenk, M. Georgiadis, E.N. Pistikopoulos, Optimal design and control of a reactive distillation system, Engineering Optimization 37 (2005) 733–753.

[4] V. Bansal, J.D. Perkins, E.N. Pistikopoulos, A case study in simultaneous design and control using rigorous, mixed-integer dynamic optimization models, Industrial & Engineering Chemistry Research 41 (2002) 760–778.

D

Nonlinear optimization formulation of the Feinberg Decomposition approach

In this appendix as the supplementary information for Chapter 10, we introduce the nonlinear programming (NLP) formulation of the Feinberg Decomposition (FD) approach proposed by Frumkin and Doherty [1]. The full set of this NLP model equations is given in detail by Eqs. (D.1), while a list of Nomenclature is provided in Table D.1.

Given the inlet molar flowrates (F_j) and reaction kinetics, the NLP problem (D.1) determines the optimum objective value (e.g., product flowrate, system temperature) considering the following optimization variables: molar fractions ($x_{j,i}$), temperatures (T_i), and pressures (P_i) in each CFSTR. In this case for olefin metathesis, 2 CFSTRs are employed since there exists 1 independent reaction (and another reverse reaction). Specifically, Eq. (D.1a) defines the objective function for the FD formulation. Eq. (D.1b) describes the overall mass balance taking the perfect separator and the CFSTRs as an integrated system. Eq. (D.1c) calculates reaction volume based on reactive holdups. Reaction rates in each CFSTR are determined by Eqs. (D.1d) and (D.1e). The mass balance around each CFSTR is depicted by Eq. (D.1f). Eq. (D.1g) ensures that the molar fractions in each CFSTR sum to unity. Eqs. (D.1h) and (D.1i) calculate inlet molar fractions for each CFSTR. Eqs. (D.1j)–(D.1o) respectively constrains molar fractions, volumes, temperatures, pressures, flowrates within certain bounds. Finally, Eq. (D.1p) constrains the molar flowrate with a ratio term α to avoid unrealistic large flowrates. The resulting NLP problem can be solved using deterministic optimization approaches and commercial solvers.

$$\max_{x_{j,i}, V_i, T_i, F_i} P_{prod} \tag{D.1a}$$

$$\text{s.t.} \quad F_j - P_j + \sum_{i=1}^{R+1} r_{j,i}(x)H_i = 0, \qquad j = 1, 2, ..., C \tag{D.1b}$$

$$V_i = H_i \sum_{j=1}^{C} \frac{x_{j,i}}{\rho_j}, \qquad i = 1, 2, ..., R+1 \tag{D.1c}$$

$$r_{j,i} = \sum_{i=1}^{R} v_{j,r}\hat{r}_{r,i}, \qquad i = 1, 2, ..., R+1, \quad j = 1, 2, ..., C \tag{D.1d}$$

$$\hat{r}_{1,i} = k_f(x_{C_5H_{10},i}^2 - \frac{x_{C_4H_8,i}x_{C_6H_{12},i}}{K_{eq}}), \qquad i = 1, 2, ..., R+1 \tag{D.1e}$$

$$P_i = F_i + \sum_{j=1}^{C} r_{j,i} H_i, \qquad i = 1, 2, ..., R+1 \tag{D.1f}$$

$$\sum_{j=1}^{C} x_{j,i} = 1, \qquad i = 1, 2, ..., R+1 \tag{D.1g}$$

$$x_{j,i}^0 = \frac{x_{j,i}[F_i + \sum_{j=1}^{C} r_{j,i}(x)H_i] - r_{j,i}(x)H_i}{F_i}, \quad j = 1, 2, ..., C-1, \quad i = 1, 2, ..., R+1 \tag{D.1h}$$

$$x_{C,i}^0 = 1 - \sum_{j=1}^{C-1} x_{j,i}^0, \qquad i = 1, 2, ..., R+1 \tag{D.1i}$$

$$0 \le x_{j,i}, x_{j,i}^0 \le 1, \qquad i = 1, 2, ..., R+1, \quad j = 1, 2, ..., C \tag{D.1j}$$

$$\sum_{i=1}^{R+1} V_i \le V_{max}, \qquad i = 1, 2, ..., R+1 \tag{D.1k}$$

$$T_{min} \le T_i \le T_{max}, \qquad i = 1, 2, ..., R+1 \tag{D.1l}$$

$$Pres_{min} \le Pres_i \le Pres_{max}, \qquad i = 1, 2, ..., R+1 \tag{D.1m}$$

$$P_i, F_i \ge 0, \qquad i = 1, 2, ..., R+1 \tag{D.1n}$$

$$P_j \ge 0, \qquad j = 1, 2, ..., C \tag{D.1o}$$

$$F_i \le \alpha F, \qquad i = 1, 2, ..., R+1 \tag{D.1p}$$

Appendix D • Nonlinear optimization formulation of the FD approach   285

Table D.1   Nomenclature for FD NLP formulation.

Variable	Physical meaning
$\hat{r}$	reaction rate
C	number of components
F	inlet molar flowrate
H	reactive molar holdup
k_f	reaction rate constant
K_{eq}	reaction equilibrium constant
P	product effluent molar flowrate
P_{prod}	product effluent molar flowrate
$Pres$	pressure
R	number of reactions
r	reaction rate for specific component
T	temperature
V	reaction volume
x	molar fraction
C_4H_8	butene
C_5H_{10}	pentene
C_6H_{12}	hexene
i	CFSTR number
j	component index
r	reaction number
α	flowrate constraint ratio
ν	stoichiometric coefficient
ρ	density

References

[1] J.A. Frumkin, M.F. Doherty, Ultimate bounds on reaction selectivity for batch reactors, Chemical Engineering Science 199 (2019) 652–660.

E

Degrees of freedom analysis and controller design in modular process intensification systems

In this appendix as the supplementary information for Chapter 13, we present in detail:

1. Degrees of freedom analysis for the olefin metathesis reactive distillation process versus reactor-distillation-recycle process.
2. Controller tuning parameters for the olefin metathesis reactive distillation process and reactor-distillation-recycle process.

E.1 Degrees of freedom analysis

To highlight the impact of design on operability and control, we distinguish hereafter three types of DOFs as defined in Nikačević et al. [1]:

  i. *Thermodynamic DOFs* – which give the number of independent intensive system properties such as pressure or temperature

  ii. *Design DOFs* – which are the number of independent geometrical properties available for process design

  iii. *Operational DOFs* – which identify the number of independent process variables that can be manipulated for process control or operation.

E.1.1 High fidelity dynamic modeling

In this section, the DOF comparison in the reactive distillation process and the reactor-distillation-recycle process is performed using high fidelity dynamic modeling with sufficient accuracy in describing the physical process as well as the correlations between design and operation (Appendix C). A generalized process system is considered, consisting of NC components, 1 feed stream, and 2 product streams. Each of the (reactive) distillation column has N_{tray} column trays. The detailed analyses of DOFs for RD and reactor-distillation-recycle are respectively presented in Tables E.1 and E.2, where the additional DOFs available in the conventional process are highlighted. It can be observed that RD has much less DOF compared to its conventional process counterpart. The general rules which results in the loss of DOFs are as summarized in Section 13.1.

288 Synthesis and Operability Strategies for Computer-Aided Modular PI

Table E.1 DOFs of reactive distillation based on dynamic modeling. (Adapted from Pistikopoulos et al. [3].)

	Variables	Number of DOFs
Thermodynamic DOFs	Reboiler pressure	1
	Condenser pressure	1
Design DOFs	Diameter, weir height, tray spacing	3
	Reflux drum diameter & length	2
	Reboiler diameter & length	2
Operational DOFs	Reflux ratio, Boilup ratio	choose 2
	Bottoms rate, Distillate rate,	
	Reboiler duty, Condenser duty	
Sum		11

Table E.2 DOFs of reactor-distillation-recycle based on dynamic modeling. (Adapted from Pistikopoulos et al. [3].)

	Variables		Number of DOFs
Thermodynamic DOFs	Distillation 1 & 2:	Reboiler pressure	1×2
		Condenser pressure	1×2
	Reactor:	Temperature, Pressure	2
Design DOFs	Distillation 1 & 2:	Diameter, weir height, tray spacing	3×2
		Reflux drum diameter & length	2×2
		Reboiler diameter & length	2×2
	Reactor:	Height & Diameter	2
Operational DOFs	Distillation 1 & 2:	Reflux ratio, Boilup ratio	choose 2×2
		Bottoms rate, Distillate rate,	
		Reboiler duty, Condenser duty	
	Reactor:	Outlet flowrate	1
	Flowsheet:	Recycle ratio	1
Sum			26

E.1.2 Steady-state modeling

As shown in Tables E.3 and E.4, steady-state modeling of the intensified reactive distillation process also has less DOFs comparing to that of the conventional reactor-distillation-recycle process. The reasons for loss of degrees of freedom are consistent with that summarized for dynamic modeling. Moreover, to briefly comment on the differences between steady-state modeling and dynamic high-fidelity modeling: steady-state modeling normally provides a more flexible description for a certain unit by considering its structural design via discrete (or binary) design variables, while dynamic modeling is mostly used for fixed design configurations. Take the (reactive) distillation column modeling as an example, the steady-state model proposed by Viswanathan and Grossmann [2] also enables the selection of feed tray location and reflux tray location (i.e., the DOFs of Feed/Reflux tray

Appendix E • DOF analysis and controller design in modular PI systems   289

structure). While attempts have been made to incorporate these discrete design decisions into dynamic modeling for the benefits of simultaneous design and control optimization, the computational tractability for mixed-integer dynamic optimization algorithms still leave it a better solution to determine discrete design variables at steady-state. Another difference is that steady-state modeling has less considerations of variable interactions due to equipment internal design. Again using the (reactive) distillation modeling as an example, the column tray pressures in dynamic modeling are calculated via pressure driving force when vapor flow passing the tray with certain geometry, while in steady-state modeling the pressure profile is pre-assigned. These all result in more DOFs, in other words a larger design space, in steady-state modeling versus dynamic modeling.

Table E.3 DOFs of reactive distillation based on steady-state modeling. (Adapted from Pistikopoulos et al. [3].)

	Variables	Number of DOFs
Thermodynamic DOFs	Stage pressures	$Ntray$
	Reboiler pressure	1
	Condenser pressure	1
Design DOFs	Feed tray structure	$Ntray - 1$
	Reflux tray structure	$Ntray - 1$
Operational DOFs	Reflux ratio, Boilup ratio	choose 2
	Bottoms rate, Distillate rate,	
	Reboiler duty, Condenser duty	
Sum		$3Ntray + 2$

Table E.4 DOFs of reactor-distillation-recycle based on steady-state modeling. (Adapted from Pistikopoulos et al. [3].)

		Variables	Number of DOFs
Thermodynamic DOFs	Distillation 1 & 2:	Stage pressures	$Ntray \times 2$
		Reboiler pressure	$1 \times 2T$
		Condenser pressure	1×2
	Reactor:	Temperature, Pressure	2
Design DOFs	Distillation 1 & 2:	Feed tray structure	$(Ntray - 1) \times 2$
		Reflux tray structure	$(Ntray - 1) \times 2$
	Reactor:	Volume	1
Operational DOFs	Distillation 1 & 2:	Reflux ratio, Boilup ratio	choose 2×2
		Bottoms rate, Distillate rate,	
		Reboiler duty, Condenser duty	
	Reactor:	Outlet flowrate	1
	Flowsheet:	Recycle ratio	1
Sum			$6Ntray + 9$

E.1.3 Superstructure-based synthesis modeling

Herein, we take the Generalized Modular Representation Framework (GMF) as an example of these PI synthesis strategies and analyze the gain or loss of DOFs in such phenomena-based representation of intensified and conventional process systems. The GMF modular representations for the reactive distillation and reactor-distillation-recycle are depicted in Fig. E.1. In Fig. E.1a, the GMF-based reactive distillation column comprises – from the top to the bottom – a pure heat exchange module where cooling water ("CW") exchanges heat with inlet vapor stream ("V"), three mass/heat exchange modules where reactive separation takes place between contacting liquid ("L") and vapor streams ("V"), and another pure heat exchange module where heating steam ("ST") exchanges heat with inlet liquid stream ("L"). Fig. E.1b can be interpreted in a similar way, just an addition of the leftmost "L-L" module represents a mass/heat exchange module with pure reaction task in analogy to the actual reactor.

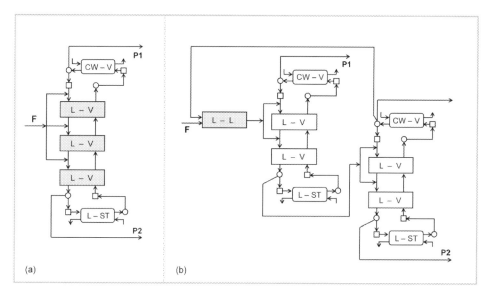

FIGURE E.1 GMF modular representation (reproduced from Pistikopoulos et al. [3]). (a) Intensified reactive distillation, (b) Conventional reactor-distillation-recycle.

At the level of full superstructure network, the number of DOFs are much surpassing the other two types of model due to the combinatorial decisions. For example, if 5 mass/heat exchange modules ($NE = 5$) are utilized to represent a process system involving 3 components ($NC = 3$), a total number of 182 degrees of freedom are available. However, many of these DOFs will appear or disappear with the selection of integer (binary) variables. As the (optimal) results of GMF synthesis, intensified reactive distillation process or conventional reactor-distillation-recycle process can be generated as shown in Fig. E.1. The DOF analyses on these configurations become consistent with that of steady-state and dynamic modeling, i.e.

Appendix E • DOF analysis and controller design in modular PI systems   291

- DOFs of GMF-based intensified reactive distillation:
  - Module temperatures and pressures ($2NE$, $NE = 5$ as the number of GMF modules)
  - Reflux ratio, Boilup ratio, Bottoms rate, Distillate rate, Reboiler duty, Condenser duty (choose 2) (2)
- DOFs of GMF-based conventional reactor-distillation-recycle:
  - Module temperatures and pressures, including mass/heat exchange modules and pure heat exchange modules ($2NE$, $NE = 9$)
  - Reflux ratio, Boilup ratio, Bottoms rate, Distillate rate, Reboiler duty, Condenser duty (choose 2) (2×2)
  - Reaction module outlet flowrate (1)
  - Recycle ratio (1)

E.2 Controller tuning for olefin metathesis case study

In what follows, we provide the key parameters used in Chapter 13 for the open-loop dynamic analyses, PI controller design, mp-MPC controller design, etc.

E.2.1 Open-loop response to step changes in manipulated variables

A series of open-loop analyses are performed for the RD process and the reactor-distillation-recycle process. For each analysis, a step change is introduced to one of the DOFs (or manipulated variables) as listed in Table E.5. The dynamic responses in these two processes are illustrated in Fig. 13.2.

Table E.5 Open-loop analysis for reactive distillation and reactor-distillation-recycle (adapted from Pistikopoulos et al. [3]).

	DOF	Nominal value	Step change	Step size
Reactive distillation	Reflux ratio	2.98	2.39	−20 %
	Distillate rate	50	49	−2 %
Reactor-distillation-recycle	Column 1 distillate rate	50	49	−2 %
	Column 1 reflux ratio	5.0	4.9	−2 %
	Column 2 reflux ratio	0.46	0.45	−2 %
	Column 2 boilup ratio	8.34	8.17	−2 %
	Valve stem position	0.70	0.72	+2 %
	Recycle ratio	1	0.995	−0.5 %

E.2.2 Open-loop response to step changes in feed flowrate

A step change at +1 kmol/h is introduced to the pentene feed flowrate (nominal value at 100 kmol/h). The dynamic responses in the RD process and the reactor-distillation-recycle process are presented in Fig. 13.7.

E.2.3 PI controller design for RD and reactor-distillation-recycle

For the PI control of RD, two controllers are designed in which reflux ratio and distillate rate are selected as manipulated variables to respectively control the top and bottom product purity as two single-input single-output (SISO) systems. The controller tuning parameters are given in Table E.6.

Table E.6 PI controller tuning parameters for RD. (Adapted from Pistikopoulos et al. [3].)

Manipulated Variable	Control Variable	K_p	K_i
Reflux ratio	Top product purity	1e-2	1e2
Distillate rate	Bottom product purity	1e-2	1e0

For the PI control of reactor-distillation-recycle, two pairing schemes are selected to control the top and bottom product purity with SISO PI controllers, which are shown in Table E.7.

Table E.7 PI controller tuning parameters for reactor-distillation-recycle. (Adapted from Pistikopoulos et al. [3].)

	Manipulated Variable	Control Variable	K_p	K_i
Pairing Scheme 1	COL1 distillate flowrate	Top product purity	1e-1	1e2
	COL2 boilup ratio	Bottom product purity	1e-1	1e3
Pairing Scheme 2	COL1 distillate flowrate	Top product purity	1e-1	1e2
	Flowsheet recycle ratio	Bottom product purity	1e-3	1e-4

E.2.4 PI controller design for modular RD units

As detailed in Section 13.3, two PI control pairing schemes are designed for the modular RD units. The pairing scheme and controller tuning parameters are presented in Table E.8.

Table E.8 PI controller tuning parameters for modular RD units. (Adapted from Pistikopoulos et al. [3].)

	Manipulated Variable	Control Variable	K_p	K_i
Pairing Scheme 1	Feed split ratio	RD1 top product purity	1e-2	1e3
	RD1 distillate rate	RD1 bottom product purity	1e-2	1e-2
	RD2 distillate rate	RD2 top product purity	1e-2	1e0
	RD2 reflux ratio	RD2 bottom product purity	1e-2	1e2
Pairing Scheme 2	Feed split ratio	Top product purity after mixing	1e-1	1e-1
	RD2 reflux ratio	Bottom product purity after mixing	1e-2	1e2
	RD2 distillate rate	RD2 top product purity	5e2	1e-2

Appendix E • DOF analysis and controller design in modular PI systems   293

E.2.5 mp-MPC controller design for RD

An explicit/multi-parametric model predictive controller (mp-MPC) is designed for the RD system following the PAROC framework (Chapter 8). The olefin metathesis RD column is managed as a multi-input multi-output (MIMO) process system using the reflux ratio and distillate rate as manipulated variables to control top and bottom product purity. Considering a model predictive control problem described by Eq. (E.1), where QR_k, $R1_k$ are the weights of the controller, P is derived from the solution of the discrete time Riccati equation, OH and CH are the output and control horizons respectively and ϵ takes into account the mismatch between the process and the developed approximate model. The tuning parameters for the mp-MPC controller are shown in Table E.9.

$$\min_{u} \quad J = x_N^T P x_N + \sum_{k=1}^{OH-1} \left((y_k - y_k^R)^T Q R_k \left(y_k - y_k^R \right) \right)$$

$$+ \sum_{k=0}^{CH-1} \Delta u_k^T R_k \Delta u_k$$

$$\text{s.t.} \quad x_{k+1} = A x_k + B u_k + C d_k \tag{E.1}$$

$$y_k = D x_k + \epsilon$$

$$x_{min} \leq x_k \leq x_{max}$$

$$u_{min} \leq u_k \leq u_{max}$$

$$\Delta u_{min} \leq \Delta u_k \leq \Delta u_{max}$$

$$y_{min} \leq y_k \leq y_{max}$$

Table E.9 Explicit MPC controller tuning parameters for RD.
(Adapted from Pistikopoulos et al. [3].)

MPC Parameters	Value
OH	3
CH	2
QR	[1E5, 0; 0, 1E5]
$R1$	[1E1, 0; 0, 1E1]
u_{min}	[2.4, 40]
u_{max}	[4.4, 60]
Δu_{min}	[−1, −1]
Δu_{max}	[1, 1]
y_{min}	[0, 0]
y_{max}	[1, 1]

References

[1] N.M. Nikačević, A.E. Huesman, P.M. Van den Hof, A.I. Stankiewicz, Opportunities and challenges for process control in process intensification, Chemical Engineering and Processing: Process Intensification 52 (2012) 1–15.

[2] J. Viswanathan, I.E. Grossmann, Optimal feed locations and number of trays for distillation columns with multiple feeds, Industrial & Engineering Chemistry Research 32 (1993) 2942–2949.

[3] E.N. Pistikopoulos, Y. Tian, R. Bindlish, Operability and control in process intensification and modular design: Challenges and opportunities, AIChE Journal 67 (2021) e17204.

F

MTBE reactive distillation model validation and dynamic analysis

In this appendix as the supplementary information for Chapter 14, we discuss: (i) MTBE reactive distillation model validation with Aspen Plus simulator, and (ii) Steady-state and dynamic analyses on the selection of manipulated variable for the MTBE reactive distillation column.

F.1 MTBE reactive distillation model validation with commercial Aspen simulator

In this section, we compare the steady-state values of the dynamic MTBE reactive distillation model built in gPROMS (hereafter referred as "gPROMS") with that obtained via Aspen simulation using the RADFRAC module for reactive distillation modeling (referred as "Aspen"). The kinetic model, thermodynamic model, column design and operating parameters are kept consistent as detailed in Chapter 14 for Operable Design 1. For the tray numbering scheme, condenser is counted as the first tray while reboiler the last tray.

The total energy consumption values are respectively 36.8 MW by Aspen simulator and 36.0 MW by the gPROMS custom model, featuring a relative error of 2%. The profiles for column temperature, vapor and liquid molar compositions are illustrated in Fig. F.1. The results indicate a good agreement between the proposed model with the Aspen simulator.

F.2 Steady-state and dynamic analyses on the selection of manipulated variable for MTBE reactive distillation

The common choices for manipulated variables in reactive distillation column include: reflux ratio, feed to reflux ratio, distillate flowrate, bottom flowrate, condenser heat duty for pressure control, reboiler heat duty for manipulation of column temperature, and feed flowrate (if applicable). For the Operable Design 1 of interest for MTBE production in Chapter 14, it is assumed to operate at a constant pressure of 8 atm with a perfectly controlled bottom product flowrate at 197 mol/s as per the product specification. Therefore,

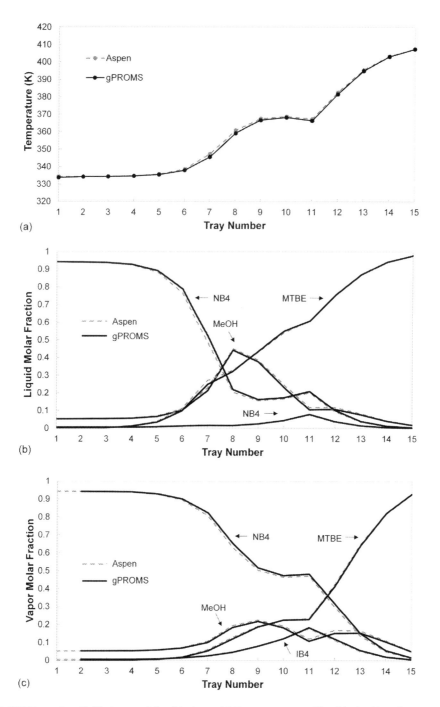

FIGURE F.1 MTBE reactive distillation model validation – (a) Temperature profile, (b) Liquid molar compositions, (c) Vapor molar compositions (adapted from Tian et al. [1]).

the remaining degrees of freedom are: (i) reflux ratio (or analogously feed to reflux ratio), (ii) reboiler heat duty, and (iii) vapor butenes feed flowrate.

The reactive distillation column is considered as a single-input single-output system to maintain the MTBE product purity at a desired setpoint despite process disturbances. To determine the proper manipulated variable, we first perform an open-loop dynamic analysis by imposing a step change at time = 0 to each of the three manipulated variable candidates. The open-loop step responses are shown in Fig. F.2. As can be noted, the MTBE product purity shows clearly a nonlinear open-loop response to the step change of reboiler duty or reflux ratio. This may give a misleading signal to the linear explicit model predictive controller for decision making. On the other hand, the vapor butenes flowrate shows a monotonic behavior on predicting the change of the MTBE product purity.

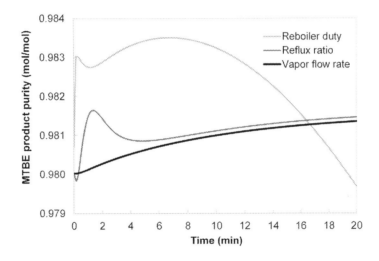

FIGURE F.2 Reactive distillation open-loop dynamic response (adapted from Tian et al. [1]).

A second analysis is performed to investigate the steady states of MTBE product purity under different values of reflux ratio, reboiler heat duty, and vapor feed flow rate. As illustrated in Fig. F.3, the butene vapor feed flowrate has the most influence in MTBE product purity. Moreover, the product purity increases monotonically with the increase of vapor feed flowrate. However, steady-state gain inversion is observed with respect to the other two manipulated variables, i.e. reflux ratio and reboiler heat duty. Given these analyses, the butene vapor feed flowrate is selected as the manipulated variable in this case study.

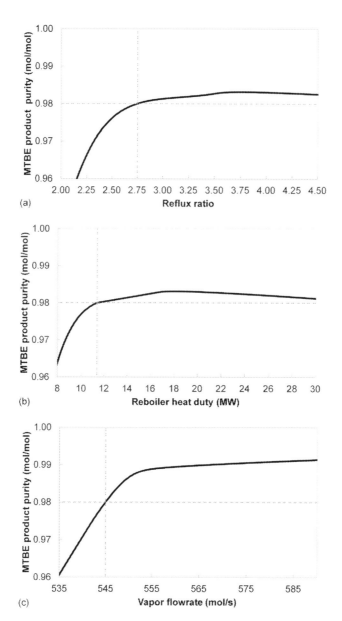

FIGURE F.3 Steady-state analysis of MTBE product purity under varying conditions. (a) Reflux ratio, (b) Reboiler heat duty, (c) Vapor butenes feed flowrate. (Adapted from Tian et al. [1].)

References

[1] Y. Tian, I. Pappas, B. Burnak, J. Katz, E.N. Pistikopoulos, Simultaneous design & control of a reactive distillation system – A parametric optimization & control approach, Chemical Engineering Science 230 (2021) 116232.

Index

A

Achievable output set (AOS), 31
Advanced
　operational strategies, 5
　optimization approach, 136
Air separation units (ASU), 34
Application Programming Interface (API), 248
Aspen
　optimization, 201, 230, 232
　simulation, 178, 195, 200–202, 229, 231, 295
　　flowsheets, 200
　　results, 184, 193, 237
　　setup, 201
Aspen Plus, 191, 200, 202, 228, 232, 236, 238, 247, 248, 255
　RADFRAC module, 196
　simulation, 177
　simulator, 295
Atmospheric pressure, 86, 89, 212
Attainable region (AR)
　approaches, 163, 164
　bounds, 163
　theory, 163
Attainable region calculation, 55
Autocatalytic reaction, 119
Available input set (AIS), 31
Azeotropic
　distillation, 13
　separation systems, 181

B

Batch operations, 15, 32
Bed reactors, 5, 15
Binary variables, 46, 52, 62, 71–73, 79, 83, 136, 155, 176, 193, 197, 229, 243, 253
Bottom
　flowrate, 295
　MTBE molar fraction, 242
　product
　　flowrate, 295
　　purity, 210, 211, 292, 293
　　purity specifications, 213
　rate, 291
　separation module, 230
　stream, 179, 190, 202
Box constraints, 131, 133, 134
Butene
　separation performance, 56
　vapor feed flowrate, 297
Butene (BUT), 53

C

Capital cost, 10, 14, 22, 175, 177, 229, 230, 232
　investment, 185
　reduction, 10
Catalyst mass, 51
Catalytic
　membrane reactor, 31
　reaction, 223
Chemical
　process, 25, 34, 47, 48, 60, 148
　　components, 22
　　control, 29
　　control basics, 123
　　designs, 45
　　functions, 45
　　industry, 14, 111, 112
　　systems, 59
　processing, 155
　reactions, 10, 116, 118
　separation operations, 7
Chemical Exposure Index (CEI), 111
　calculation, 111, 112
　guide, 111
Chromatographic
　reactors, 15
　separators, 15

299

300　Index

Cold
 process streams, 105, 106
 side stream, 82
 streams, 82, 157
Comprehensive process safety assessment,
  111
Conceptual synthesis, 20
Condenser, 22, 74, 89, 178, 194, 201, 227, 230,
  235, 236, 255, 295
 duty, 191, 201
 heat duty, 295
 heat duty optimal, 90
Condenser/reboiler duty, 201, 230
Consecutive reactions, 11
Constant
 system pressure, 229
 temperature, 49
Constituent subsystems, 220
Constrained Linear Quadratic Regulator
  (CLQR) problem, 126, 128, 129, 137
Constraints
 driving force, 25, 48–53, 59, 62, 65, 81, 149,
  163, 170, 180, 228
 feasibility, 83
 material/energy balance, 27
 nonconvex, 30
 operation, 213
 operational, 95
 process, 36, 125, 151, 153, 207, 211
 process design, 31
 relevant, 29
Continuous
 microreactor operation, 13
 operation, 14
 process, 13
 variables, 71, 73, 79, 104, 153, 176, 197, 229,
  253
Continuous Flow Stirred Tank Reactor (CFSTR)
 equivalence principle, 163, 164
 principle, 163, 164
 temperatures, 165
Control
 actions, 34, 123, 124, 139, 215, 243
 analysis, 29, 247

 approaches, 37, 249
  metrics, 37
 characteristics, 207
 concerns, 211
 considerations, 147, 248, 249
 criterion, 249
 decisions, 220
 efficiency, 219
 horizon, 158, 293
 inputs, 219
 intensified process, 29
 investigations, 239
 law, 123, 125, 129, 137, 138
 layers, 32
 measures, 115
 metrics, 36, 220
 objective, 140, 141, 157, 242
 optimal, 125, 128, 129, 135, 137, 138, 148,
  242
 optimization, 31, 34, 150, 219, 223, 237, 243,
  289
 optimization problem, 138
 output, 215, 241
 performance, 138, 157, 217, 245, 255, 259
 PI, 123, 124, 144, 211, 213, 292
 problem, 138, 239
 problem formulation, 243
 process, 34, 123, 207, 287
 schemes, 215
 strategies, 135, 219, 242
 structure, 219
 suite, 247, 248, 255
 system, 116, 123
 target, 147
 variables, 31, 95, 123, 125, 136, 137, 141, 147,
  151, 158, 219, 239, 241
Controller, 123, 137, 158, 259, 293
 design, 29, 31, 136, 141, 157, 209, 219, 239,
  242
 outputs, 124
 performances, 243
 tuning, 291
 tuning parameters, 142, 243, 287, 292

Index 301

Conventional
  distillation columns, 202
  PID control, 34
  process, 22, 207, 211, 214, 215, 219, 287
    counterpart, 36, 207, 287
    solution, 255
    synthesis, 25
    systems, 290
  reactors, 14
  unit operations, 25, 29
Cooling water (CW), 290
Cost
  optimality, 178
  optimization, 177
Critical Region (CR), 131–135, 138
CSTR
  integrated, 54
  reactor model, 55
Cyclic distillation, 14

D

Decanter, 193, 195, 199
  dewatered organic stream, 197, 199, 202
  dewatered organic stream split ratio, 200
  flowsheet, 191
  module, 191, 193, 200
  temperature, 201
Degrees of freedom (DOF)
  analysis, 147, 208, 287
  loss of, 207, 288, 290
  operational, 208, 209
Design feasibility, 175
Desired input set (DIS), 31
Desired output set (DOS), 31
Dewatered
  organic stream, 191, 197
  organic stream decanter, 197, 199, 202
  stream, 191, 202
Distillate
  rate, 74, 191, 202, 211, 291–293
  stream, 176, 179, 199, 202
Distillation, 7, 8, 10, 45, 47
  column, 8, 14, 20, 46, 53, 69, 74, 75, 89, 141,
    179, 183, 188, 213, 232, 255, 287
    modeling, 288

  modules, 236
  optimization, 75
  representation, 24
  sequences, 86, 87
  setup, 200
  superstructure, 88
  module RADFRAC, 229
  operating window, 212
  operations, 13, 213
  processes, 3, 12, 13, 26, 213
  reactive, 3, 5, 6, 8, 10, 22, 34, 47, 48, 144, 165,
    168, 207, 208, 212, 213, 227, 229, 231,
    238, 248, 255, 290, 291, 295
  systems, 47, 177
  tray, 52, 53, 154, 178, 180, 185
Distribution network (DN), 25
Disturbance, 36, 123, 129, 136, 147, 150, 157,
    158, 207, 211, 215, 226, 239, 249, 255
  process, 29, 239, 243
  profiles, 144
  rejection, 138, 215, 219
Dividing wall column (DWC), 6–8, 187
  design, 188
  design optimization, 187
  systems, 187
Downstream separation equipment, 10
Driving force constraints, 25, 48–53, 59, 62, 65,
    81, 149, 163, 170, 180, 228

E

Elevated operating temperatures, 10
Emergency equipment, 114
Endothermic processes, 114
Energy
  balances, 62, 64, 79, 81, 84, 85, 105, 135, 141
  consumption, 7, 14, 22, 36, 75, 90, 149, 181,
    184, 188, 219
  consumption rates, 185
Enhanced
  flexibility, 150
  safety, 33
  safety performance, 33
Enriched superstructure representation, 228

302   Index

Equipment
   assignment, 23
   design, 180
   failure data, 157
   failure frequency, 33, 120, 149, 153, 226, 233
   intensified, 23, 155, 173, 207
   intensified process, 219
   optimal, 148
   optimization, 173
   process, 5, 33, 150
   sizes, 4, 5
   sizing, 68, 103
   translation, 178
   types, 25, 153
   units, 68
   volume reactive, 170
Ethylene glycol (EG), 139, 174–176, 178, 179
   production, 11
   solvent, 182
   solvent flowrate, 140, 144
Excessive pentene reactant, 167
Exothermic
   chemical reactions, 114
   reactions, 10, 16, 32
Expected disturbance set (EDS), 31
Explosive substances, 156
Extractive
   distillation column, 139, 140, 178, 182, 185
   separation, 53, 79, 181, 185
   separation processes, 173, 175
   separation systems, 173

F
Feasibility, 29, 98, 232
   analysis, 157
   constraints, 83
   design, 175
   dynamic, 31
   function, 31
   heat exchange, 47
   mass/heat exchange, 25
   operational, 248
   process, 29

Feed
   flowrate, 188, 213–215, 291, 295
   flowrate disturbances, 218, 249
   preheating, 16
   stream, 16, 61, 74, 75, 82, 86, 89, 139, 140,
      147, 173, 175, 193, 194, 196, 249, 287
   temperature, 74
   tray location, 288
Feinberg Decomposition (FD), 163–165
   approach, 163, 164, 171
   boundaries, 170, 171
   design boundaries, 170, 171
   optimization formulation, 164
   theory, 163, 165
Flammability, 33, 34, 113, 118, 147, 156, 226
Flammability risk, 233
Flash column, 55, 182, 185, 202
   process, 54, 55
   process reactive, 54
   reactive, 55–57
Flash distillation, 12
Flexibility
   analysis, 29, 31, 95, 99, 105, 150, 233, 235,
      239, 248
   considerations, 103, 239
   index, 97, 99, 102, 107, 109
      calculation, 107
      problem, 29, 95, 97, 98, 101, 102, 107
   operational, 34, 219
   performance, 153
   problems, 31
   process, 95, 147
   programming problem nonconvex, 30
   requirement, 97, 151
   steady-state, 150, 151
   target, 104, 147
   test, 96, 98, 99, 101, 103, 149, 151, 152, 235,
      239
   test problem, 29, 95, 98, 100, 104, 151
   text, 101
   text problem, 97
   vertex problem, 98
Flowrate
   bottom, 295
   feed, 188, 213–215, 291, 295

Index 303

product, 200, 227
vapor, 69, 79, 228
Flowsheet synthesis, 22, 28
Flowsheet synthesis process, 28, 29

G

Gas separation, 11, 14
Gas to liquids (GTL) process, 13
General Algebraic Modeling System (GAMS), 229
Generalized Bender Decomposition (GBD), 70, 73, 229
  algorithm, 71
  algorithm workflow, 71
  convergence criteria, 73
  decomposition strategy, 73
  method, 70
  solution optimal, 73
  solution strategy, 70
Generalized Modular Representation Framework (GMF), 25, 45, 148, 163, 290
  design, 153, 167, 170, 235
  heat exchange, 157
  intermediate process solutions, 179
  mass/heat exchange module, 50, 52, 68, 70, 167, 180, 185, 229
  modular
    building blocks, 149, 193
    configuration, 227
    designs, 150, 236
    layout, 229
    optimization, 54
    representation, 227
    structure, 193
  module, 54, 68, 70, 79, 150, 154–156, 179, 227, 231, 249, 291
  module selection, 181
  optimal solution, 169
  optimal structure, 229
  optimization, 253
  process intensification synthesis, 83
  process stream, 61
  reaction module, 56
  reactive separation module, 57

representation, 48, 52, 55, 56, 74–76, 193, 195, 197, 231
representation capabilities, 45
synthesis, 52, 75, 79, 86, 87, 90, 151, 153, 176, 178, 187, 193, 197, 200, 226, 231, 290
  model, 60, 62, 69, 71, 79, 82, 150, 153, 228, 230, 234
  optimization, 197
  optimization problem, 53
  problem, 61, 70, 87, 89
  representation, 73
  results, 150, 237, 255
  strategy, 190
  structure, 202
gPROMS, 138, 243, 247, 248, 255, 295
  custom model, 295
  environment, 255
  ModelBuilder, 136, 141
  process, 158
Guaranteed operability, 150, 249
Guaranteed operability performance, 37

H

Hazard Distance (HD), 112
Hazardous process failure, 32
Heat
  capacity, 74, 105
  duties, 87, 105, 151
  exchange, 47, 61, 68, 75, 83, 106, 157, 158, 255, 290, 291
    area, 104
    feasibility, 47
    module, 25, 47, 48, 60, 74, 75, 79, 82, 83, 148, 151, 153, 169, 176, 180, 181, 194, 197–199, 227–229, 248, 253, 290
  integration feasibility constraints, 86
  reaction, 10, 118
  recovery, 106
  synthesis, 25
  transfer, 45, 47, 82, 173
    coefficient, 157
    feasibility, 82
    feasibility constraints, 106
    loads, 66, 68
    performances, 25

304 Index

Heat exchanger (HE), 82, 105, 106, 153, 157, 200
    areas, 158
Heat exchanger network (HEN), 105, 157
    problem, 108
    synthesis, 34, 157
    toxicity risk, 157
Heat Integration (HI), 82, 89
    block, 82, 83, 89
    considerations, 27, 83, 89
Heating
    duty, 191, 201, 202
    energy consumption, 185
    module, 230
Hexene (HEX), 53
Homogeneous azeotropic separation, 48
Hot utility (HU), 157

I

Inactive
    constraints, 129, 130, 132, 134
    inequality constraints, 131, 132, 134
Indication number, 121, 154
    approach, 121, 153
    for eac GMF module, 154
Indirect control systems, 219
Inequality constraints, 83, 95, 96, 107, 130, 131, 134, 151, 176
Inequality constraints process, 98
Infeasibility, 98
Infeasibility areas, 30
Infinite DimEnsionAl State Space (IDEAS) approach, 25
Inherent process
    physics, 34
    safety, 13
Inherent safety, 223, 235, 239
    analysis, 33, 151, 157
    assessment, 33
    considerations, 32, 33
    criteria, 148
    index, 33
    indicator, 239
    level, 239

    performance, 111, 115, 147, 149, 151, 153, 233, 249
    process, 31, 33
    targets, 157
Inherently safer
    design, 32, 156, 157, 234
    process, 156
    process systems, 111
Inlet stream
    conditions, 167
    liquid, 180
    vapor, 180
Inlet/outlet
    temperatures, 105
    vapor flowrate, 69
Instructive process alternatives, 46
Intensification synthesis, 247
Intensification synthesis process, 20, 22, 37, 45, 79, 111, 148, 171, 193, 202, 223, 227, 231, 247, 249, 252
Intensified
    chemical processes synthesis, 27
    equipment, 23, 155, 173, 207
    equipment safety performance, 32
    GMF modular structures, 150
    modular production process, 207
    process, 5, 15, 20, 25, 28, 29, 45, 125, 150, 207, 209, 247, 249
        control, 29
        designs, 249, 255
        equipment, 219
        systems, 19, 29, 32, 35, 45, 52, 207, 248
    reaction, 255
    reactive distillation process, 288
    reactive separation systems, 249
    systems, 20, 28, 34, 36, 48, 215, 219
Interconnecting
    streams, 66, 193, 197, 208, 217, 236
    vapor, 183
Ionic liquids, 173–175, 181, 185

K

Karush-Kuhn-Tucker (KKT) active set strategy, 130

Index   305

L

Limit value, 121, 122, 154, 155
  for explosive substances, 122
  for toxicity, 121
Liquid
  activity coefficient, 51, 53, 224
  composition, 231
  composition profiles, 195
  ethanol, 139, 175, 181
  flashing, 112
  flowrates, 81
  holdup, 69, 70, 231
  hydraulic calculation, 141
  inlet, 49, 170, 185
    flowrate, 230
    stream, 180
  level, 154
  mixture, 167
  molar
    compositions, 295
    fraction profiles, 53
    holdup, 69
  MTBE, 224
  outlet, 154, 155, 185
  outlet stream, 53, 54, 177, 180, 182, 193, 199, 200
  phase, 50, 53, 165, 167, 194, 223, 249
  phase reaction, 69
  pool evaporation, 121
  reactive mixture, 165
  stream, 65, 183, 201, 253
  stream outlet, 197
  temperatures, 57
  water, 181

M

Manipulated variables, 123, 125, 136, 137, 147, 158, 209, 211, 230, 241, 291–293, 295, 297
  for control, 209
  in reactive distillation column, 295
  optimal, 125
Mass
  balances, 62, 63, 83, 85, 141
  conservation, 50

densities, 154
exchange, 69
exchange module, 153
fractions, 155
transfer, 45, 47, 49, 50, 52, 53, 62, 65, 69, 116, 153, 194
transfer feasibility, 47, 50, 55, 149, 180
vapor, 141
Mass/energy integration, 207
Mass/heat
  exchange, 255
    feasibility, 25
    module, 25, 47–49, 60, 64, 69, 79, 82, 83, 148, 150, 151, 153, 154, 166, 169, 170, 176, 177, 194, 197–199, 227, 228, 248, 253
  transfer, 48, 91, 148, 181
  transfer phenomena, 36
  utilities, 147
Material factor (MF), 113, 115
Material/energy balance constraints, 27
Maximum
  butene production rates, 56
  pressure, 32, 68
Membrane
  distillation, 6, 7, 11
  reactors, 3, 5, 6, 8, 10, 11, 27
Metathesis reaction, 167, 249
Methanol
  contamination, 179
  feed flowrate, 226, 235, 239
  liquid feed flowrate, 239
Methyl methacrylate (MMA)
  feed stream, 197, 199, 202
  product
    flowrate specification, 200
    purity, 190, 202
    specifications, 200
    stream, 190, 194, 199, 202
  purification process, 190, 193, 197
  recovery flowrate, 202
Methyl methacrylate (MMA) purification, 187
Methyl tert-butyl ether (MTBE), 11, 223, 224, 233
  liquid, 224

306 Index

molar composition, 239
product purity, 297
production, 223, 228, 231, 241, 295
   process, 226, 227, 233
   systems, 235, 238, 239, 245
   task, 227, 245
production process systems, 223
reactive distillation, 238, 295
reactive distillation column, 295
Microreactors, 5, 6, 8, 13, 36, 219
Minimum temperature, 82, 83
Mixed-integer dynamic optimization (MIDO)
   problem, 31
Mixed-integer nonlinear programming
   (MINLP)
  model, 176
  problem, 149, 229
Mixed-integer programming (MIP) problem,
   72, 101
Model predictive control (MPC), 34, 125
  control, 34
  online optimization, 128
  problem, 293
  schemes, 137
Model predictive controller, 138
Modular
  building blocks, 47, 60, 148
  building blocks GMF, 149, 193
  characteristics, 6
  chemical process intensification, 25
  configuration GMF, 227
  design, 3, 20, 207, 219
  flowsheet, 219
  flowsheet features, 217
  intensified process, 220
  layout GMF, 229
  mass/heat exchange, 59
  operation, 19
  optimization, 53, 54
  plant design, 31
  process, 34
  process equipment, 14
  process intensification, 3, 7, 19, 20, 25, 29, 45
   control optimization systems, 220
   operation systems, 37

   opportunities, 37
   synthesis, 148, 249
   systems, 19, 31, 32, 95, 147
   systems operation, 37
   technologies, 6, 8, 10, 47
  processing systems, 219
  reactors, 13
  structure GMF, 193
  systems, 29, 34, 37, 219
  units, 219
Modularization, 36, 216, 217, 219
Module
  cost, 68
  decanter, 191, 193, 200
  design parameters, 79
  diameters, 235
  flowsheet, 253
  heat exchange, 48, 75, 82, 83, 148, 176, 180,
   185, 229, 248, 290
  heating, 230
  height, 154
  holdup, 68
  mass transfer capability, 57
  mass/heat exchange, 48, 150, 151, 154, 166,
   177, 194, 227, 228, 248
  outlet streams, 52, 180
  pressure, 75
  reaction, 55, 291
  reactive, 166, 168
  reactive separation, 55, 57, 167, 230, 234–236
  separation, 53, 55, 56, 167, 229, 230
  volume, 154
Molar flowrate vapor, 69, 239
Monolith reactors, 7
Multi-input multi-output (MIMO)
  control system, 141
  process system, 293
  system, 140
Multi-parametric model predictive control
  (mp-MPC), 128, 141, 239, 255
Multi-parametric quadratic optimization
  (mp-QP) problem, 129, 130, 137

Multiperiod
  GMF synthesis, 239
  optimization problem, 103
  synthesis, 104
Multistage separation, 79

N
Network operation, 151
Nonconvex
  constraints, 30
  flexibility programming problem, 30
  systems, 153
Nonideal
  liquid behavior, 139
  reactor network synthesis, 26
Nonlinear
  optimization basics, 130
  process dynamics, 125
Nonlinear programming problem (NLP), 164

O
Olefin metathesis, 164, 165, 209, 212, 213, 217, 247, 293
  process, 166
  reactive distillation process, 287
Operability
  analysis, 36, 148, 151, 220
  metrics, 36
  performances, 211, 235
  PI, 248
  process, 247, 248, 255
  study, 33
Operable
  designs, 235, 236, 238, 239, 241, 242, 295
  GMF modular, 235
  MTBE production systems, 223, 227
  process, 241
    design, 31
    intensification, 150, 223, 245
    intensification synthesis, 247
    intensification synthesis systems, 148
Operation
  agileness, 3, 36
  characteristics, 164
  constraints, 213

continuous, 14
design, 231
dynamic, 34, 148
feasible, 95–98
fully automated, 14
infeasible, 96
information, 178
mode, 14
modular, 19
nodes, 23
parameters, 209
period, 103
periodic, 5, 14, 15
process, 47
reactive distillation, 230
space, 212
temperatures, 166
vacuum pressure, 181
Operational
  analysis, 19, 231
  analysis methods, 36
  aspects, 36
  challenges, 14, 36
  characteristics, 37
  constraints, 95
  DOFs, 208
  feasibility, 248
  flexibility, 34, 219
  information, 46
  objectives, 219
  optimization, 46, 135
  optimization approaches, 219
  performances, 36, 150
  perspective, 231
Optimal
  condenser heat duty, 90
  configuration, 230
  control, 125, 128, 129, 135, 137, 138, 148, 242
  control law, 138
  design, 87, 103, 149, 150, 177–179, 235, 238, 255
    parameters, 232
    solution, 230
  dynamic operation strategies, 249
  equipment, 148

308   Index

explicit model predictive controller design, 150
GBD solution, 73
GMF
   configuration, 181
   modular process, 228
   modular process alternatives, 149
   process, 182
   solution, 75, 200
grassroots design, 197
heat duties, 90
manipulated variables, 125
mass separating agent, 173
modular solution, 230
operating conditions, 75, 89
performance, 228
process, 52
   design, 57, 59, 87, 90, 135, 150, 173, 190, 252, 259
   solution, 89, 190, 248, 253
reboiler heat duty, 90
retrofit design, 197
separation, 176
solution, 130–133, 177, 197, 198, 230
solvent choice, 174
structure, 230, 235
TAC value, 177
thermal coupling process scheme, 91
Optimality, 59, 75, 129, 138, 175
gap, 229
process, 62
Optimization
algorithms, 187
Aspen, 201, 230, 232
capabilities, 59
control, 31, 34, 150, 219, 223, 237, 243, 289
environment, 33
equipment, 173
format, 127
GMF, 253
GMF modular, 54
GMF synthesis, 197
methods, 20
modular, 53, 54
objective, 75, 87, 243

objective function, 74
operational, 46, 135
problem, 22, 68, 96, 98, 100, 101, 104, 128, 138, 149, 153, 166, 231, 243, 248
procedure, 228
steps, 136
variables, 59, 89, 229
Organic stream, 197, 202
Orthogonal Collocation (OC), 79, 173
Outlet
liquid, 154, 155, 185
streams, 49, 60, 79, 81, 167, 170, 180, 193, 199
vapor, 185

P
Parametric optimization, 243
PAROC (PARametric Optimization and Control) framework, 34, 135, 141, 144, 150, 158, 215, 239, 255, 293
Partial Differential Algebraic Equation (PDAE) model, 158
Pentene (PEN), 53
feed flowrate, 291
feed stream, 168, 169
metathesis reaction, 165, 249
Pertinent process units, 112, 114
Pervaporation (PV), 12
plants, 13
vapor, 12
Petlyuk column, 8, 48, 90, 91, 191, 195, 198, 200, 201
Petlyuk column reactive, 236
Phase contact (PC), 28
Phase separation (PS), 28, 45, 47
Phase transition (PT), 28
Phenomena building block (PBB), 27
Phenomenological process intensification synthesis, 163
Pool evaporation, 112
Pool evaporation liquid, 121
Predictive control, 139, 151, 237, 239, 249
Prefractionator, 8, 179, 195, 197, 200, 202
column, 8
section, 194, 197, 199, 201
streams mix, 197

Index   309

Pressure
  effects, 118, 120
  module, 75
  profile, 289
  relief, 111
  vapor, 118, 181, 188
  vessel, 68
Pressure swing adsorption (PSA), 14, 34
  process, 14, 15
Pressure swing adsorptive reactor (PSAR), 15
Process
  alternatives, 29, 111, 163, 178, 202
  bottlenecks, 20, 28, 48
  boundaries, 163
  conditions, 121, 154
  constraints, 36, 125, 151, 153, 207, 211
  continuous, 13
  control, 34, 123, 207, 287
  conventional, 22, 207, 211, 214, 215, 219, 287
  cost, 46
  description, 74, 86, 105, 139, 157, 165, 188, 223
  design, 11, 33, 46, 97, 103, 111, 116, 147, 171, 173, 177, 182, 187, 197, 248, 249, 255, 287
    constraints, 31
    intensification, 181
    parameters, 255
    stages, 208
    variables, 111
  disturbances, 29, 239, 243
  dynamics, 14, 36
  efficiencies, 5
  equipment, 5, 33, 150
  feasibility, 29
  flexibility, 95, 147
  flexibility analysis, 29
  flowsheet design, 111
  flowsheet synthesis, 28, 29
  gPROMS, 158
  hazards, 114, 116
  improvements, 46, 163
  inequality constraints, 98
  information, 111
  inherent safety, 31, 33

  inherent safety performance, 33, 153, 233
  integration, 4, 28, 173
  intensified, 5, 15, 20, 25, 28, 29, 45, 125, 150, 207, 209, 247, 249
  metrics, 45
  model, 33, 125, 138, 139, 231
  model formulation, 102
  modeling constraints, 211
  MTBE production, 226, 227, 233
  operability, 247, 248, 255
  operable, 241
  operation, 47
  optimal, 52
  optimal GMF, 182
  optimality, 62
  performances, 5, 11
  reactive distillation, 13, 26, 211, 287, 290
  risk, 153, 239
  route, 33
  safety, 29, 32, 115
    analysis, 111
    performance, 120
  separation, 53, 56, 175
  specifications, 74, 95, 148, 150, 213, 215, 230
  stream, 23, 33, 59, 62, 65, 82, 83, 105
  stream integration, 47
  synthesis, 20, 22, 28, 33, 45, 59, 163, 173, 223
  synthesis problem, 23, 28
  systems, 33, 99, 125
  tasks, 47, 53
  temperature, 113, 121–123, 155, 190
  unit hazards factor, 114
  unites, 112, 234
  units, 114
Process graph (P-graph), 23
Process intensification (PI), 3, 6, 22, 27, 32, 79, 207, 217, 231, 247, 248, 255
  control, 123, 124, 144, 211, 213, 292
  controller, 211, 215, 218
  controller design, 291, 292
  objectives, 20
  operability, 248
  synthesis, 20, 22, 36, 37, 45, 79, 111, 148, 171, 193, 202, 223, 227, 231, 247, 249, 252
  synthesis suite, 247, 249

310 Index

Process operators (OP), 25
Process Route Index (PRI), 33
Process Stream Index (PSI), 33
Process Systems Engineering (PSE), 5, 19, 187
Process Unit Hazards Factor, 115
Product
  flowrate, 200, 227
  purity, 10, 123, 140, 190, 249, 252, 297
  separation, 5
  specifications, 175, 178, 180, 185, 190, 200,
    201, 211, 219, 224, 249
  streams, 86, 197
Proportional-Integral-Derivative (PID)
  control, 34, 123–125
  control conventional, 34

Q

Quadratic Program (QP), 128
  problem, 128
Quadratic programming (QP) problem, 130,
  131, 134
Quantitative risk analysis (QRA), 33, 153
Quantitative Risk Assessment (QRA), 120

R

RADFRAC module, 178
Reaction
  catalysts, 59
  engineering, 7
  equilibrium, 55, 170
  for RD operations, 10
  heat, 10, 118
  intensified, 255
  kinetics, 53, 54, 165, 166, 224, 238
  mixture, 11, 170
  module, 55, 291
  network, 24
  processes, 15, 52
  products, 10
  rate, 32, 51, 53, 65, 224, 230
  schemes, 59
  stoichiometry, 252
  systems, 7, 11, 25
  temperature, 165, 166
  temperature range, 11

  time, 16
  zone, 168, 227, 229, 230
Reaction/Separation Simulation function, 252
Reactive
  absorption, 48, 53
  column, 236
  equipment volume, 170
  flash column, 55–57
  flash column process, 54
  holdup, 165, 170
  module, 166, 168
  Petlyuk column, 236
  separation, 7, 10, 36, 47, 48, 52, 69, 154, 168,
    227, 229, 232, 253, 290
    module, 55, 57, 167, 230, 234–236
    process design, 247
    synthesis systems, 249
    systems, 157
  separator, 55, 163
  tray location, 231
  volume, 164–166
  volume minimization, 168
  zone, 236
Reactive distillation (RD), 3, 5, 6, 8, 10, 22, 34,
    47, 48, 144, 165, 168, 207, 208, 212, 213,
    227, 229, 231, 238, 248, 255, 290, 291,
    295
  column, 47, 209, 213, 229, 231, 235, 236, 238,
    255, 297
  column in Aspen Plus, 230
  in Aspen Plus, 229
  model, 209
  operation, 230
  operations, 10
  process, 13, 26, 210, 211, 287, 290, 291
  synthesis, 25
  systems, 25
  units, 27
Reactor
  design, 7, 16, 163
  effluent, 22
  network synthesis, 36, 163
  outlet flowrate, 209
  outlet stream, 208, 213
  volumes, 10

Reactor-mixer-separator (RMS) system, 163, 164
Reboiler, 22, 74, 89, 178, 194, 201, 227, 230, 235, 236, 255, 295
 duty, 179, 191, 291, 297
 heat duty, 140, 295, 297
 heat duty optimal, 90
Reboiler/condenser duty, 178
Reflux
 ratio, 74, 191, 297
 tray location, 288
Regulatory control, 34
Relative cost optimality, 197
Retrofit optimization, 200
Reverse osmosis (RO), 11
 membranes, 12
 process, 12
Risk analysis, 149–151, 153, 233, 239, 248
 for inherent safety, 233
 for inherent safety assessment, 239
Rotating
 packed bed reactors, 36
 reactors, 5

S

Safe operation region, 211
Safety
 analysis, 20, 112
 aspects, 29
 assessment, 150
 considerations, 126, 150, 157, 235, 237, 239
 evaluation, 111
 evaluation methods, 33
 factors, 33
 indicator, 33
 indices, 33, 120, 150
 metrics, 33
 performance, 32, 150
 process, 29, 32, 115
 properties, 33
 target, 147
 weighted hazard index, 115
Saturation pressure, 121
Seasonal disturbances, 34

Selective separation, 45, 47
Separation
 energy consumption, 169, 170
 module, 53, 55, 56, 167, 229, 230
 operations, 7
 operations chemical, 7
 optimal, 176
 performance, 180
 phenomena, 155, 229, 231
 problem, 174
 process, 53, 56, 175
  for olefin metathesis, 171
  for pentene metathesis, 54
  synthesis problem, 86
  systems, 36
 product, 5
 reactive, 7, 10, 36, 47, 48, 52, 69, 154, 168, 227, 229, 232, 253, 290
 systems, 53, 68, 181, 187, 255
 target, 139, 175, 181
 tasks, 11, 48, 169, 228, 253
 tasks from heat, 52
 techniques, 7, 11
 technology, 7
Sequential
 reaction, 56
 separation steps, 177
Single-input single-output (SISO)
 PI controllers, 211, 215, 292
 systems, 211, 241, 292
Sinusoidal disturbance, 215
Solvay process, 11
Sorption Enhanced Reaction Process (SERP), 15
Specialized equipment, 185
Spill control, 114
Stage reaction, 55
State variables, 95, 102, 106, 125, 127, 129, 151, 241, 242, 259
State-Equipment-Network (SEN), 22
State-Task-Network (STN), 22
Steady-state
 flexibility, 150, 151
 synthesis, 157

312   Index

Stochastic flexibility
  analysis, 95
  index, 31
Stream
  bottom, 179, 190, 202
  conditions, 185
  connections, 187
  feed, 16, 61, 74, 75, 82, 86, 89, 139, 140, 147, 173, 175, 193, 196, 249, 287
  flowrates, 65, 157
  flows, 62, 65, 68, 104, 153, 213
  inlet temperature, 158
  mixture, 86
  MMA product, 190, 194, 199, 202
  outlet temperature, 158
  property, 82
  property variables, 79
  temperature, 167
  toxicity, 157
  variables, 180
Stripping
  section, 74, 168, 200, 229, 230
  trays, 230
Substance
  boiling point, 121
  phase, 121
Superstructure
  optimization, 223, 227, 228
  optimization problem, 157
  representation, 22–25, 187, 227
Synchronized operation, 219
SYNOPSIS framework, 148, 150, 157, 223, 249
Synthesis
  capabilities, 28
  conventional process, 25
  framework, 36, 173
  function, 255
  GMF, 52, 75, 79, 86, 87, 90, 151, 153, 176, 178, 187, 193, 197, 200, 226, 231, 290
  heat, 25
  methods, 163
  model, 34, 193
  model GMF, 60, 62, 69, 71, 79, 82, 150, 153, 228, 230, 234
  modular process intensification, 148, 249

  multiperiod, 104
  objective, 190
  PI, 36, 37
  problem, 62, 79, 149, 157, 252
  problem GMF, 61, 87, 89
  process, 22, 28, 33, 45, 59, 163, 173, 223
  reactive distillation, 25
  representation, 25, 149
  representation GMF, 73
  results, 201
  results GMF, 150, 237, 255
  stage, 28
  strategy GMF, 190
  structure GMF, 202
  suite, 247–249
Systematic process intensification, 25
Systems
  chemical process, 59
  conventional process, 290
  design, 46
  distillation, 47, 177
  intensified, 20, 28, 34, 36, 48, 215, 219
  intensified process, 19, 29, 32, 35, 45, 52, 207, 248
  modular, 29, 34, 37, 219
  modular PI, 19, 32
  modular process intensification, 31, 95
  MTBE production, 235, 238, 239, 245
  nonconvex, 153
  process, 33, 99, 125
  reaction, 7, 11, 25
  reactive separation, 157
  separation, 53, 68, 181, 187, 255
  separation process, 36

T
Tailed operation, 219
Tailored operability, 37
Temperature
  approximation, 106
  combinations, 83
  decanter, 201
  difference, 83
  effects, 118, 119
  feed, 74

gradient, 47, 82
process, 113, 121–123, 155, 190
profile, 75, 91, 219, 230
range, 165
reaction, 165, 166
stream, 167
swing, 15
Ternary
separation, 8, 89
separation system, 86
systems, 175
Thermally coupled distillation, 8
Thermally coupled distillation systems, 25
Total annualized cost (TAC), 174
Toxicity, 33, 34, 120, 147, 155, 156, 226
risk, 158, 233
stream, 157
Tray
distillation, 52, 53, 154, 178, 180, 185
location feed, 288
numbering scheme, 295
reactive distillation column, 231
solvent recovery, 185
Tutorial example, 105, 126, 130, 137

U
Unconventional process flowsheets, 25

V
Vacuum pressure operation, 181
Vapor
butenes feed flowrate, 242, 297
butenes flowrate, 241, 297
compression distillation, 12
connect stream flowrates, 191

density, 120
feed flowrate, 297
flow, 229
flowrate, 69, 79, 228
fugacity coefficient, 51, 53
holdup, 155
inlet, 185, 199
inlet stream, 180
inlet stream conditions, 53
mass, 141
density, 69
holdups, 154
mixture, 69
molar
density, 69
flowrate, 69, 239
fraction profiles, 196
outlet, 185
outlet streams, 52, 55, 177, 180, 200
pervaporation, 12
phase, 50, 53, 56, 155, 194
pressure, 118, 181, 188
streams, 65, 154, 155, 170, 181, 182
temperatures, 57
velocity, 69
volumetric flowrate, 69
Vapor permeation (VP), 12
Vaporization, 10, 74, 197

W
Water
decanter, 188, 190, 191, 201, 202
decanter process, 190
recovery stream, 191, 193, 202

Printed in the United States
by Baker & Taylor Publisher Services